AF787516

Advance in Polymer Composites: Fire Protection and Thermal Management

Advance in Polymer Composites: Fire Protection and Thermal Management

Fei Xiao
Fubin Luo
Ke Sun

MDPI

Basel • Beijing • Wuhan • Barcelona • Belgrade • Novi Sad • Cluj • Manchester

Fei Xiao
School of Safety Science and Emergency Management
Wuhan University of Technology
Wuhan
China

Fubin Luo
Engineering Research Center of polymer Green Recycling of Ministry of Education
Fujian Normal University
Fuzhou
China

Ke Sun
Institute of Intelligent Manufacturing Technology
Shenzhen Polytechnic University
Shenzhen
China

Editorial Office
MDPI AG
Grosspeteranlage 5
4052 Basel, Switzerland

This is a reprint of articles from the Special Issue published online in the open access journal *Polymers* (ISSN 2073-4360) (available at: www.mdpi.com/journal/polymers/special_issues/8TB1S0J029).

For citation purposes, cite each article independently as indicated on the article page online and using the guide below:

Lastname, A.A.; Lastname, B.B. Article Title. *Journal Name* **Year**, *Volume Number*, Page Range.

ISBN 978-3-7258-1914-0 (Hbk)
ISBN 978-3-7258-1913-3 (PDF)
https://doi.org/10.3390/books978-3-7258-1913-3

Cover image courtesy of Fei Xiao

© 2024 by the authors. Articles in this book are Open Access and distributed under the Creative Commons Attribution (CC BY) license. The book as a whole is distributed by MDPI under the terms and conditions of the Creative Commons Attribution-NonCommercial-NoDerivs (CC BY-NC-ND) license (https://creativecommons.org/licenses/by-nc-nd/4.0/).

Contents

About the Editors

Fei Xiao

Dr. Fei Xiao obtained his Ph.D. at Centrale Lille Institut, France in 2021, where he addressed numerous challenges related to the flame retardancy of polymer composites. In 2022, he joined the School of Safety Science and Emergency Management, Wuhan University of Technology, China as a lecture. Throughout his research career, Dr. Xiao has published more than 30 peer-reviewed journal papers. Currently, he is serving as a guest editor for *Polymers* and *Frontiers in Chemistry*. He is also serving as the topical advisory panel member for the *Journal of Composites Science*. His research interests lie in the thermal stability and fire retardancy of polymer composites.

Fubin Luo

Dr. Fubin Luo works at the Engineering Research Center of polymer Green Recycling of Ministry of Education, Fujian Normal University, China. Throughout his research career, Dr. Luo has published over 50 original research articles in internationally recognized, peer-reviewed journals, including *Materials Horizons, Journal of Materials Chemistry A, Composites Science & Technology*, and *Composites Part B: Engineering*. Currently, he is serving as a guest editor of a Special Issues in MDPI's *Polymers*. His research focuses on thermally conductive composites and flame retardant composites.

Ke Sun

Dr. Ke Sun works at the Institute of Intelligent Manufacturing Technology, Shenzhen Polytechnic University. His research focuses on photopolymerization, photoinitiators, and stimuli-responsive materials.

Preface

Most synthetic polymers are derived from petroleum, which makes them highly flammable. Flame retardancy and thermal management are common methods to reduce fire hazards. With the rapid development of materials science, more and more highly flame-retardant and highly thermally conductive polymer materials are being developed for various applications, ranging from the electrical and electronic (E&E) sector and transportation to energy materials and aerospace.

The purpose of this reprint is to collect the latest advances in polymer composites, including research on various aspects related to fire protection, flame retardancy, thermal stability, and thermal management. We sincerely thank all the authors for their enthusiasm and significant contributions to this reprint. These papers emphasize the use of different flame retardancy and thermal management methods to improve the fire resistance of polymers, which is of great significance for reducing fire accidents and the resulting loss of life and property.

We also thank Ms. Luna Shen, the editor-in-chief of the MDPI department, for her generous and effective help throughout the publishing process.

This reprint is dedicated to Professor Serge Bourbigot, my beloved mentor, on the occasion of his 60th birthday.

Fei Xiao, Fubin Luo, and Ke Sun
Editors

polymers MDPI

Article

Exploration of the Fire-Retardant Potential of Microencapsulated Ammonium Polyphosphate in Epoxy Vitrimer Containing Dynamic Disulfide Bonds

Wenlong Shao [1,†], Tongbing Li [2,†], Fei Xiao [1,*], Fubin Luo [3,*], Yong Qiu [4], Yanyan Liu [1], Bihe Yuan [1] and Kaiyuan Li [1]

1 School of Safety Science and Emergency Management, Wuhan University of Technology, Wuhan 430070, China; shaowenlong@whut.edu.cn (W.S.); yanyan.liu@whut.edu.cn (Y.L.); bhyuan@whut.edu.cn (B.Y.)
2 Guangdong Advanced Thermoplastic Polymer Technology Co., Ltd., Dongguan 523125, China; rd03@atpchem.com
3 Engineering Research Center of Polymer Green Recycling of Ministry of Education, College of Environmental and Resource Sciences, Fujian Normal University, Fuzhou 350007, China
4 China Light Industry Engineering Technology Research Center of Advanced Flame Retardants, Beijing Technology and Business University, Beijing 100048, China; yongqiu@btbu.edu.cn
* Correspondence: fei.xiao@whut.edu.cn (F.X.); luofubin@fjnu.edu.cn (F.L.)
† These authors contributed equally to this work.

Abstract: Epoxy vitrimers appear as a promising alternative to common epoxy thermoset composites. Nevertheless, the possibilities of applying these materials are limited due to their high flammability which may cause high fire risks. To date, the flame-retardant epoxy vitrimer systems reported in the literature almost all rely on intrinsic flame retardancy to achieve high fire safety; however, the complex and expensive synthesis process hinders their large-scale application. In this work, disulfide-based epoxy vitrimer (EPV) was fabricated with 4, 4′-dithiodianiline as the curing agent, and microencapsulated ammonium polyphosphate (MFAPP) was employed as a potential additive flame retardant to improve their fire retardancy. As a comparative study, common epoxy (EP) composites were also prepared using 4,4′-diaminodiphenylmethane as the curing agent. The results showed that the introduction of dynamic disulfide bonds led to a reduction in the initial thermal decomposition temperature of EPV by around 70 °C compared to EP. Moreover, the addition of 7.5 wt.% of MFAPP endowed EP with excellent fire performance: the LOI value was as high as 29.9% and the V-0 rating was achieved in the UL-94 test (3.2 mm). However, under the same loading, although EPV/MFAPP7.5% showed obvious anti-dripping performance, it did not reach any rating in the UL-94 test. The flame-retardant mechanisms in the condensed phase were evaluated using SEM-EDS, XPS, and Raman spectroscopy. The results showed that the residue of EPV/MFAPP7.5% presented numerous holes during burning, which failed to form a continuous and dense char layer as a physical barrier resulting in relatively poor flame retardancy compared to EP/MFAPP7.5%.

Keywords: fire retardant; epoxy vitrimers; thermal stability; microencapsulated ammonium polyphosphate

check for updates

Citation: Shao, W.; Li, T.; Xiao, F.; Luo, F.; Qiu, Y.; Liu, Y.; Yuan, B.; Li, K. Exploration of the Fire-Retardant Potential of Microencapsulated Ammonium Polyphosphate in Epoxy Vitrimer Containing Dynamic Disulfide Bonds. *Polymers* **2023**, *15*, 2839. https://doi.org/10.3390/polym15132839

Academic Editor: Paul Joseph

Received: 8 June 2023
Revised: 25 June 2023
Accepted: 26 June 2023
Published: 27 June 2023

Copyright: © 2023 by the authors. Licensee MDPI, Basel, Switzerland. This article is an open access article distributed under the terms and conditions of the Creative Commons Attribution (CC BY) license (https://creativecommons.org/licenses/by/4.0/).

1. Introduction

Epoxy vitrimers are constantly emerging. They are a special class of polymers that integrate the unique characteristics of both thermoplastics and thermosets [1–4]. The presence of abundant dynamic exchangeable bonds in their structure endows epoxy vitrimers with high self-healing (reparable) [5,6], recyclable [7,8], and reprocessing [9,10] capabilities, showing the potential to extend their lifetime and reduce maintenance costs in many application fields [11]. While epoxy vitrimers have been developed in recent years, the main bottleneck persists in their intrinsic flammability [12], which releases a large amount of heat and smoke during burning, resulting in high fire risk during their application.

To address this issue, efforts have been made to explore the fire retardancy of epoxy vitrimers. Most studies focused on the incorporation of flame-retardant elements into crosslinked EP networks. In this regard, some phosphorus-/nitrogen-containing functional groups, such as 9,10-dihydro-9-oxa-10-phosphazene-10-oxide (DOPO) derivatives [13,14], 2-(bis(2-hydroxyethyl)amino)ethyl diphenylphosphinate [15], cyclolinear cyclotriphosphazene [16], cyclophophazene [17,18], and β-ketoester-containing phosphonate [19], were introduced into the crosslinked network of EP by modifying epoxy monomers or curing agents. For instance, Chen et al. prepared intrinsically flame-retardant epoxy vitrimers by introducing dynamic phosphorus-containing ester linkages into the crosslinking networks. The resulting epoxy vitrimers exhibited desirable thermal stability and excellent fire retardancy with a V-0 rating in the UL-94 test (3.2 mm) [13]. Intrinsic flame retardants are connected to the EP matrix through chemical bonds; they participate in curing and become a component of the cured system structure [20], ensuring the uniformity of EP vitrimers and preventing the precipitation of flame retardants [21]. However, this inevitably increases the complexity of the synthesis and curing process of epoxy vitrimers which hinders its commercial application. Comparatively, additive flame retardants do not participate in the curing reaction and are the most ideal industrial solution due to their relative cheapness and convenience [22–24]. Moreover, inspired by the recycling of carbon fibers in epoxy vitrimers [25], additive flame retardants and epoxy monomers are expected to be recycled separately.

In this work, a common epoxy vitrimer (EPV) was prepared based on the exchangeable disulfide bonds using bisphenol A diglycidyl ether as the epoxy monomer and 4, 4′-dithiodianiline (DTDA) as the curing agent [26]. To endow the EPV with flame retardancy, melamine–formaldehyde resin microencapsulated ammonium polyphosphate (MFAPP) was incorporated as a promising flame retardant due to the excellent compatibility, flame retardancy, and water resistance, benefiting from the presence of the organic MF coating on the surface of APP [27–29]. As a comparative study, the fire performance of epoxy composites was also studied with 4,4′-methylenedianiline (DDM) as the curing agent and MFAPP as the flame retardant. Overall, the details of fire behaviors, flame retardancy, and the mechanism of actions of flame-retardant EP and EPV composites were investigated to assess the flame-retardant potential of MFAPP for epoxy vitrimers.

2. Experimental Setup

2.1. Materials

Diglycidyl ether of bisphenol-A (DGEBA, E-51, WSR618) with an epoxide equivalent weight of 184–200 g/eq was supplied by Nantong Xingchen Synthetic Material Co., Ltd., Nantong, China. 4, 4′-Diaminodiphenylmethane (DDM, 99%) and 4, 4′-dithiodianiline (DTDA, 98%), supplied by Shanghai Aladdin Biochemical Technology Co., Ltd., Shanghai, China, were used as the curing agents for the curing of the epoxy matrix. Melamine–formaldehyde resin microencapsulated ammonium polyphosphate (MFAPP, TF-MF201, ammonium polyphosphate with crystalline form II) was purchased from Guangzhou Zhanpu Chemical Co., Ltd., Guangzhou, China.

2.2. Preparation of the Cured Epoxy Composites

For epoxy resin (EP) composites, DGEBA was added to a beaker with magnetic stirring at 95 °C for 15 min. Subsequently, various contents of MFAPP were slowly added into DGEBA and adequately mixed for 20 min to form a uniform mixed system. Then, DDM was incorporated into the mixture and continuously stirred for another 20 min. After degassing at 105 °C for 5 min, the blend was poured into a pre-heated mold and cured at 120 °C for 2 h and then post-cured at 150 °C for 2 h. Finally, the cured specimens were naturally cooled to room temperature. Epoxy vitrimer (EPV) composites were fabricated by the same method, except that 4, 4′-dithiodianiline was used as a curing agent. Moreover, the pre-curing time was increased to 40 min to avoid the precipitation of flame retardants. The detailed formulas are presented in Table 1.

Table 1. Formulation of EP and EPV composites.

Sample	Composition				
	EP (g)	**DDM (g)**	**DTDA (g)**	**MFAPP (g)**	**FRs (%)**
EP	100	25.3	/	0	0
EP/MFAPP2.5%	100	25.3	/	3.21	2.5
EP/MFAPP5.0%	100	25.3	/	6.59	5
EP/MFAPP7.5%	100	25.3	/	10.16	7.5
EP/MFAPP10%	100	25.3	/	13.92	10
EPV	100	/	31.6	0	0
EPV/MFAPP2.5%	100	/	31.6	3.37	2.5
EPV/MFAPP5.0%	100	/	31.6	6.92	5
EPV/MFAPP7.5%	100	/	31.6	10.67	7.5
EPV/MFAPP10.0%	100	/	31.6	14.62	10

2.3. Characterization

The thermogravimetric analysis (TGA) was performed to evaluate the thermal stability of the samples using a simultaneous thermogravimetric analyzer (NETZSCH STA 449 F3, NETZSCH, Bayern, Germany). Approximately 5 mg of samples were heated from 25 °C to 800 °C at a heating rate of 10 °C/min under nitrogen atmospheres. Limiting oxygen index (LOI) values of the samples were measured using an AOI LOI apparatus (Motis Fire Technology Co., Ltd., Kunshan, China) standardized as ASTM D2863 with dimensions of 130 × 7.0 × 3.2 mm^3. The UL-94 vertical burning test was carried out using an M607 horizontal-vertical combustion apparatus (Qingdao Shang-fang Instrument Co., Ltd., Qingdao, China) according to ASTM D3801-2020 standards. The dimensions of the samples were 127 × 12.7 × 3.2 mm^3. A Thermo Scientific K-Alpha X-ray photoelectron spectrometer (XPS) (Thermo Fisher Scientific, Waltham, MA, USA) was used to determine the elemental composition and oxidation states of elements at the surface of the residues after the UL-94 test, and Al Ka radiation (1361 eV) was used as the excitation source. The LabRAM Odyssey high-speed and high-resolution confocal microscopic Raman spectrometer (HORIBA, Gières, France) was applied to investigate the graphitization degree of the char residue. The Raman spectra of residual chars were obtained at room temperature under the excitation line of a 532 nm laser. The morphology and element mappings of char residue after the UL-94 test were performed using a Gemini 300 scanning electron microscope (SEM, ZEISS, Oberkochen, Germany), which was integrated with an energy-dispersive X-ray (EDX) microanalyzer for elemental analysis.

3. Results and Discussion

3.1. Thermal Stability

TGA was performed to investigate the thermal performance of EP and EPV composites under a nitrogen atmosphere, and the TGA/DTG curves and the relative data are presented in Figure 1 and Table 2. It is clear that the incorporation of MFAPP into the epoxy matrix inevitably affects the thermal stability of EP and EPV composites. For EP composites, it is evidently observed from DTG curves that a neat EP displays one-step degradation behavior as well as EP/MFAPP composites, occurring approximately from 300 to 500 °C. The initial decomposition temperature ($T_{5\%}$) of neat EP is 375 °C, and the thermal degradation rate rapidly reaches the maximum at 388 °C (T_{max}). There is a 19.7 wt.% residue formed at 800 °C in the EP matrix, indicating the presence of abundant aromatic/polyaromatic structures. It is worth noting that with the increase in MFAPP loading (2.5–10.0 wt.%), the $T_{5\%}$ values of EP composites with MFAPP gradually decrease and stabilize at around 340 °C, and the char residues gradually increase to approximately 29%. This phenomenon is attributed to the weak bonds of P-O-C and P-N in the structure of MFAPP [30] and is probably ascribed to the decreased cross-linking density of EP due to the addition of

MFAPP [31]. The TGA and DTG curves of EPV composites exhibit a similar behavior to those of EP composites. Nevertheless, there is a remarkable reduction (~67 °C) in the $T_{5\%}$ for EPV compared to EP, which might be attributed to the instability of disulfide bonds in the crosslinked networks [2]. The incorporation of MFAPP does not significantly reduce the thermal stability ($T_{5\%}$ and T_{max}) of EPV, and char residues at 800 °C increase with the increase in MFAPP loading. In conclusion, although the incorporation of MFAPP leads to a certain reduction in the thermal stability of EP and EPV composites, the char yields of those composites are significantly improved, which is beneficial to the flame retardancy of EP.

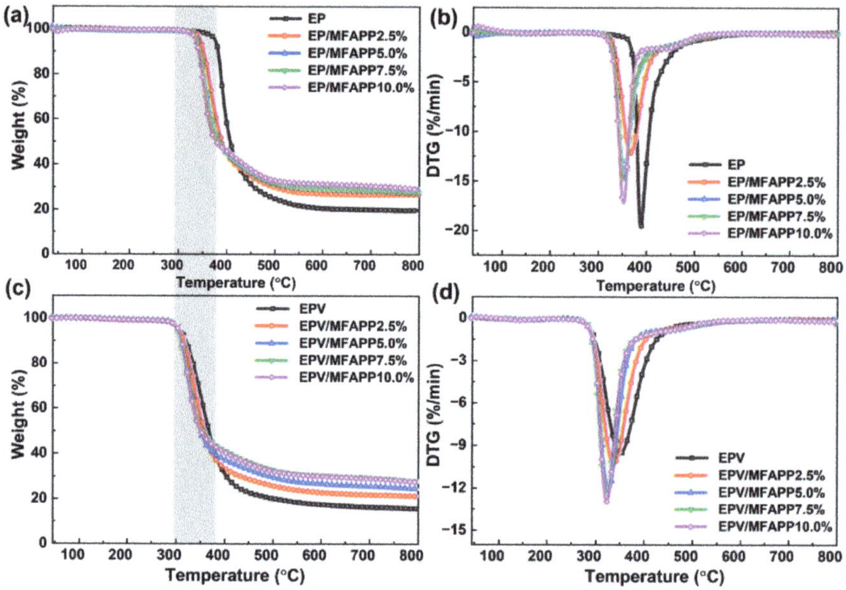

Figure 1. TGA and DTG curves of EP (**a**,**b**) and EPV composites (**c**,**d**).

Table 2. TGA data from EP and EPV composites.

Sample	$T_{5\%}$ (°C)	$T_{50\%}$ (°C)	T_{Max} (°C)	Residues at 800 °C (wt%)
EP	375	406	388	19.7
EP/MFAPP2.5%	346	391	369	26.7
EP/MFAPP5.0%	338	385	355	28.6
EP/MFAPP7.5%	340	387	360	28.1
EP/MFAPP10.0%	337	380	354	29.1
EPV	308	367	350	15.5
EPV/MFAPP2.5%	306	359	339	21.3
EPV/MFAPP5.0%	304	351	328	24.6
EPV/MFAPP7.5%	303	353	324	27.8
EPV/MFAPP10.0%	305	352	324	27.6

3.2. Flame Retardancy

The flammability, fire retardancy, and anti-dripping behavior of EP and EPV composites were evaluated using LOI and UL-94 vertical burning tests [32–34], and the corresponding results are presented in Table 3 and Figures 2 and 3. It is seen that the LOI value of EP is only 23.4%, and it burns violently after ignition with severe dripping fire behavior, which is considered as a significant fire hazard [35]. With the incorporation of MFAPP at a low loading (2.5%), EP/MFAPP2.5% still burns out (no rating) in the UL-94 test but no dripping occurred (Figure 2). By further increasing the loading of MFAPP to 5.0 wt.%, 7.5 wt.%, and

10.0 wt.%, the flame retardancy of EP/MFAPP composites gradually improved, and when the loading reached 7.5 wt.%, a V-0 rating was achieved for EP/MFAPP7.5% with an LOI value of 29.7%.

Table 3. Related data from EP and EPV composites from LOI and UL-94.

Sample	LOI (%)	UL-94 Test			
		t_1 (s)	t_2 (s)	Dripping	Rating
EP	23.4	206.2	/	Yes	NR
EP/MFAPP2.5%	23.6	149.8	/	No	NR
EP/MFAPP5.0%	26.3	57.6	1.4	No	NR
EP/MFAPP7.5%	29.9	2.8	0.2	No	V-0
EP/MFAPP10%	35.5	0.2	0	No	V-0
EPV	19.9	165.8	/	Yes	NR
EPV/MFAPP2.5%	22.2	151.4	/	No	NR
EPV/MFAPP5.0%	23.1	149.6	/	No	NR
EPV/MFAPP7.5%	24.3	136.4	/	No	NR
EPV/MFAPP10.0%	25.3	142.2	/	No	NR

t_1 and t_2 refers to the average times of first and second ignitions; NR: no rating.

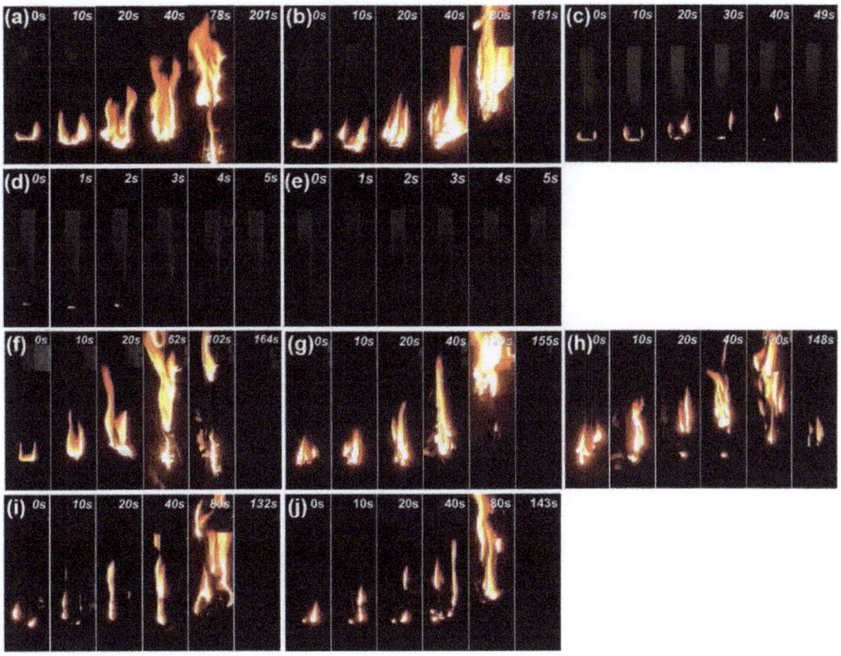

Figure 2. Fire behaviors of (**a–e**) EP and (**f–j**) EPV composites after the first ignition (0, 2.5 wt.%, 5 wt.%, 7.5 wt.%, and 10 wt.% MFAPP loading).

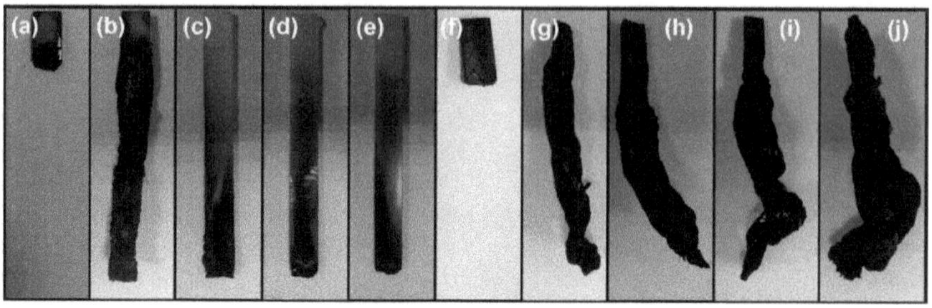

Figure 3. Residues of (**a–e**) EP and (**f–j**) EPV composites after the UL-94 tests (0, 2.5 wt.%, 5.0 wt.%, 7.5 wt.%, and 10.0 wt.% MFAPP loading).

Regarding EPV composites, neat EPV is very flammable, showing an LOI value of 19.9% and no rating (NR) in the UL-94 test. As the loading of MFAPP increases to 10 wt.%, the LOI value gradually increases to 25.3%, which is far lower than the EP composite with the same content (Table 3). However, in terms of the UL-94 ratings, EPV composites showed poor results relative to the EP composites (Figure 2f–j). Although all EPV composites exhibited significant enhancements in anti-dripping behavior, the UL-94 ratings of EPV/MFAPP systems seemed to not largely improve with the increasing loading of MFAPP, as all EPV composites failed to pass any rating in the UL-94 test. It can be concluded that the addition of MFAPP does not significantly improve the flame retardancy of EPV.

Figure 3 presents the char residues of EP and EPV composites after UL-94 tests. The residual chars of neat EP and EPV almost burn out, which illustrates the weak char formation capacity and high fire risk of neat EP and EPV matrices. It is observed that increasing the content of MFAPP causes the EP composites to self-extinguish over a shorter burning distance, resulting in the formation of a dense carbon layer. On the other hand, when MFAPP is introduced into the EPV matrix, the yields of char residues for the EPV composites rise significantly; however, all EPV composites do not extinguish spontaneously within 60 s after the first ignition. Visual observation shows that intumescent char residues are formed due to the presence of MFAPP, which does not retard the burning of the matrix underneath.

3.3. *Morphology and Chemical Structure of Char Residues*

3.3.1. SEM-EDS Analysis

On the basis of the above analysis of LOI and UL-94 tests, it can be concluded that EP/MFAPP composites possess better flame retardancy than EPV/MFAPP composites under the same loading. To further investigate its relevant mechanisms in the condensed phase, the micro-morphologies of residual chars of EP/MFAPP7.5% and EPV/MFAPP7.5% after the UL-94 test were analyzed by SEM-EDS.

It is observed in Figure 4a,c that the surface residues of neat EP and EPV exhibit very loose and porous morphologies. In contrast, due to the addition of MFAPP, the surface of char residues of EP/MFAPP7.5% become compact, smooth, and homogeneous, implying the improvement of the char quality (Figure 4c) [36]. The continuous and dense char layer serves as an effective physical barrier to inhibit the transfer of heat and prevent the internal pyrolysis products from being released into the combustion zone [37]. For EPV/MFAPP7.5%, there are numerous holes observed on the surface of char residues (Figure 4d). These holes allow combustible pyrolysis products to be released from the internal matrix into the external environment, resulting in poor flame retardancy.

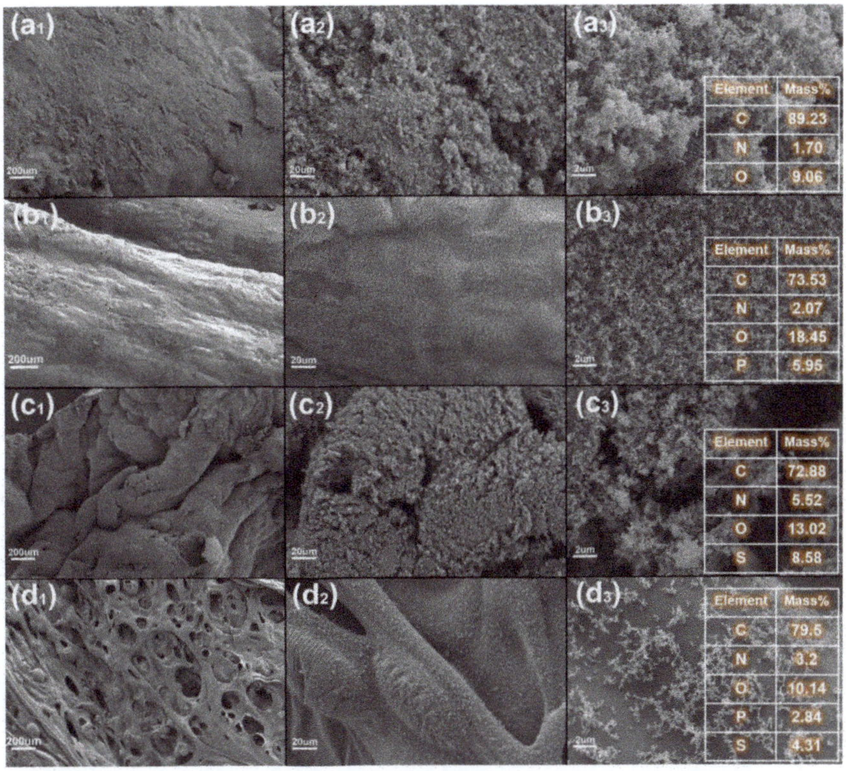

Figure 4. SEM-EDS images of the exterior surface of the char residues after the UL-94 test: (**a**$_1$–**a**$_3$) neat EP, (**b**$_1$–**b**$_3$) EP/FMAPP7.5%, (**c**$_1$–**c**$_3$) neat EPV, (**d**$_1$–**d**$_3$) EPV/FMAPP7.5%.

EDS analysis was performed to study the chemical composition of the char residues, as shown in Figure 4. In the EDS spectrum of EP/MFAPP7.5%, the main composition of the residue is carbon (73.53%), oxygen (18.45%), nitrogen (2.07%), and phosphorus (5.96%). The relatively high phosphorus content mainly comes from the phosphorus-containing species formed by the decomposition of MFAPP, which are mainly orthophosphoric acid and phosphoric acid, according to previous work [38,39]. Moreover, the oxygen content is also high (18.45 %), implying that the formation of a continuous and compact char layer is related to the increase in phosphorus and oxygen content [31]. It is concluded that the introduction of MFAPP in the EP matrix promotes the formation of a char layer with a more phosphorus-crosslinked structure, improving the quality of the char layer. For the EDS spectrum of EPV/MFAPP7.5%, it is found that the phosphorus content (2.84%) on the char layer from the char residue is much lower than that of the EP/MFAPP7.5%. Sulfur emerges in the char residue of EPV/MFAPP7.5% due to the presence of DTDA as the curing agent.

3.3.2. Elemental Mappings

The distribution of species in the residual chars is very important when explaining the mechanism in the condensed phase. Figure 5a shows the elemental mapping images of C, O, N, and P elements, respectively, of the surface char of EP/MFAPP7.5%. The results provide visible evidence that MFAPP decomposes to form phosphorus-containing species, which combine with oxygen (see the O and P elemental mappings in Figure 5) to form phosphoric acid/orthophosphates/pyrophosphates and the related analogues during burning. The phosphorus-containing species derived from MFAPP promote the dehydration and carbonization of the epoxy matrix and form a continuous and dense carbonaceous char layer. It is found that the aggregated phosphate species are embedded on the surface of the carbon layer, which enhances the cohesion and resistance of the char residues, thereby improving its flame retardancy. For EPV/MFAPP7.5%, the mapping images confirm the presence and dispersion of C, O, N, P, and S elements, which indicates those elements are uniformly distributed on the surface of char.

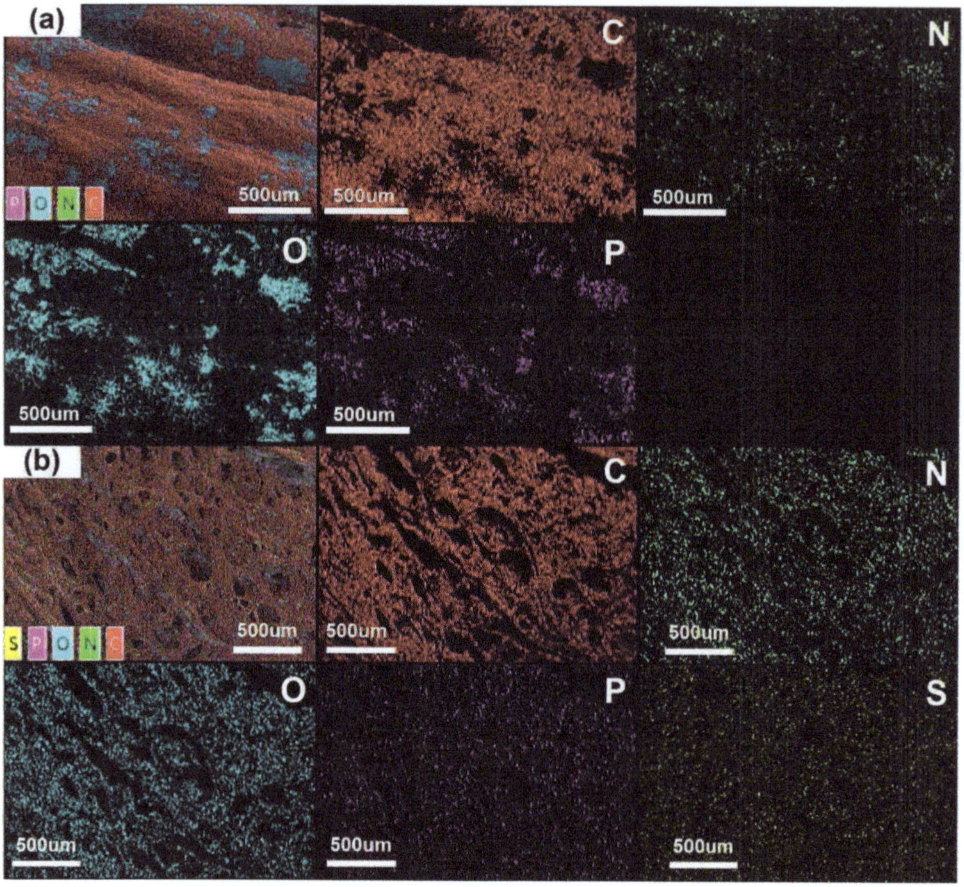

Figure 5. EDS mapping images of total elements, C, N, O, P, and S elements of the exterior char residues after the UL-94 tests: (**a**) EP/FMAPP7.5%, (**b**) EPV/FMAPP7.5%.

3.3.3. XPS Analysis

To further explore the effect of MFAPP on the charring of EP and EPV, XPS analysis was conducted for the chars of EP/MFAPP7.5% and EPV/MFAPP7.5%. Figure 6 shows the C_{1s}, O_{1s}, N_{1s}, P_{2p} and S_{2p} spectra of the external char of EP/MFAPP7.5% and EPV/MFAPP7.5%. In the C_{1s} spectra, the peak at 284.8 eV is assigned mainly to the C-H, C-C, and C=C in the aliphatic and aromatic species in char, while the peak at 286.5 eV corresponds to C-O, C-N, C-P, and C=O linkages [40]. The high-resolution O_{1s} spectra consist of two characteristic peaks corresponding to the double-bonded oxygen (C=O and P=O groups) in phosphate and carbonyl compounds at a binding energy of 531.3 eV and single-bonded oxygen (-O-) in O-C-O, C-O-P, O-P=O, and P-O-P at around 533.0 eV [41]. In addition, in the N_{1s} spectrum of EP/MFAPP7.5% (Figure 6), the broad bands centered at 399.8 and 401.5 eV are assigned to C=N group and C-N/N-H groups, respectively, while the peak intensity at 401.0 eV for EPV/MFAPP7.5% decreases compared to EP/MFAPP7.5%, indicating that less nitrogen-containing aromatic heterocyclic cross-linking structures are formed in the char residue. The deconvoluted P_{2p} region spectrum of EPV/MFAPP7.5% shows two signals at 134.0 eV attributed to O-P-C, O-P-O, O-P=O groups formed during the decomposition of MFAPP, while the peak at 134.8 eV is assigned to the structure of pyrophosphates and metaphosphates in P-O-P and PO_3, respectively [42]. Similar linkages are also found for the P_{2p} survey of EP/MFAPP7.5% char, except that the linkages of O-P=O can be fitted in an independent peak at a binding energy of 133.4 eV. The high-resolution S_{2p} spectrum is detected for EPV/MFAPP7.5% as demonstrated earlier by the EDS results. The signal can be fitted into four peaks, where the binding energies at 163.6 and 164.2 eV are assigned to the S-C and S=C, thiols, or sulfur ethers [43,44], while the peaks observed at 164.8 and 166.3 eV are attributed to the oxidized sulfur species (SO_x) and sulfoxide sulfur due to the burning [45,46]. Compared to EPV/MFAPP7.5%, the N_{1s} and P_{2s} spectra of EP/MFAPP7.5% showed broader bands, suggesting that a more complex heterocyclic carbonaceous structure containing phosphorus and nitrogen elements was formed. These results verify that the introduction of MFAPP is conducive to the formation of crosslinking structures containing P-O-C, P-O-P, C-N, and C-O-C groups during burning, endowing the resulting char with better barrier effects and stability, effectively insulating the heat and fuel transfer between the gas and condensed phases.

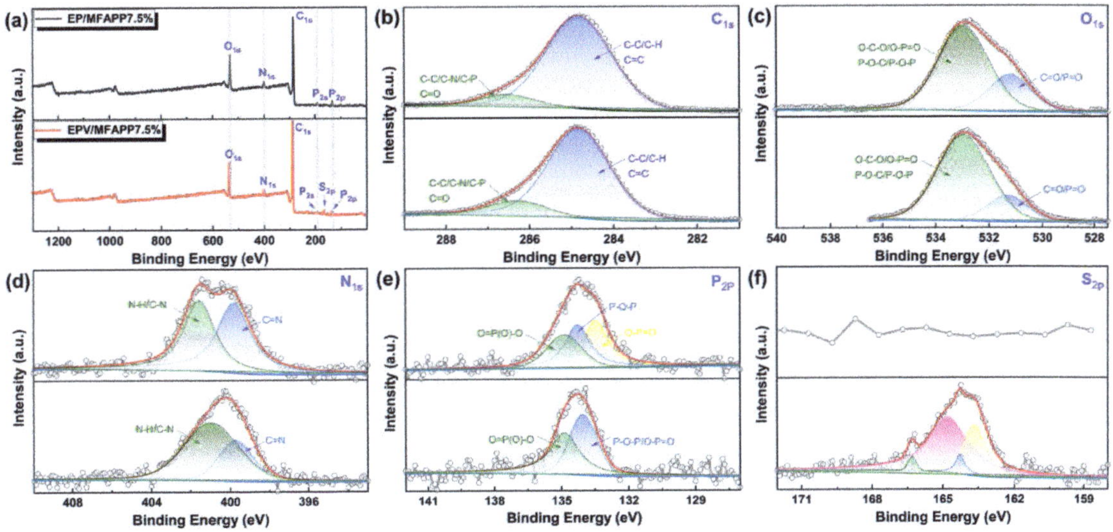

Figure 6. (**a**) XPS survey spectra and high-resolution XPS spectra of (**b**) C_{1s}, (**c**) O_{1s}, (**d**) N_{1s}, (**e**) P_{2s}, and (**f**) S_{2s} of the exterior char residues of EP/FMAPP7.5% and EPV/FMAPP7.5% after the UL-94 tests.

3.3.4. Raman Analysis

To investigate the effect of MFAPP on the structure of the char layers of EP and EPV composites, laser Raman spectroscopy was conducted to study the graphitization level of the char of EP/MFAPP7.5% and EPV/MFAPP7.5%, as presented in Figure 7. Two strong characteristic peaks at approximately 1370 and 1590 cm^{-1} are assigned to the D band (disorder band or defect band, indicating the presence of some disorder in the carbon aromatic structure) and G band (tangential vibration mode along the graphitic plane of the tube), respectively. Generally, the ratio of the band intensity of the D to G band (I_D/I_G) is an important measure of the order to estimate the graphitization degree of the char residues, and a lower value corresponds to a higher graphitization degree. The I_D/I_G value of EP/MFAPP7.5% is 3.74 and this value increases to 4.45 for EPV/MFAPP7.5%. The phenomenon indicates that compared to EP/MFAPP7.5%, the residue of EPV/MFAPP7.5%

presents more defects after burning. It demonstrates that a more graphitized structure in residual char is formed during the burning of EP/MFAPP7.5%, which improves the barrier effect of the heat and fuel transfer.

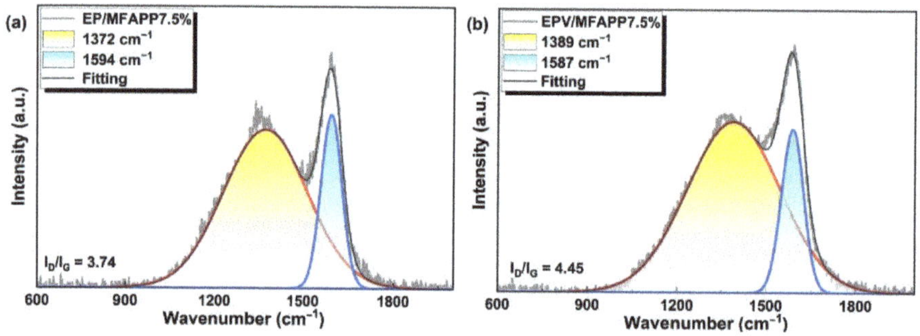

Figure 7. SEM-EDS images of the exterior char residues after the UL-94 tests: (**a**) EP/FMAPP7.5%, (**b**) EPV/FMAPP7.5%.

3.4. Flame-Retardant Mechanisms

Based on the above analyses, the possible flame-retardant mechanisms of EP/MFAPP and EPV/MFAPP are proposed as shown in Figure 8. In the condensed phase, phosphonic acid and its derivatives can be formed during the early decomposition of MFAPP and are then converted to pyrophosphate and polyphosphate species with viscous natures [47]. These phosphorus-containing acids react with the decomposed epoxy matrix through dehydration and esterification, promoting char formation [48]. Moreover, the viscous pyrophosphate and polyphosphate species tightly cover the surface of residual char (see Figure 5). These factors lead to the formation of a compact and phosphorus-rich char layer with polyaromatic/heteroaromatic structures, which acts as a physical barrier to isolate the underneath matrix from heat and oxygen. For EPV/MFAPP composites, the presence of disulfide bonds in the crosslinking networks leads to the degradation of the EPV composite at a lower temperature as presented in the TGA. They fracture during combustion to form sulfur-containing gases [45], which are released into the gas phase together with the decomposition gases of MFAPP, resulting in the formation of numerous pores on the surface of the char. The porous surface char layer cannot insulate the transfer of heat and oxygen, exhibiting relatively poor flame retardancy.

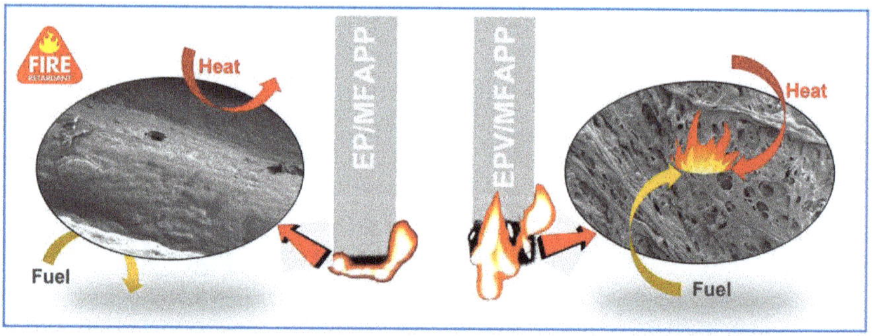

Figure 8. Possible flame-retardant mechanisms of EP/MFAPP and EPV/MFAPP composites.

4. Conclusions

This work performed a comparative study on the flame-retardant properties of EP/MFAPP and EPV/MFAPP composites to evaluate the flame-retardant potential of MFAPP in epoxy vitrimers. The results showed that the addition of 7.5 wt.% MFAPP endowed EP with excellent fire performance, including that the LOI value was as high as 29.9% and a V-0 rating was achieved in the UL-94 test (3.2 mm). Unfortunately, with the same loading (7.5 wt.%), although EPV/MFAPP7.5% showed an obvious anti-dripping performance, it did not reach any rating in the UL-94 test. SEM-EDS, XPS, and Raman were employed to assess the flame-retardant mechanism in the condensed phase. The results showed that the residue of EPV/MFAPP7.5% presented numerous holes during burning, which failed to form a continuous and dense char layer as a physical barrier resulting in relatively poor flame retardancy compared to EP/MFAPP7.5%. In summary, these results indicate that the disulfide-based epoxy vitrimers showed relatively high flammability, and it was difficult to improve their flame retardancy by adding small amounts of additive flame retardants (MFAPP). This study provides a perspective on the application of additive flame retardants in epoxy vitrimers and can contribute to future research on the improvement of flame retardancy in epoxy vitrimers.

Author Contributions: Conceptualization, F.X. and K.L.; methodology, F.X. and F.L.; software, F.X.; validation, W.S., T.L. and Y.L.; formal analysis, B.Y.; investigation, W.S.; resources, K.L.; data curation, W.S.; writing—original draft preparation, F.X. and W.S.; writing—review and editing, F.X. and W.S.; visualization, T.L.; supervision, F.X. and K.L.; project administration, F.X.; funding acquisition, Y.Q. and T.L. All authors have read and agreed to the published version of the manuscript.

Funding: This research was funded by the Open Project Program (No. BTBUFR22-3) of China Light Industry Engineering Technology Research Center of Advanced Flame Retardants, Beijing Technology and Business University, China, and the Fundamental Research Funds for the Central Universities (WUT: 2023IVA071).

Institutional Review Board Statement: Not applicable.

Data Availability Statement: The raw/processed data required to reproduce these findings cannot be shared at this time as the data also forms part of an ongoing study.

Conflicts of Interest: The authors declare no conflict of interest.

References

1. Schenk, V.; Labastie, K.; Destarac, M.; Olivier, P.; Guerre, M. Vitrimer composites: Current state and future challenges. *Mater. Adv.* **2022**, *3*, 8012–8029. [CrossRef]
2. Li, B.; Zhu, G.; Hao, Y.; Ren, T. An investigation on the performance of epoxy vitrimers based on disulfide bond. *J. Appl. Polym. Sci.* **2022**, *139*, 51589. [CrossRef]
3. Wang, M.; Gao, H.; Wang, Z.; Mao, Y.; Yang, J.; Wu, B.; Jin, L.; Zhang, C.; Xia, Y.; Zhang, K. Rapid self-healed vitrimers via tailored hydroxyl esters and disulfide bonds. *Polymer* **2022**, *248*, 124801. [CrossRef]
4. Chong, K.; Lai, J.; Rahman, R.; Adrus, N.; Al-Saffar, Z.; Hassan, A.; Lim, T.; Wahit, M. A review on recent approaches to sustainable bio-based epoxy vitrimer from epoxidized vegetable oils. *Ind. Crop. Prod.* **2022**, *189*, 115857. [CrossRef]
5. Wu, P.; Liu, L.; Wu, Z. A transesterification-based epoxy vitrimer synthesis enabled high crack self-healing efficiency to fibrous composites. *Compos. Part A Appl. Sci. Manuf.* **2022**, *162*, 107170. [CrossRef]
6. Lorwanishpaisarn, N.; Srikhao, N.; Jetsrisuparb, K.; Knijnenburg, J.T.N.; Theerakulpisut, S.; Okhawilai, M.; Kasemsiri, P. Self-healing Ability of Epoxy Vitrimer Nanocomposites Containing Bio-Based Curing Agents and Carbon Nanotubes for Corrosion Protection. *J. Polym. Environ.* **2022**, *30*, 472–482. [CrossRef]
7. Xu, Y.; Dai, S.; Bi, L.; Jiang, J.; Zhang, H.; Chen, Y. Catalyst-free self-healing bio-based vitrimer for a recyclable, reprocessable, and self-adhered carbon fiber reinforced composite. *Chem. Eng. J.* **2022**, *429*, 132518. [CrossRef]
8. Tratnik, N.; Tanguy, N.; Yan, N. Recyclable, self-strengthening starch-based epoxy vitrimer facilitated by exchangeable disul-fide bonds. *Chem. Eng. J.* **2023**, *451*, 138610. [CrossRef]
9. Yang, W.; Ding, H.; Zhou, W.; Liu, T.; Xu, P.; Puglia, D.; Kenny, J.M.; Ma, P. Design of inherent fire retarding and degradable bio-based epoxy vitrimer with excellent self-healing and mechanical reprocessability. *Compos. Sci. Technol.* **2022**, *230*, 109776. [CrossRef]
10. Azcune, I.; Odriozola, I. Aromatic disulfide crosslinks in polymer systems: Self-healing, reprocessability, recyclability and more. *Eur. Polym. J.* **2016**, *84*, 147–160. [CrossRef]
11. Gao, W.; Bie, M.; Quan, Y.; Zhu, J.; Zhang, W. Self-healing, reprocessing and sealing abilities of polysulfide-based polyurethane. *Polymer* **2018**, *151*, 27–33. [CrossRef]
12. Li, X.; Zhang, J.; Zhang, L.; de Luzuriaga, A.R.; Rekondo, A.; Wang, D.-Y. Recyclable flame-retardant epoxy composites based on disulfide bonds: Flammability and recyclability. *Compos. Commun.* **2021**, *25*, 100754. [CrossRef]

13. Chen, J.-H.; Lu, J.-H.; Pu, X.-L.; Chen, L.; Wang, Y.-Z. Recyclable, malleable and intrinsically flame-retardant epoxy resin with catalytic transesterification. *Chemosphere* **2022**, *294*, 133778. [CrossRef] [PubMed]
14. Tian, P.-X.; Li, Y.-D.; Weng, Y.; Hu, Z.; Zeng, J.-B. Reprocessable, chemically recyclable, and flame-retardant biobased epoxy vitrimers. *Eur. Polym. J.* **2023**, *193*, 112078. [CrossRef]
15. Ren, Q.-R.; Gu, S.; Liu, J.-H.; Wang, Y.-Z.; Chen, L. Catalyst-free reprocessable, degradable and intrinsically flame-retardant epoxy vitrimer for carbon fiber reinforced composites. *Polym. Degrad. Stab.* **2023**, *211*, 110315. [CrossRef]
16. Zhou, L.; Zhang, G.; Feng, Y.; Zhang, H.; Li, J.; Shi, X. Design of a self-healing and flame-retardant cyclotriphosphazene-based epoxy vitrimer. *J. Mater. Sci.* **2018**, *53*, 7030–7047. [CrossRef]
17. Peng, J.; Xie, S.; Liu, T.; Wang, D.; Ou, R.; Guo, C.; Wang, Q.; Liu, Z. High-performance epoxy vitrimer with superior self-healing, shape-memory, flame retardancy, and antibacterial properties based on multifunctional curing agent. *Compos. Part B Eng.* **2022**, *242*, 110109. [CrossRef]
18. Zhang, X.; Eichen, Y.; Miao, Z.; Zhang, S.; Cai, Q.; Liu, W.; Zhao, J.; Wu, Z. Novel Phosphazene-Based flame retardant polyimine vitrimers with Monomer-Recovery and high performances. *Chem. Eng. J.* **2022**, *440*, 135806. [CrossRef]
19. Markwart, J.C.; Battig, A.; Urbaniak, T.; Haag, K.; Koschek, K.; Schartel, B.; Wurm, F.R. Intrinsic flame retardant phosphonate-based vitrimers as a recyclable alternative for commodity polymers in composite materials. *Polym. Chem.* **2020**, *11*, 4933–4941. [CrossRef]
20. Dai, X.; Li, P.; Sui, Y.; Zhang, C. Thermal and flame-retardant properties of intrinsic flame-retardant epoxy resin containing biphenyl structures and phosphorus. *Eur. Polym. J.* **2021**, *147*, 110319. [CrossRef]
21. Feng, X.; Li, G. Versatile Phosphate Diester-Based Flame Retardant Vitrimers via Catalyst-Free Mixed Transesterification. *ACS Appl. Mater. Interfaces* **2020**, *12*, 57486–57496. [CrossRef] [PubMed]
22. Bekeshev, A.; Mostovoy, A.; Shcherbakov, A.; Zhumabekova, A.; Serikbayeva, G.; Vikulova, M.; Svitkina, V. Effect of phospho-rus and chlorine containing plasticizers on the physicochemical and mechanical properties of epoxy composites. *J. Compos. Sci.* **2023**, *7*, 178. [CrossRef]
23. Mostovoy, A.; Yakovlev, A.; Lopukhova, M. Directional control of physico-chemical and mechanical properties of epoxide composites by the addition of graphite-graphene structures. *Polym. Plast. Technol. Mater.* **2020**, *59*, 874–883. [CrossRef]
24. Bao, X.; Wu, F.; Wang, J. Thermal Degradation Behavior of Epoxy Resin Containing Modified Carbon Nanotubes. *Polymers* **2021**, *13*, 3332. [CrossRef]
25. Memon, H.; Wei, Y.; Zhang, L.; Jiang, Q.; Liu, W. An imine-containing epoxy vitrimer with versatile recyclability and its appli-cation in fully recyclable carbon fiber reinforced composites. *Compos. Sci. Technol.* **2020**, *199*, 108314. [CrossRef]
26. de Luzuriaga, A.; Martin, R.; Markaide, N.; Rekondo, A.; Cabañero, G.; Rodríguez, J.; Odriozola, I. Epoxy resin with exchange-able disulfide crosslinks to obtain reprocessable, repairable and recyclable fiber-reinforced thermoset composites. *Mater. Horiz.* **2016**, *3*, 241–247. [CrossRef]
27. Wu, K.; Zhang, Y.; Zhang, K.; Shen, M.; Hu, Y. Effect of microencapsulation on thermal properties and flammability perfor-mance of epoxy composite. *J. Anal. Appl. Pyrolysis* **2012**, *94*, 196–201. [CrossRef]
28. Luo, F.; Wu, K.; Zhu, H.; Li, X.; Lu, M. Intrinsically flame retarded foams based on melamine—Formaldehyde condensates: Thermal and mechanical properties. *Polym. Int.* **2017**, *66*, 779–786. [CrossRef]
29. Luo, F.; Wu, K.; Guo, H.; Zhao, Q.; Liang, L.; Lu, M. Effect of cellulose whisker and ammonium polyphosphate on thermal prop-erties and flammability performance of rigid polyurethane foam. *J. Therm. Anal. Calorim.* **2015**, *122*, 717–723. [CrossRef]
30. Zhang, K.; Shen, M.; Wu, K.; Liu, H.; Zhang, Y. Comparative study on flame retardancy and thermal degradation of phospho-rus- and silicon-containing epoxy resin composites. *J. Polym. Res.* **2011**, *18*, 2061–2070. [CrossRef]
31. Ai, Y.-F.; Pang, F.-Q.; Xu, Y.-L.; Jian, R.-K. Multifunctional Phosphorus-Containing Triazolyl Amine toward Self-Intumescent Flame-Retardant and Mechanically Strong Epoxy Resin with High Transparency. *Ind. Eng. Chem. Res.* **2020**, *59*, 11918–11929. [CrossRef]
32. Jian, R.-K.; Ai, Y.-F.; Xia, L.; Zhao, L.-J.; Zhao, H.-B. Single component phosphamide-based intumescent flame retardant with potential reactivity towards low flammability and smoke epoxy resins. *J. Hazard. Mater.* **2019**, *371*, 529–539. [CrossRef]
33. Kempel, F.; Schartel, B.; Marti, J.M.; Butler, K.M.; Rossi, R.; Idelsohn, S.R.; Oñate, E.; Hofmann, A. Modelling the vertical UL 94 test: Competition and collaboration between melt dripping, gasification and combustion. *Fire Mater.* **2015**, *39*, 570–584. [CrossRef]
34. Xiao, F.; Fontaine, G.; Bourbigot, S. Improvement of flame retardancy and antidripping properties of intumescent polybutyl-ene succinate combining piperazine pyrophosphate and zinc borate. *ACS Appl. Polym. Mater.* **2022**, *4*, 1911–1921. [CrossRef]
35. Sun, P.; Jia, Y.; Zhang, X.; Huang, X. Fire risk of dripping flame: Piloted ignition and soaking effect. *Fire Saf. J.* **2021**, *122*, 103360. [CrossRef]
36. Wang, Y.; Ma, L.; Yuan, J.; Zhu, Z.; Liu, X.; Li, D.; He, L.; Xiao, F. Furfural-based P/N/S flame retardant towards high-performance epoxy resins with flame retardancy, toughness, low dielectric properties and UV resistance. *Polym. Degrad. Stab.* **2023**, *212*, 110343. [CrossRef]
37. Xiao, F.; Fontaine, G.; Bourbigot, S. Recent developments in fire retardancy of polybutylene succinate. *Polym. Degrad. Stab.* **2020**, *183*, 109466. [CrossRef]
38. Hu, C.; Bourbigot, S.; Delaunay, T.; Collinet, M.; Marcille, S.; Fontaine, G. Poly(isosorbide carbonate): A 'green' char forming agent in polybutylene succinate intumescent formulation. *Compos. Part B Eng.* **2019**, *184*, 107675. [CrossRef]

39. Jimenez, M.; Bellayer, S.; Revel, B.; Duquesne, S.; Bourbigot, S. Comprehensive study of the influence of different aging scenar-ios on the fire protective behavior of an epoxy based intumescent coating. *Ind. Eng. Chem. Res.* **2013**, *52*, 729–743. [CrossRef]
40. Yan, L.; Xu, Z.; Wang, X.; Deng, N.; Chu, Z. Preparation of a novel mono-component intumescent flame retardant for enhanc-ing the flame retardancy and smoke suppression properties of epoxy resin. *J. Therm. Anal. Calorim.* **2018**, *134*, 1505–1519. [CrossRef]
41. Ali, W.; Zilke, O.; Danielsiek, D.; Salma, A.; Assfour, B.; Shabani, V.; Caglar, S.; Phan, H.M.; Kamps, L.; Wallmeier, R.; et al. Flame-retardant finishing of cotton fabrics using DOPO functionalized alkoxy- and amido alkoxysilane. *Cellulose* **2023**, *30*, 2627–2652. [CrossRef]
42. Guan, Y.-H.; Huang, J.-Q.; Yang, J.-C.; Shao, Z.-B.; Wang, Y.-Z. An Effective Way to Flame-Retard Biocomposite with Ethanolamine Modified Ammonium Polyphosphate and Its Flame Retardant Mechanisms. *Ind. Eng. Chem. Res.* **2015**, *54*, 3524–3531. [CrossRef]
43. Yang, Z.; Yao, Z.; Li, G.; Fang, G.; Nie, H.; Liu, Z.; Zhou, X.; Chen, X.; Huang, S. Sulfur-Doped Graphene as an Efficient Metal-free Cathode Catalyst for Oxygen Reduction. *ACS Nano* **2012**, *6*, 205–211. [CrossRef]
44. A Howell, B.; Daniel, Y.G. The impact of sulfur oxidation level on flame retardancy. *J. Fire Sci.* **2018**, *36*, 518–534. [CrossRef]
45. Li, W.; Xiao, L.; Wang, Y.; Huang, J.; Liu, Z.; Chen, J.; Nie, X. Thermal-induced self-healing bio-based vitrimers: Shape memory, recyclability, degradation, and intrinsic flame retardancy. *Polym. Degrad. Stab.* **2022**, *202*, 110039. [CrossRef]
46. Wang, H.; Qiu, X.; Wang, W.; Jiang, L.; Liu, H. Iron Sulfide Nanoparticles Embedded into a Nitrogen and Sulfur Co-doped Carbon Sphere as a Highly Active Oxygen Reduction Electrocatalyst. *Front. Chem.* **2019**, *7*, 855. [CrossRef] [PubMed]
47. Bourbigot, S.; Le Bras, M.; Delobel, R.; Amoureux, J.-P.; Decressain, R. Synergistic effect of zeolite in an intumescence process: Study of the carbonaceous structures using solid-state NMR. *J. Chem. Soc. Faraday Trans.* **1996**, *92*, 149–158. [CrossRef]
48. Wang, P.; Cai, Z. Highly efficient flame-retardant epoxy resin with a novel DOPO-based triazole compound: Thermal stability, flame retardancy and mechanism. *Polym. Degrad. Stab.* **2017**, *137*, 138–150. [CrossRef]

Disclaimer/Publisher's Note: The statements, opinions and data contained in all publications are solely those of the individual author(s) and contributor(s) and not of MDPI and/or the editor(s). MDPI and/or the editor(s) disclaim responsibility for any injury to people or property resulting from any ideas, methods, instructions or products referred to in the content.

polymers MDPI

Article

Constructing Heterostructured MWCNT-BN Hybrid Fillers in Electrospun TPU Films to Achieve Superior Thermal Conductivity and Electrical Insulation Properties

Yang Zhang †, Shichang Wang †, Hong Wu * and Shaoyun Guo

The State Key Laboratory of Polymer Materials Engineering, Polymer Research Institute of Sichuan University, Chengdu 610065, China; 17844626708@163.com (Y.Z.); 15166793239@163.com (S.W.); nic7702@scu.edu.cn (S.G.)
* Correspondence: wh@scu.edu.cn
† These authors contributed equally to this work.

Abstract: The development of thermally conductive polymer/boron nitride (BN) composites with excellent electrically insulating properties is urgently demanded for electronic devices. However, the method of constructing an efficient thermally conductive network is still challenging. In the present work, heterostructured multi-walled carbon nanotube-boron nitride (MWCNT-BN) hybrids were easily prepared using an electrostatic self-assembly method. The thermally conductive network of the MWCNT-BN in the thermoplastic polyurethane (TPU) matrix was achieved by the electrospinning and stack-molding process. As a result, the in-plane thermal conductivity of TPU composite films reached 7.28 W m^{-1} K^{-1}, an increase of 959.4% compared to pure TPU films. In addition, the Foygel model showed that the MWCNT-BN hybrid filler could largely decrease thermal resistance compared to that of BN filler and further reduce phonon scattering. Finally, the excellent electrically insulating properties (about 10^{12} Ω·cm) and superior flexibility of composite film make it a promising material in electronic equipment. This work offers a new idea for designing BN-based hybrids, which have broad prospects in preparing thermally conductive composites for further practical thermal management fields.

Keywords: electrostatic self-assembly; MWCNT-BN hybrids; thermal conductivity; thermal resistances; electrical insulation

check for updates

Citation: Zhang, Y.; Wang, S.; Wu, H.; Guo, S. Constructing Heterostructured MWCNT-BN Hybrid Fillers in Electrospun TPU Films to Achieve Superior Thermal Conductivity and Electrical Insulation Properties. *Polymers* **2024**, *16*, 2139. https://doi.org/10.3390/polym16152139

Academic Editor: Alexey L. Iordanskii

Received: 2 July 2024
Revised: 23 July 2024
Accepted: 25 July 2024
Published: 27 July 2024

Copyright: © 2024 by the authors. Licensee MDPI, Basel, Switzerland. This article is an open access article distributed under the terms and conditions of the Creative Commons Attribution (CC BY) license (https://creativecommons.org/licenses/by/4.0/).

1. Introduction

With the increased integration and high power density of electronic devices, military equipment, and communication instruments, heat accumulation continues to be a critical limiting factor in the narrow space [1–3]. Highly efficient heat dissipation is essential to ensure the service life and stability of the electronic components. Polymer-based thermally conductive materials have been extensively investigated due to their cost-effectiveness, lightweight, and ease of processing [4–6]. Nevertheless, the polymer materials are limited in practical applications because they possess a low intrinsic thermal conductivity (TC, <0.50 W m^{-1} K^{-1}) [7–9]. In the past decades, aluminum nitride [10], alumina [11], boron nitride [12–15], silicon carbide [16], metal [17], carbon nanotube [18,19], and graphene [20,21] have served as thermally conductive fillers to enhance the TC of polymer materials. Particularly, BN has attracted tremendous attention in the development of thermally conductive yet electrically insulating composites because of its extremely high thermal conductivity (400 W m^{-1} K^{-1}) and relatively wide band gap (5.5–6.4 eV) [22–24]. However, due to its chemical inertness, the improvement of the obtained TC is usually limited due to the undesirable dispersion of BN in polymer composites [25].

Interfacial optimization has become a trend through the surface treatment of BN to improve the TC of polymer-based materials. The common way is to modify the BN surface through covalent bonding and noncovalent interactions, which could promote the BN dispersion and the phonon transfer from the BN to the polymer matrix, and thus enhance the TC of polymer composites [26–29]. For example, Liu et al. prepared (3-aminopropyl) triethoxysilane (APTES) functionalized BN, and the epoxy/BN composites reached 5.86 W m^{-1} K^{-1} at a filler content of 40 wt% [30]. Additionally, Liu et al. functionalized a noncovalent BN with polydopamine, which could improve interfacial compatibility and uniform dispersion of BN in the polymer matrix [31]. Although the thermal conductivity can be effectively enhanced by optimizing the dispersion of BN, it is still a challenge to further increase thermal conductivity by constructing an efficient thermally conductive network of polymer composites.

Previous studies have demonstrated that hybrid thermally conductive fillers prepared with different dimensions and morphologies have been widely used to fabricate multifunctional polymer composites, which also provide a practical method for constructing efficient thermally conductive networks of polymer composites [32–37]. Silver nanoparticles (AgNPs) are introduced onto the surface of BNNS to construct efficient thermally conductive pathways, resulting in a higher TC of the composite [38]. Similarly, Qian et al. developed a novel hetero-structured BN/ZC hybrid filler through in-situ growth of CNTs onto the surface of BNNSs, and thus the TC of epoxy composites reached 3.21 W m^{-1} K^{-1} at a filler content of 9.86 vol%, which is increased by about 13.4 times compared to that of pure epoxy [39]. As a result, it is found that the hybrid thermally conductive fillers can construct more thermal conduction pathways, ultimately improving the TC of thermally conductive composites.

Furthermore, it is known to all that the TC of the composite can be further enhanced by the orientation of single or multi-hybrid fillers with anisotropic properties among the polymer matrix [40–43]. The commonly used strategies are spin-coating [30], solvent-casting [44,45], hot-pressing [46,47], vacuum filtration [48,49], electrospinning [50,51], etc. Recently, advanced functional composites based on electrospinning have attracted much attention. It has also been found that the anisotropic thermally conductive fillers tend to be oriented along the nanofiber, which can help to achieve higher TC. Shen et al. also aligned BNNS in a horizontal direction by electrospinning technology to build efficient thermal conduction pathways, and the obtained composite possessed a high in-plane TC of 10.9 W m^{-1} K^{-1} [52]. Accordingly, it can be concluded that combining the orientation and the hybrid thermally conductive fillers can enhance the TC of the composites effectively.

In this study, one-dimensional polyethyleneimine (PEI) modified multi-walled carbon nanotubes (MWCNTs) were successfully introduced onto the polydopamine (PDA) functionalized BN via the electrostatic self-assembly method. Then, the obtained MWCNT-BN hybrid fillers were oriented along the fiber during the electrospinning and hot-pressing processes. The MWCNT-BN hybrid fillers significantly construct efficient thermal conduction pathways compared to single BN fillers, displaying superior TC enhancement in TPU composite films. Furthermore, it is worth noting that the addition of a low concentration of MWCNTs can not only greatly enhance the TC of the composite but also maintain electrical insulation properties. The excellent TC, electrical insulation, and mechanical properties of composite ensure its great potential applications in electronic devices, such as computers and drones.

2. Materials and Methods

2.1. Materials

TPU (Elastogran S85A) was bought from BASF Co., Ltd. (Shanghai, China). MWCNTs-COOH were obtained from Chengdu Organic Chemicals Co., Ltd. (Chengdu, China). The MWCNTs-COOH (MWCNT) had the following features: a length of 10−30 μm and a diameter of 5–15 nm. Hexagonal BN (average particle size of 1 μm) was obtained from Yingkou Tianyuan Chemical Research Institute Co., Ltd. (Hubei, China). Tris (hydroxymethyl) aminomethane (Tris), dopamine hydrochloride, and PEI (Mw = 1800) were purchased from Adamas Reagent Co., Ltd. (Shanghai, China). Ethanol, N,N-dimethylformamide (DMF), and tetrahydrofuran (THF) were purchased from Chengdu Kelong Chemicals Co. Ltd. (Chengdu, China).

2.2. *Preparation of MWCNT-BN Hybrids*

2.2.1. Preparation of PDA Modified BN (PDA-BN)

Typically, 2.0 g of BN powder was dispersed in 400 mL of tris-buffer solution (10 mM, pH 8.5) by ultrasonic for 1 h, then 0.8 g of dopamine hydrochloride was added and stirred for 12 h at ambient temperature, followed by vacuum-assisted filtration, washing, and drying to obtain BN-PDA.

2.2.2. Preparation of PEI Functionalized MWCNT (PEI-MWCNT)

Firstly, 5 g of PEI was added to deionized water (10 mg/mL) and this was stirred at room temperature until the PEI was completely dissolved. Subsequently, 0.5 g of MWCNTs were added into the PEI solution and it was ultrasonically processed for 1 h and stirred at 80 °C for another 12 h. Finally, the PEI-MWCNT was filtered, washed five times, and dried at 80 °C overnight.

2.2.3. Preparation of MWCNT-BN Hybrids

According to Figure 1a, 1.2 g of BN-PDA was dispersed in 100 mL of deionized water for 0.5 h by ultrasound, and, at the same time, a certain mass of PEI-MWCNT was dispersed in 50 mL of deionized water. Then, the PEI-MWCNT dispersion was gradually added to the BN-PDA dispersion and it was stirred at room temperature for 12 h. Finally, the product was centrifuged, washed, and dried to obtain MWCNT-BN hybrids.

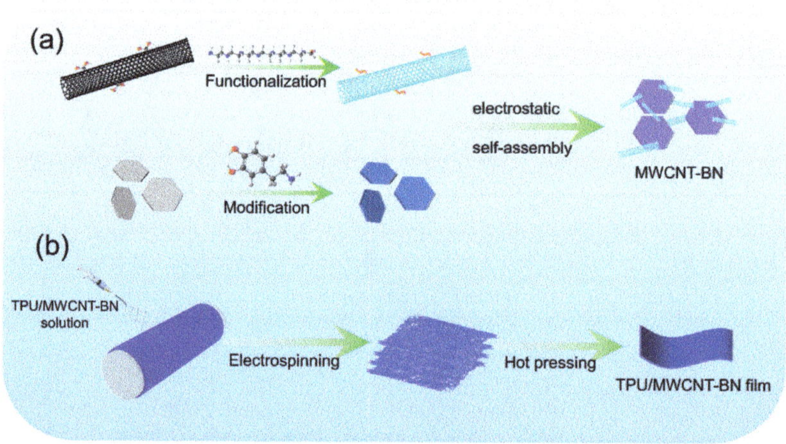

Figure 1. (**a**) Schematic illustration of fabrication for (**a**) hetero-structured MWCNT−BN fillers and (**b**) thermally conductive TPU/MWCNT−BN composite films.

2.3. *Fabrication of the TPU/MWCNT-BN Composite Films*

According to Figure 1b, a certain mass of MWCNT-BN hybrids was dispersed in a mixed solvent of DMF and THF (1:1 by weight) under an ultrasonic process for 0.5 h. TPU was added to the above solution until it was completely dissolved. Then, the electrospinning process was carried out according to our previous work. Finally, the TPU/MWCNT-BN electrospun fibers were hot pressed at 170 °C and 20 MPa for 10 min to obtain TPU/MWCNT-BN composite films. For comparison, the pure TPU film and TPU/BN composite films were also fabricated through the same procedures.

2.4. Characterization

The surface morphology of BN, PDA-BN, MWCNT-BN, and the electrospun fibers and the cross-section of the composite films were observed by scanning electron microscopy (SEM, FEI Quanta 250, Hillsboro USA). The morphologies of the TPU fibers, TPU/BN fibers, and TPU/MWCNT-BN fibers were observed by transmission electron microscopy (TEM, FEI Tecnai G2 F20, Hillsboro USA) at 200 kV. Atomic force microscopy (AFM, AFM+, Anasys Instruments, Hillsboro USA) images were recorded to measure the thickness of BN. Fourier transform-infrared (FT-IR) spectroscopies of MWCNT and PEI-MWCNT were obtained using the FT-IS10 spectrometer (FT-IR, Thermo Fisher Scientific, Massachusetts USA) over a frequency range of 400–4000 cm^{-1}. Raman spectroscopies of BN, PDA-BN, MWCNT, PEI-MWCNT, and MWCNT-BN were obtained using a micro-Raman spectrometer (InVia Reflex, London UK) with laser light focusing at a wavelength of 532 nm. The crystal structures of composite films were characterized by X-ray diffraction (XRD, Ultima IV, Rigaku, Tokyo Japan) in the 2θ range of 2–90°. The chemical compositions of BN, PDA-BN, MWCNT, PEI-MWCNT, and MWCNT-BN were determined by X-ray photoelectron spectroscopy (XPS, AXIS Ultra DLD (Kratos) Kyoto Japan). Thermal gravimetric (TG) analyses were performed using TG209F1 equipment (Selb Germany) under a nitrogen atmosphere at a heating rate of 10 °C/min. A Zetasizer Nano ZS analyzer (Malvern Instruments Ltd., Malvern, UK) was used to characterize the zeta potentials of BN, PDA-BN, MWCNT, PEI-MWCNT, and MWCNT-BN at pH = 7.0. The in-plane or through-plane TC of composite films was calculated by multiplication of thermal diffusivity (α, mm^2/s), density (ρ, g/cm^3), and specific heat capacity (Cp, J/(g·K)), i.e., TC = $\alpha \times Cp \times \rho$. A laser flash apparatus (LFA, NETZSCH LFA 467, Selb Germany) was used to measure the α of composite films at 25 °C. The water displacement method was performed to determine the ρ of composite films. Differential scanning calorimetry (DSC, TA Instruments, New Castle, DE, USA) was used to measure the Cp of composite films at 25 °C. An infrared camera (T620, FLIR Systems Inc., Boston, MA, USA) recorded the variations in the surface temperature of the LED device. A high resistance meter (KEITHLEY, 6487 Oregon, OR, USA) was carried out to measure the volume electrical resistivity of the composite films. The tensile strength at the breaking points of the composite samples was determined using an INSTRON 5966 (Shanghai China) electronic tensile strength meter with a tensile rate of 200 mm/min.

3. Results and Discussion

3.1. Characterization of MWCNT-BN Hybrids

Bulk BN displays a similar multi-layered structure to graphite and is also a promising thermally conductive filler [50], as shown in Figure 2a. First, BN is easily and successfully modified through dopamine chemistry, which effectively improves the chemical activity of inert BN. As seen in Figure S1, the color of the BN aqueous solution changes from white to gray after modification because of the formation of a PDA layer on the BN surface. At the same time, the dispersibility of PDA-BN is also significantly improved compared to BN after modification. Figure 2b shows the SEM image of PDA-BN with an average size of 0.47 μm (Figure S2), and the dimension has not changed compared to the pure BN. The MWCNT-BN hybrid is obtained by the electrostatic self-assembly method. The MWCNT shows the uniform size and high aspect ratio in Figure 2f. The AFM image and the corresponding thickness curve of BN are shown in Figure 2d,e, and the thickness is about 10 nm. Thermally conductive pathways can be easily constructed using the electrostatic self-assembly method due to the different dimensions and morphologies of BN and MWCNT. As can be seen from Figure 2c, the MWCNT with a high aspect ratio is anchored on the surface of BN without agglomerating and can act as an effective "bridge" to significantly reinforce the thermal conduction pathways (Figure S3).

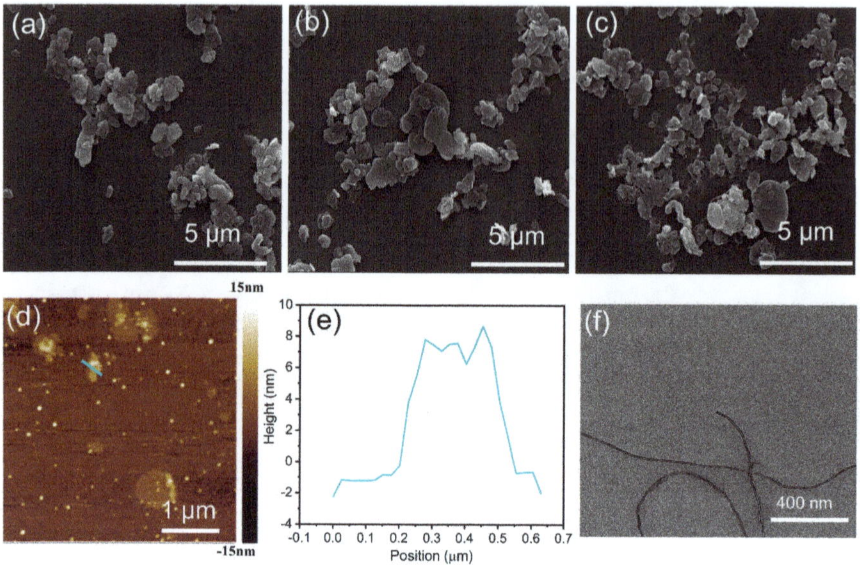

Figure 2. SEM images of (**a**) BN, (**b**) PDABN, and (**c**) MWCNTBN. (**d**) AFM image and (**e**) height profile of the BN. (**f**) TEM image of MWCNT.

FTIR measurement is carried out to prove effective modification of BN and MWCNT. Compared to MWCNT, a small new peak at 1671 cm^{-1} is assigned to O=C-NRR′C=O stretching vibration in the amide I band or C=C skeletal vibration. The absorbance peaks at 3440 cm^{-1} are attributed to the -NH and -OH groups, which are much stronger than MWCNT, indicating that PEI is successfully grafted onto the surface of MWCNT (Figure 3a,b). Obviously, the dispersibility of MWCNT and PEI-MWCNT is significantly improved compared with CNT after modification (Figure S4). As shown in Figure 3c, the absorption peaks at 811 and 1376 cm^{-1} are attributed to the in-plane stretching vibration and out-of-plane bending vibration of the B—N bond, respectively [51]. After PDA modification, the peak corresponding to oxygen-containing groups becomes stronger, suggesting the successful preparation of PDA-BN. The successfully modified BN and MWCNT are further validated using XPS analysis (Figure 3f–i). From Figure 3f, the peak intensities of C 1s and O 1s for PDA-BN are remarkably increased compared with pristine BN, demonstrating that the PDAs are successfully coated on the surface of BN (Table S1). Furthermore, the N 1s spectrum of BN is deconvoluted into 398.25 eV, corresponding to the N-B bond (Figure S5). In comparison, a new signal of 399.8 eV occurs in the N 1s spectrum of PDA-BN, which is assigned to the existence of N-C, further indicating that DA is successfully polymerized on the surface of BN (Figure 3g). The two characteristic peaks at 532.2 eV (O 1s) and 284.9 eV (C 1s) are detected for MWCNT, while a new peak of N 1s for PEI-MWCNT exists at 399.3 eV, which is fitted to 399.5 eV (N-H) and 401.0 eV (N-C) due to the PEI being grafted onto the surface of MWCNT (Figure 3h). As presented in Figure 3i, it is also observed that the N 1s peak of MWCNT-BN is deconvoluted into three peaks, N-C (400.8 eV), N-H (399.1 eV), and N-B (397.9 eV), demonstrating the successful preparation of MWCNT-BN.

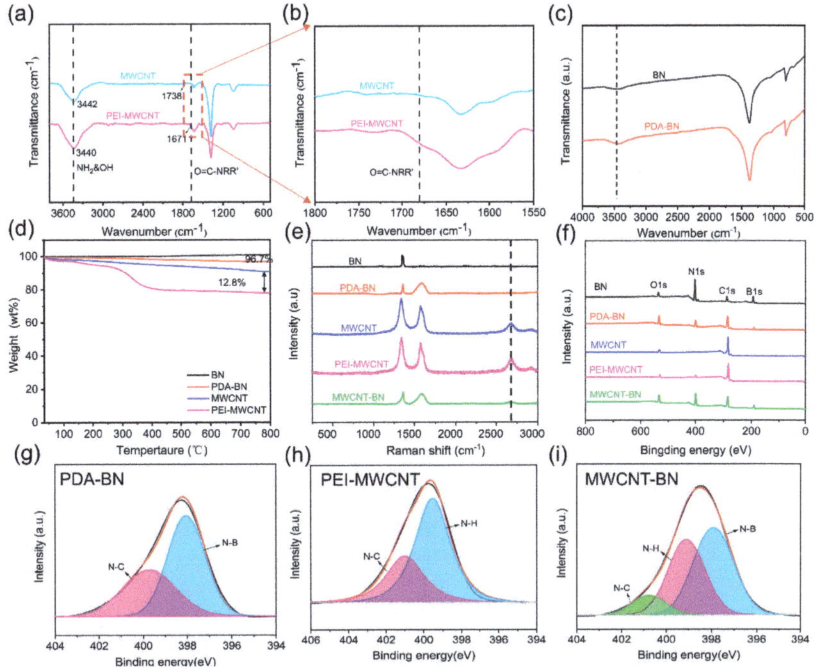

Figure 3. (**a**,**b**) FT-IR spectra of MWCNT and PEI−MWCNT. (**c**) FT-IR spectra of BN and PDA−BN. (**d**) TGA spectra of BN, PDA−BN, MWCNT, and PEI−MWCNT. (**e**) Raman results of BN, PDA−BN, MWCNT, PEI−MWCNT, and MWCNT-BN. (**f**) XPS patterns of BN, PDA-BN, MWCNT, PEI−MWCNT, and MWCNT−BN. (**g**–**i**) N1s XPS core level of PDA−BN, PEI−MWCNT, and MWCNT−BN.

Furthermore, the TGA curves in Figure 3d show that the content of the PDA layer is calculated at 3.3%. Additionally, the PEI grafted on the surface of MWCNTs is about 12.8%. As shown in Figure S6, the zeta potential of BN is estimated to be −32.8 mV due to the hydroxyl groups attached to the surface of BN, resulting in a negatively charged state. Meanwhile, the zeta potential of PDA-BN is about −37.2 mV, mainly because more hydroxyl groups are generated on the surface of BN, which is consistent with the results of FTIR measurement. In addition, the zeta potential of MWCNT in deionized water is −41.0 mV, indicating the impossible self-assembly with the PDA-BN that has a negative charge. Thus, the PEI is successfully grafted on the surface of MWCNT and the resultant PEI-MWCNT displays a zeta potential value of 33.3 mV. The photographs of PDA-BN before and after mixing with PEI-MWCNT are shown in Figure S7. The PDA-BN exhibits a homogeneous dispersion in deionized water, while the PDA-BN fillers quickly precipitate to the bottom due to the strong electrostatic adsorption between them. Raman spectra analyses are performed to further confirm the surface chemical compositions of fillers (Figure 3e). The characteristic peak at 1358 cm^{-1} is ascribed to high-frequency intralayer E2g tangential mode for BN. A new signal at 1587 cm^{-1} occurs after functionalization, which is attributed to the deformation of catechol from PDA. Furthermore, due to the PDA layer attached to the BN surface, the intensity of the BN peak at 1358 cm^{-1} is merely weakened. The MWCNT has three characteristic peaks, i.e., D peak (1343 cm^{-1}), G peak (1575 cm^{-1}), and G′ peak (2678 cm^{-1}). The G peak is caused by the vibration of the sp_2 carbon atom in the carbon nanotubes, and its location and intensity are closely related to the structure and chirality of the carbon nanotubes. The intensity of the G peak is high, generally occurring at about 1500 cm^{-1}. The D peak is due to disordered vibrations in carbon nanotubes and usually occurs at about 1300 cm^{-1}. The intensity of the D peak is low, but it is related to the diameter, length, and chirality of the carbon nanotubes. The G′ is caused by the Van der Waals force between the layers of carbon nanotubes, which occurs at about 2000 cm^{-1}. The G′ peak intensity is low and sensitive to the chirality, diameter, and number of layers of carbon nanotubes. Meanwhile, the PEI-MWCNT has the same peak and intensity as MWCNT, indicating that the PEI has little impact on the crystal structure of MWCNT. Due to the small amount of PEI-MWCNT in the MWCNT-BN hybrid filler, the D and G peaks of PEI-MWCNT are not easily observed, while a weak G′ peak can be observed in the Raman spectra of MWCNT-BN, further proving the successful preparation of MWCNT-BN.

3.2. Preparation and Morphology of TPU/MWCNT-BN Composite Films

The SEM images of pure TPU and TPU/MWCNT-BN electrospun fibers are shown in Figure 4. The pure TPU fibers exhibit a smooth and uniform surface, illustrating that TPU can be completely dissolved in the DMF/THF solution (Figure 4a). The average diameter of TPU fibers is about 1.29 μm. With the MWCNT-BN concentrations rising, the average diameter of the electrospun fibers increases gradually. The primary reason is that the introduction of MWCNT-BN fillers increases the viscosity of the electrospinning solution. Meanwhile, due to the instability of the jet, the uniformity of the fiber diameter decreases and the MWCNT-BN also occurs on the surface of electrospun TPU/MWCNT-BN fibers. Figure S8 shows the TEM images of TPU, TPU/40 BN, and TPU/40 MWCNT-BN electrospun fibers, respectively. There are no fillers observed in pure TPU fiber, while the BN is better dispersed in the TPU fibers. Furthermore, the heterostructured MWCNT-BN is also oriented in the TPU fibers, which is better for forming thermal conduction pathways than BN. The cross-sectional morphologies of the TPU/MWCNT-BN composite film after hot-pressing are shown in Figure S9. It can be clearly observed that the MWCNT-BN exhibits obvious orientation along the in-plane direction. To visually demonstrate the flexibility of TPU/40MWCNT-BN composite films, Figure 4f shows the optical photographs of TPU/MWCNT-BN composite films, which could be bent, folded, curled, and fixed into the shape of a small windmill without obvious cracks. Meanwhile, the composite film could withstand the pulling of 500 g without any breakages, demonstrating its application prospects in electronic equipment.

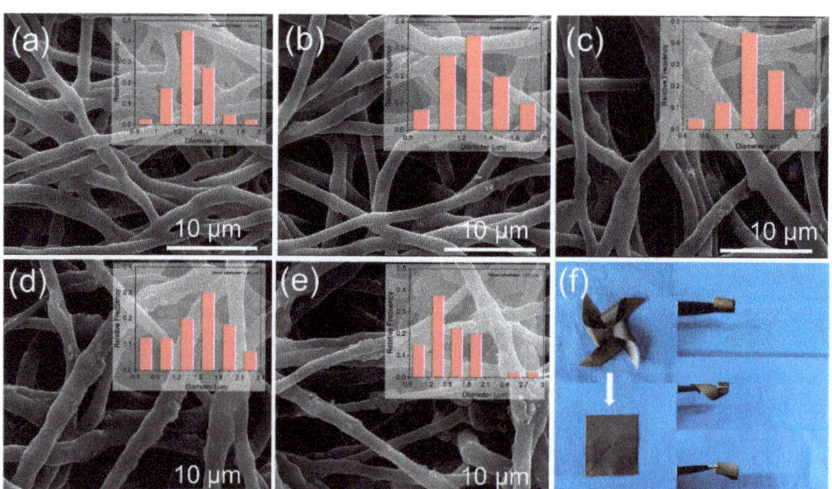

Figure 4. (**a**–**e**) SEM images of TPU/MWCNT-BN fibers with different filler content of 0 wt%,10 wt%, 20 wt%, 30 wt%, and 40 wt%. The insets in (**a**–**e**) represent the diameter distribution of fibers. (**f**) Photographs of the TPU/MWCNT-BN composite films.

3.3. Properties of the TPU/MWCNT-BN Composite Films

Figure 5 shows the in-plane TC and out-of-plane TC of TPU/BN and TPU/MWCNT-BN composite films with different content, respectively. From Figure 5a, pure TPU film shows a low in-plane TC of 0.68 W m^{-1} K^{-1} due to phonon scattering between the interfaces, defects, and impurities [52]. It is evident that the introduction of BN can sharply enhance the in-plane TC of TPU/BN composite film. Furthermore, as the BN content increases, the in-plane TC of TPU/BN composite films shows a notably increasing tendency. The TPU/40 BN composite film has an excellent in-plane TC of 6.04 W m^{-1} K^{-1}, which is higher than that of pure TPU. This is mainly because the BNs are mostly arranged along the horizontal direction of the composite. In addition, it is worth noting that the TPU/MWCNT-BN composite films exhibit higher thermal conductivity compared with TPU/BN composite films at the same filler content, which is primarily due to the fact that the heterogeneous MWCNT-BN is more prone to constructing continuous thermally conductive pathways than BN. One can see that the TPU/40 MWCNT-BN composite film shows ultrahigh in-plane TC of 7.28 W m^{-1} K^{-1}, which is 20.5% higher than the TPU/40 BN composite film. Each composite film displays a much lower out-of-plane TC, as shown in Figure 5b. The through-plane TC of TPU/40 MWCNT-BN composite film increases to 0.54 W m^{-1} K^{-1} when the mass fraction of MWCNT-BN is 40 wt%, which is 3.86 times larger than that of pure TPU film (0.14 W m^{-1} K^{-1}). Similarly, the through-plane TC of TPU/MWCNT-BN composite films is also higher than that of TPU/BN composite films, the through-plane TC value of TPU/40 BN composite film is as high as 0.44 W m^{-1} K^{-1}.

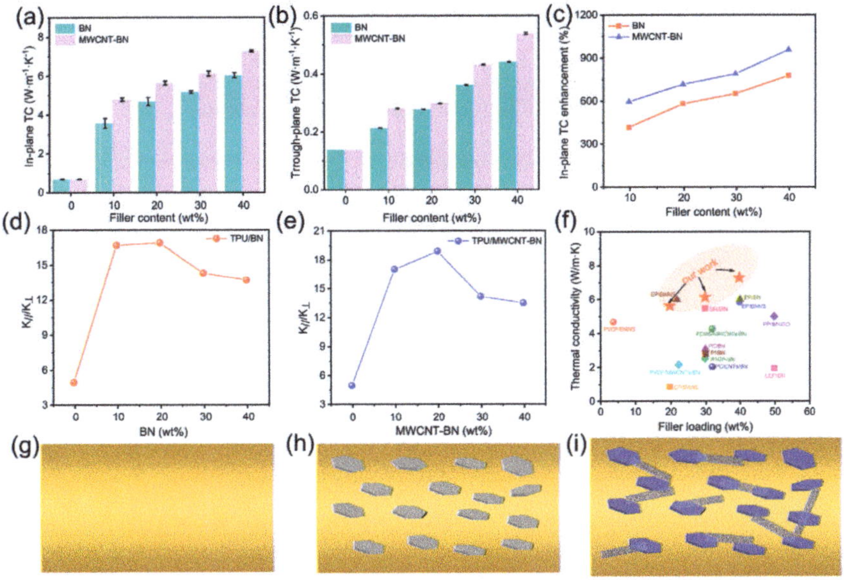

Figure 5. (**a**) In-plane TC and (**b**) through-plane TC of TPU/BN and TPU/MWCNT−BN composite films with different filler content. (**c**) The in-plane TC enhancement of TPU/BN and TPU/MWCNT−BN composite films with different filler content. (**d**,**e**) Thermal conductivity anisotropy of TPU/BN and TPU/MWCNT−BN composite films, respectively. (**f**) Comparison of in-plane TC of TPU/MWCNT−BN composite films and other thermally conductive composite reported previously. (**g**–**i**) Schematic diagram of thermally conductive pathways of TPU film, TPU/BN composite film and TPU/MWCNT−BN composite film.

Due to the typical anisotropic feature of h-BN, the orientation of BN in the TPU matrix can be studied by the XRD investigation, which is a vital factor in improving the TC of the composite. The (002) plane and (100) plane of BN are assigned to the BN oriented in in-plane and through-plane directions, respectively. Therefore, the in-plane orientation degree of BN in the TPU matrix can be expressed by the intensity ratio I (002)/I (100). As presented in Figure S10, the peak intensities of (002), (100), and (004) increase as the MWCNT-BN content increases. The high I (002)/I (100) peak ratio was obtained in the TPU/MWCNT-BN composite films, indicating that the MWCNT-BN displays a well-oriented structure in the TPU matrix because of the electrospinning and hot-pressing methods, which could greatly enhance the in-plane TC of TPU composite films.

In addition, the TC enhancement (TCE) is calculated to compare the in-plane TC of TPU composite films with that of pure TPU films, which is defined as follows:

$$TCE = \frac{TC - TC_0}{TC_0} \tag{1}$$

where TC and TC_0 are the in-plane TC of TPU composite films with different filler content and pure TPU film, respectively. As shown in Figure 5c, the TCE value of TPU/40BN composite film is as high as 778.6% when the BN content reaches 40%. In contrast, at the same filler content, the TCE value is greatly enhanced to 959.4% for TPU/40MWCNT-BN, indicating that the MWCNT-BN shows better TC enhancement than BN. Furthermore, the anisotropic TC of the composite films is also studied. It is found that the in-plane TC of TPU/BN and TPU/40MWCNT-BN is higher than that of the through-plane TC, displaying a high anisotropic index (AI, the ratio values of $K_{\parallel}$ and $K_{\perp}$), as shown in Figure 5d,e. This is due to the well-dispersed BN and MWCNT-BN which are highly oriented along the in-plane direction because of the synergy effect of electrospinning and hot pressing method, reducing the phonon scattering in this direction. Figure 5f and Table S2 summarize the previously reported in-plane TC of BN-based thermally conductive materials. Compared with other polymer materials, TPU/MWCNT-BN composite films exhibit considerable competitiveness. The in-plane TC of TPU/40MWCNT-BN composite film is up to 7.28 W m^{-1} K^{-1}, which is greater than most thermally conductive materials previously reported. The superior in-plane TC is due to the formation of heterostructured MWCNT-BN hybrid fillers as well as the oriented microstructure of the composite films. To further make sense of the possible mechanism of enhanced thermal conductivity, the thermal conduction pathways in the fibers are graphically displayed in Figure 5g–i.

Due to the severe phonon scattering between the interfaces, defects, and impurities, thermal conduction pathways are difficult to construct in pure TPU, which exhibits a low value of TC. When BN is dispersed into the TPU matrix, the stacked BN in the TPU fibers is conducive to phonon transfer, however, it is not enough to construct continuous thermal conduction pathways. Nevertheless, heterostructured MWCNT-BN hybrid fillers more easily form highly efficient phonon transmission pathways. Therefore, phonons can be efficiently transported in the TPU fibers, resulting in a significantly improved TC.

The Agari model is employed to elucidate the effects of MWCNT on constructing thermal conduction pathways in TPU composite films. Generally, the in-plane TC of composite films can be effectively predicted by the following equation:

$$\log K_C = V_f C_2 \log K_f + (1 - V_f)\log(C_1 K_m) \tag{2}$$

where K_c, K_m, and K_f are the TCs of the composite films, matrix, and fillers, respectively; K_f is the volume fraction of fillers, C_1 represents that the fillers influence the polymer crystallinity; C_2 suggests that the ability of the fillers to form thermal conduction pathways, $0 \le C_2 \le 1$. Figure 6a,b show the fitted curves of TPU/BN and TPU/MWCNT-BN composite films, respectively. It is worth noting that the obtained C_2 value of TPU/BN composite films is 0.248, while a higher C_2 value is obtained (0.287) for TPU/MWCNT-BN composite films, indicating that the MWCNT more easily forms efficient heat conduction pathways (Table S3). In addition, the Foygel model is further fitted to calculate the interfacial thermal resistances (R) of TPU composite films. The fitting curves and calculation results of the TPU composite films are shown in Figure S11 and Table S4, and the calculation process is exhibited in detail in the supporting information. The interfacial thermal resistances of TPU/BN and TPU/MWCNT-BN are 1.17×10^6 K/W and 7.43×10^6 K/W, showing a decrease of 36.5% due to the addition of MWCNT. Therefore, the fitting results of the Foygel model illustrate that the preparation of MWCNT-BN hybrids is conducive to reducing the R-value of TPU composite films.

To further demonstrate the potential application of composite films as thermal management materials in electronic devices, pure TPU film, TPU/40 BN composite film, and TPU/40 MWCNT-BN composite film with the same size are sandwiched between a light-emitting diode (LED) chip and a heat sink, respectively (Figure 6c). Surface temperature variations of the LED chip with a working time of 120 s are recorded by the infrared thermal imager. As shown in Figure 6d, the surface temperature of the LED chip on TPU/40 BN and TPU/40 MWCNT-BN composite films increased to 79.2 °C and 76.8 °C, while that of pure TPU film reached as high as 86.7 °C. It is worth noting that the LED chip shows the slowest heating rate when the TPU/40 MWCNT-BN composite film is sandwiched between an LED chip and a heat sink (Figure 6e). The infrared thermal image results also verify a better thermal conduction ability of TPU/40 MWCNT-BN composite film than TPU film and TPU/40 BN composite film. Therefore, the flexible TPU/40 MWCNT-BN composite films have great potential in the field of electronic components to ensure service life and stability.

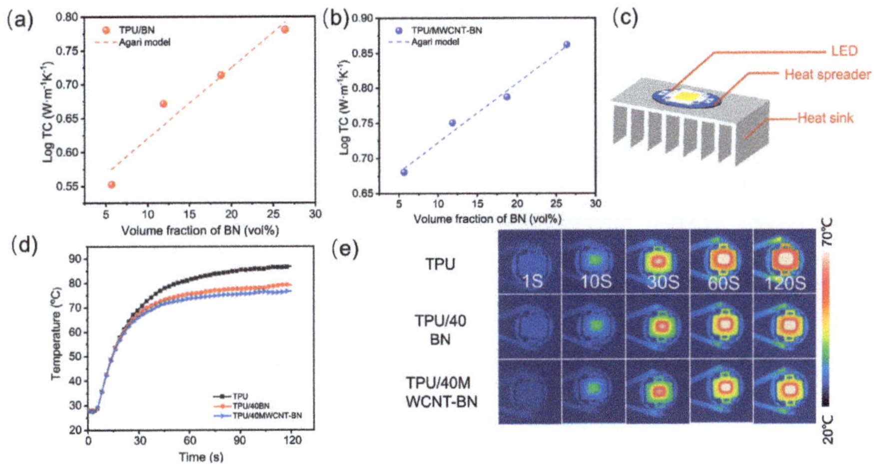

Figure 6. Agari model fitting curve of composite films: (**a**) TPU/BN and (**b**) TPU/MWCNT−BN. (**c**) Schematic illustration of the thermal conduction in an LED bulb. (**d**) Surface temperature versus time of the LED chip when TPU, TPU/40BN, and TPU/40MWCNT−BN as a heat spreader between LED chips and the heat sink and (**e**) corresponding IR images.

In addition to high TC, the excellent electrical insulation property is also important for thermal management applications to ensure the efficient operation and safety of electronic equipment. Figure 7d shows the volume resistivity results of pure TPU, TPU/BN, and TPU/MWCNT-BN composite films. The volume resistivity of TPU/20 BN composite film is higher than pure TPU film, which is mainly due to the introduction of BN with ultra-high resistivity. Nevertheless, the TPU/MWCNT-BN composite films exhibit lower volume electrical resistivities because of the addition of MWCNT. It is found that the volume resistivity of TPU/40 MWCNT-BN composite film is as high as 2×10^{12} Ω·cm, which is far beyond the standard of insulating materials (10^9 Ω·cm). The TPU/40 MWCNT-BN composite film is connected to an LED in series under a 3 V external voltage (Figure S12). It can be seen that the LED lamps failed to light up when integrated with the TPU/40 MWCNT-BN composite film, further visually proving the insulating property of the composite films. Therefore, it is believed that the composite films can be satisfied with the fields of high electrical insulation requirements. The TGA curves of TPU film and TPU/MWCNT composite films are shown in Figure S13. It is found that both the TPU film and TPU/MWCNT-BN composite films show excellent thermal stability. Moreover, with an increasing MWCNT-BN content, the decomposition temperatures of 10 wt% ($T_{10\%}$) and 5 wt% ($T_{5\%}$) are enhanced, suggesting that the thermal stability of the composite film is increased. The mechanical properties of TPU/MWCNT-BN composite films are characterized, and the results are shown in Figure 7a. The tensile strength and elongation at the break of pure TPU film are 67.4 MPa and 682.9%, respectively. It can be clearly observed that the tensile strength and elongation at the break of composite films decrease gradually with increasing MWCNT-BN loadings (Figure 7b,c). In general, more interfaces between the thermally conductive filler and the polymer matrix are formed, which act as stress concentration points, leading to the decline in the mechanical properties of composite films. A tensile strength of 25.44 MPa is obtained when the MWCNT-BN loading is up to 40 wt%. At the same time, the composite films maintain superior flexibility, which is also vital in the field of electric equipment and electronic devices.

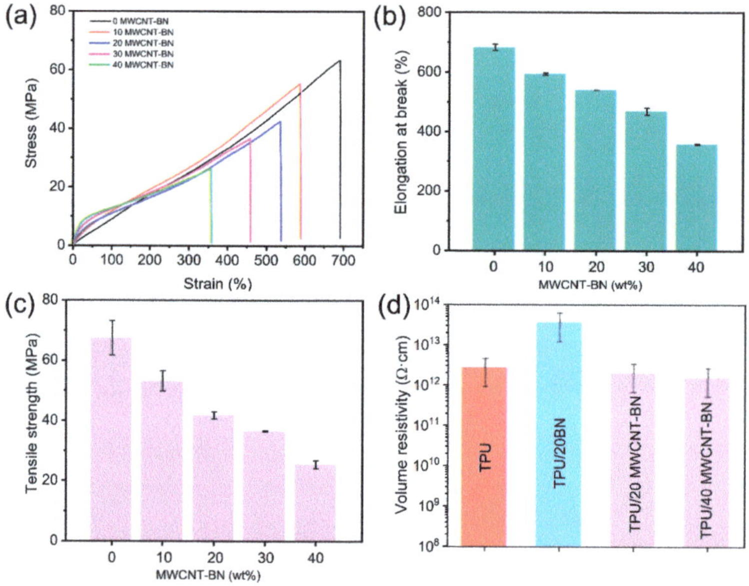

Figure 7. (**a**) Tensile stress–strain curves of the TPU film and TPU/MWCNT-BN composite films. (**b**) Elongation at break and (**c**) tensile strength of TPU/MWCNT-BN composite films. (**d**) Volume resistivity of TPU, TPU/BN, and TPU/MWCNT-BN composite films.

4. Conclusions

In conclusion, heterostructured MWCNT-BN hybrid fillers were prepared via a facile electrostatic self-assembly method, and highly thermally conductive and electrically insulating TPU/MWCNT-BN composite films were successfully fabricated by electrospinning and hot-pressing. The positively charged MWCNT was anchored on the surface of BN and served as a bridge to connect the uncontacted BN, endowing the TPU/40 MWCNT-BN composite film with a high in-plane TC of 7.28 W m^{-1} K^{-1}, which is an increase of 959.4% compared to pure TPU films. Moreover, infrared thermal images demonstrated that the TPU/40 MWCNT-BN composite film exhibited strong heat dissipation capability for LED lamps. Furthermore, the composite film possessed superior heat resistance, outstanding electrical insulation properties (about 10^{12} Ω·cm), and perfect flexibility, indicating that it can be widely used as a thermal interface material in high power density electrical equipment and electronic devices.

Supplementary Materials: The following supporting information can be downloaded at: https://www.mdpi.com/article/10.3390/polym16152139/s1, Figure S1: Optical images of BN and PDA-BN.; Figure S2: Lateral size statistics of BN, PDA-BN and MWCNT-BN. Figure S3: SEM image of MWCNT-BN. Figure S4: Optical images of CNT, MWCNT and PEI-MWCNT. Figure S5: N1s XPS core level of BN and MWCNT. Figure S6: Zeta potential of BN, PDA-BN, MWCNT, PEI-MWCNT and MWCNT-BN. Figure S7: Optical images the fabrication process of MWCNT-BN. Figure S8: TEM images of (a) TPU, (b) TPU/40BN and (c) TPU/40MWCNT-BN electrospun fibers, respectively. Figure S9: Cross-sectional morphologies of the TPU/40MWCNT-BN composite film. Figure S10: XRD patterns of TPU/MWCNT-BN composite film with different filler content. Figure S11: Foygel model fitting curves of composite fims: (a) TPU/BN; (b) TPU/MWCNT-BN. Figure S12: Insulation exhibition of TPU/40MWCNT-BN compsoite films. Figure S13: TGA curves of TPU/MWCNT-BN composite film with different filler content. Table S1: Atomic weight ratio of BN, PDA-BN, MWCNT, PEI-MWCNT and MWCNT-BN; Table S2: Comparison of in-plane TC of TPU/MWCNT-BN composite films and other thermally conductive composite reported previously. Table S3: Agari model fitting parameters of composite films. Table S4: Foygel model fitting parameters of composite films.

Author Contributions: Conceptualization, S.W. and H.W.; methodology, S.W.; investigation, Y.Z.; data curation, Y.Z.; writing—original draft preparation, Y.Z.; writing—review and editing, S.W. and Y.Z. and H.W.; visualization, Y.Z.; supervision, H.W. and S.G.; project administration, S.G.; funding acquisition, H.W. and S.G. All authors have read and agreed to the published version of the manuscript.

Funding: This work was funded by the Fundamental Research Funds for the Central Universities and State Key Laboratory of Polymer Materials Engineering.

Institutional Review Board Statement: Not applicable.

Informed Consent Statement: Not applicable.

Data Availability Statement: Data from this study are available upon request from the corresponding authors.

Conflicts of Interest: The authors declare no conflicts of interest.

References

1. Mehra, N.; Mu, L.; Ji, T.; Yang, X.; Kong, J.; Gu, J.; Zhu, J. Thermal transport in polymeric materials and across composite interfaces. *Appl. Mater. Today* **2018**, *12*, 92–130. [CrossRef]
2. Guo, Y.; Ruan, K.; Shi, X.; Yang, X.; Gu, J. Factors affecting thermal conductivities of the polymers and polymer composites: A review. *Compos. Sci. Technol.* **2020**, *193*, 108134. [CrossRef]
3. Zhang, F.; Feng, Y.; Feng, W. Three-dimensional interconnected networks for thermally conductive polymer composites: Design, preparation, properties, and mechanisms. *Mater. Sci. Eng. R Rep.* **2020**, *142*, 100580. [CrossRef]
4. Kim, G.H.; Lee, D.; Shanker, A.; Shao, L.; Kwon, M.S.; Gidley, D.; Kim, J.; Pipe, K.P. High thermal conductivity in amorphous polymer blends by engineered interchain interactions. *Nat. Mater.* **2015**, *14*, 295–300. [CrossRef]
5. Chen, J.; Huang, X.; Sun, B.; Jiang, P. Highly Thermally Conductive Yet Electrically Insulating Polymer/Boron Nitride Nanosheets Nanocomposite Films for Improved Thermal Management Capability. *ACS Nano* **2019**, *13*, 337–345. [CrossRef]
6. Chen, Y.; Zhang, H.; Chen, J.; Guo, Y.; Jiang, P.; Gao, F.; Bao, H.; Huang, X. Thermally Conductive but Electrically Insulating Polybenzazole Nanofiber/Boron Nitride Nanosheets Nanocomposite Paper for Heat Dissipation of 5G Base Stations and Transformers. *ACS Nano* **2022**, *16*, 14323–14333. [CrossRef]
7. Henry, A.; Chen, G. High thermal conductivity of single polyethylene chains using molecular dynamics simulations. *Phys. Rev. Lett.* **2008**, *101*, 235502. [CrossRef] [PubMed]
8. Pan, G.; Yao, Y.; Zeng, X.; Sun, J.; Hu, J.; Sun, R.; Xu, J.-B.; Wong, C.-P. Learning from Natural Nacre: Constructing Layered Polymer Composites with High Thermal Conductivity. *ACS Appl. Mater. Interfaces* **2017**, *9*, 33001–33010. [CrossRef] [PubMed]
9. Wang, X.; Wu, P. Fluorinated Carbon Nanotube/Nanofibrillated Cellulose Composite Film with Enhanced Toughness, Superior Thermal Conductivity, and Electrical Insulation. *ACS Appl. Mater. Interfaces* **2018**, *10*, 34311–34321. [CrossRef]
10. Zhang, Z.; Zhao, L.; Li, Z.; Hua, Y.; Jiao, T.; Lin, L.; Zhao, Y.; Han, B.; Deng, Q.; Hu, N. Construction and the performance of the combination heat transfer mode of Dots-and-Plane based on liquid metal and AlN. *Appl. Surf. Sci.* **2023**, *624*, 157166. [CrossRef]

11. Chen, Y.; Hou, X.; Liao, M.; Dai, W.; Wang, Z.; Yan, C.; Li, H.; Lin, C.-T.; Jiang, N.; Yu, J. Constructing a "pea-pod-like" alumina-graphene binary architecture for enhancing thermal conductivity of epoxy composite. *Chem. Eng. J.* **2020**, *381*, 122690. [CrossRef]
12. Tian, X.; Wu, N.; Zhang, B.; Wang, Y.; Geng, Z.; Li, Y. Glycine functionalized boron nitride nanosheets with improved dispersibility and enhanced interaction with matrix for thermal composites. *Chem. Eng. J.* **2021**, *408*, 127360. [CrossRef]
13. Jiang, F.; Song, N.; Ouyang, R.; Ding, P. Wall Density-Controlled Thermal Conductive and Mechanical Properties of Three-Dimensional Vertically Aligned Boron Nitride Network-Based Polymeric Composites. *ACS Appl. Mater. Interfaces* **2021**, *13*, 7556–7566. [CrossRef]
14. Hu, Z.; Wang, S.; Liu, Y.; Qu, Z.; Tan, Z.; Wu, K.; Shi, J.; Liang, L.; Lu, M. Constructing a Layer-by-Layer Architecture to Prepare a Transparent, Strong, and Thermally Conductive Boron Nitride Nanosheet/Cellulose Nanofiber Multilayer Film. *Ind. Eng. Chem. Res.* **2020**, *59*, 4437–4446. [CrossRef]
15. Cheng, W.-C.; Hsieh, Y.-T.; Liu, W.-R. Enhanced Thermal Conductivity of Silicone Composites Filled with Few-Layered Hexagonal Boron Nitride. *Polymers* **2020**, *12*, 2072. [CrossRef]
16. Liang, C.; Hamidinejad, M.; Ma, L.; Wang, Z.; Park, C.B. Lightweight and flexible graphene/SiC-nanowires/ poly(vinylidene fluoride) composites for electromagnetic interference shielding and thermal management. *Carbon* **2020**, *156*, 58–66. [CrossRef]
17. Zhao, H.; Huang, Y.; Han, Y.; Yun, J.; Wang, X.; Jin, L.; Zheng, Y.; Chen, L. Flexible and lightweight porous polyether sulfone/Cu composite film with bidirectional differential structure for electromagnetic interference shielding and heat conduction. *Chem. Eng. J.* **2022**, *440*, 135919. [CrossRef]
18. Zhang, X.; Wu, K.; Liu, Y.; Yu, B.; Zhang, Q.; Chen, F.; Fu, Q. Preparation of highly thermally conductive but electrically insulating composites by constructing a segregated double network in polymer composites. *Compos. Sci. Technol.* **2019**, *175*, 135–142. [CrossRef]
19. Wang, S.; Feng, D.; Guan, H.; Guo, Y.; Liu, X.; Yan, C.; Zhang, L.; Gu, J. Highly efficient thermal conductivity of polydimethylsiloxane composites via introducing "Line-Plane"-like hetero-structured fillers. *Compos. Part A Appl. Sci. Manuf.* **2022**, *157*, 106911. [CrossRef]
20. Guo, H.; Zhao, H.; Niu, H.; Ren, Y.; Fang, H.; Fang, X.; Lv, R.; Maqbool, M.; Bai, S. Highly Thermally Conductive 3D Printed Graphene Filled Polymer Composites for Scalable Thermal Management Applications. *ACS Nano* **2021**, *15*, 6917–6928. [CrossRef]
21. Wu, Z.; Xu, C.; Ma, C.; Liu, Z.; Cheng, H.M.; Ren, W. Synergistic Effect of Aligned Graphene Nanosheets in Graphene Foam for High-Performance Thermally Conductive Composites. *Adv. Mater* **2019**, *31*, e1900199. [CrossRef]
22. Cao, C.; Xue, Y.; Liu, Z.; Zhou, Z.; Ji, J.; Song, Q.; Hu, Q.; Fang, Y.; Tang, C. Scalable exfoliation and gradable separation of boric-acid-functionalized boron nitride nanosheets. *2D Mater.* **2019**, *6*, 035014. [CrossRef]
23. Ding, J.-H.; Zhao, H.-R.; Yu, H.-B. High-yield synthesis of extremely high concentrated and few-layered boron nitride nanosheet dispersions. *2D Mater.* **2018**, *5*, 045015. [CrossRef]
24. Teng, C.; Su, L.; Chen, J.; Wang, J. Flexible, thermally conductive layered composite films from massively exfoliated boron nitride nanosheets. *Compos. Part A Appl. Sci. Manuf.* **2019**, *124*, 105498. [CrossRef]
25. Morishita, T.; Okamoto, H. Facile Exfoliation and Noncovalent Superacid Functionalization of Boron Nitride Nanosheets and Their Use for Highly Thermally Conductive and Electrically Insulating Polymer Nanocomposites. *ACS Appl. Mater Interfaces* **2016**, *8*, 27064–27073. [CrossRef]
26. Sahu, M.; Narashimhan, L.; Prakash, O.; Raichur, A.M. Noncovalently Functionalized Tungsten Disulfide Nanosheets for Enhanced Mechanical and Thermal Properties of Epoxy Nanocomposites. *ACS Appl. Mater Interfaces* **2017**, *9*, 14347–14357. [CrossRef]
27. Jiang, Y.; Shi, X.; Feng, Y.; Li, S.; Zhou, X.; Xie, X. Enhanced thermal conductivity and ideal dielectric properties of epoxy composites containing polymer modified hexagonal boron nitride. *Compos. Part A Appl. Sci. Manuf.* **2018**, *107*, 657–664. [CrossRef]
28. Wang, Y.; Wu, W.; Drummer, D.; Liu, C.; Shen, W.; Tomiak, F.; Schneider, K.; Liu, X.; Chen, Q. Highly thermally conductive polybenzoxazine composites based on boron nitride flakes deposited with copper particles. *Mater. Des.* **2020**, *191*, 108698. [CrossRef]
29. Wang, X.; Sun, M.; Wang, R.; Jiao, L.; Bian, H.; Dai, H. Promoting h-BN dispersion in cellulose-based composite by lignosulfonate for regulatable effectual thermal management. *Mater. Des.* **2022**, *214*, 110379. [CrossRef]
30. Liu, Z.; Li, J.; Liu, X. Novel Functionalized BN Nanosheets/Epoxy Composites with Advanced Thermal Conductivity and Mechanical Properties. *ACS Appl. Mater. Interfaces* **2020**, *12*, 6503–6515. [CrossRef]
31. Chen, L.; Li, K.; Li, B.; Ren, D.; Chen, S.; Xu, M.; Liu, X. Enhanced thermal conductivity of benzoxazine nanocomposites based on non-covalent functionalized hexagonal boron nitride. *Compos. Sci. Technol.* **2019**, *182*, 107741. [CrossRef]
32. Zhang, S.; Wu, J.; Liu, J.; Yang, Z.; Wang, G. Ti3C2Tx MXene Nanosheets Sandwiched between Ag Nanowire-Polyimide Fiber Mats for Electromagnetic Interference Shielding. *ACS Appl. Nano Mater.* **2021**, *4*, 13976–13985. [CrossRef]
33. Fan, X.; Wang, F.; Gao, Q.; Zhang, Y.; Huang, F.; Xiao, R.; Qin, J.; Zhang, H.; Shi, X.; Zhang, G. Nature inspired hierarchical structures in nano-cellular epoxy/graphene-Fe_3O_4 nanocomposites with ultra-efficient EMI and robust mechanical strength. *J. Mater. Sci. Technol.* **2022**, *103*, 177–185. [CrossRef]
34. An, D.; Cheng, S.; Zhang, Z.; Jiang, C.; Fang, H.; Li, J.; Liu, Y.; Wong, C.-P. A polymer-based thermal management material with enhanced thermal conductivity by introducing three-dimensional networks and covalent bond connections. *Carbon* **2019**, *155*, 258–267. [CrossRef]

35. Kang, S.; Kang, T.-H.; Kim, B.S.; Oh, J.; Park, S.; Choi, I.S.; Lee, J.; Son, J.G. 2D reentrant micro-honeycomb structure of graphene-CNT in polyurethane: High stretchability, superior electrical/thermal conductivity, and improved shape memory properties. *Compos. Part B-Eng.* **2019**, *162*, 580–588. [CrossRef]
36. Xue, Y.; Li, X.; Wang, H.; Zhao, F.; Zhang, D.; Chen, Y. Improvement in thermal conductivity of through-plane aligned boron nitride/silicone rubber composites. *Mater. Des.* **2019**, *165*, 107580. [CrossRef]
37. Zhou, Y.; Yu, S.; Niu, H.; Liu, F. Synergistic Improvement in Thermal Conductivity of Polyimide Nanocomposite Films Using Boron Nitride Coated Copper Nanoparticles and Nanowires. *Polymers* **2018**, *10*, 1412. [CrossRef] [PubMed]
38. Chen, C.; Xue, Y.; Li, Z.; Wen, Y.; Li, X.; Wu, F.; Li, X.; Shi, D.; Xue, Z.; Xie, X. Construction of 3D boron nitride nanosheets/silver networks in epoxy-based composites with high thermal conductivity via in-situ sintering of silver nanoparticles. *Chem. Eng. J.* **2019**, *369*, 1150–1160. [CrossRef]
39. Li, X.; Wu, B.; Li, Y.; Alam, M.M.; Chen, P.; Xia, R.; Lin, C.T.; Qian, J. Construction of Oriented Interconnected BNNS Skeleton by Self-Growing CNTs Leading High Thermal Conductivity. *Adv. Mater. Interfaces* **2021**, *8*, 2001910. [CrossRef]
40. Zhang, H.; Ju, X.; Jiang, H.; Yang, D.; Wei, R.; Hu, W.; Lu, X.; Zhu, M. Implementation of high thermal conductivity and synaptic metaplasticity in vertically-aligned hexagonal boron nitride-based memristor. *Sci. China-Mater.* **2024**, *67*, 1907–1914. [CrossRef]
41. Niu, H.; Guo, H.; Kang, L.; Ren, L.; Lv, R.; Liu, L.; Bashir, A.; Bai, S. Highly thermally conductive and soft thermal interface materials based on vertically oriented boron nitride film. *Compos. Part B-Eng.* **2024**, *272*, 111219. [CrossRef]
42. Wang, Z.-G.; Wang, J.-Y.; Guo, Z.-H.; Wang, W.-H.; Zhou, S.-Y.; Li, Y.; Jia, L.-C.; Niu, J.-H.; Zhong, G.-J.; Xu, L.; et al. Highly and Uniformly Thermal Conductive Phase Change Composites by Constructing the Bidirectionally Oriented and Interconnected Boron Nitride Nanosheet Network. *Ind. Eng. Chem. Res.* **2024**, *63*, 9438–9446. [CrossRef]
43. Liu, Y.; Gong, W.; Liu, X.; Fan, Y.; He, A.; Nie, H. Enhancing Thermal Conductivity in Polymer Composites through Molding-Assisted Orientation of Boron Nitride. *Polymers* **2024**, *16*, 1169. [CrossRef]
44. Yue, Y.; Yang, X.; Yang, K.; Li, K.; Liu, Z.; Wang, F.; Zhang, R.; Huang, J.; Wang, Z.; Zhang, L.; et al. Highly Thermally Conductive Super-Aligned Boron Nitride Nanotube Films for Flexible Electronics Thermal Management. *Acs Appl. Mater. Interfaces* **2024**, *16*, 33971–33980. [CrossRef] [PubMed]
45. Shen, B.; Zhai, W.; Zheng, W. Ultrathin Flexible Graphene Film: An Excellent Thermal Conducting Material with Efficient EMI Shielding. *Adv. Funct. Mater.* **2014**, *24*, 4542–4548. [CrossRef]
46. Yang, X.; Guo, Y.; Han, Y.; Li, Y.; Ma, T.; Chen, M.; Kong, J.; Zhu, J.; Gu, J. Significant improvement of thermal conductivities for BNNS/PVA composite films via electrospinning followed by hot-pressing technology. *Compos. Part B-Eng.* **2019**, *175*, 107070. [CrossRef]
47. Yu, C.; Gong, W.; Tian, W.; Zhang, Q.; Xu, Y.; Lin, Z.; Hu, M.; Fan, X.; Yao, Y. Hot-pressing induced alignment of boron nitride in polyurethane for composite films with thermal conductivity over 50 $Wm^{-1} K^{-1}$. *Compos. Sci. Technol.* **2018**, *160*, 199–207. [CrossRef]
48. Wang, Y.; Zhang, Z.; Li, T.; Ma, P.; Zhang, X.; Xia, B.; Chen, M.; Du, M.; Liu, T.; Dong, W. Artificial Nacre Epoxy Nanomaterials Based on Janus Graphene Oxide for Thermal Management Applications. *ACS Appl. Mater. Interfaces* **2020**, *12*, 44273–44280. [CrossRef] [PubMed]
49. Yan, Q.; Dai, W.; Gao, J.; Tan, X.; Lv, L.; Ying, J.; Lu, X.; Lu, J.; Yao, Y.; Wei, Q.; et al. Ultrahigh-Aspect-Ratio Boron Nitride Nanosheets Leading to Superhigh In-Plane Thermal Conductivity of Foldable Heat Spreader. *ACS Nano* **2021**, *15*, 6489–6498. [CrossRef]
50. Chen, J.; Wei, H.; Bao, H.; Jiang, P.; Huang, X. Millefeuille-Inspired Thermally Conductive Polymer Nanocomposites with Overlapping BN Nanosheets for Thermal Management Applications. *ACS Appl Mater Interfaces* **2019**, *11*, 31402–31410. [CrossRef]
51. Yang, G.; Zhang, X.; Pan, D.; Zhang, W.; Shang, Y.; Su, F.; Ji, Y.; Liu, C.; Shen, C. Highly Thermal Conductive Poly(vinyl alcohol) Composites with Oriented Hybrid Networks: Silver Nanowire Bridged Boron Nitride Nanoplatelets. *Acs Appl. Mater. Interfaces* **2021**, *13*, 32286–32294. [CrossRef]
52. Wu, N.; Yang, W.; Che, S.; Sun, L.; Li, H.; Ma, G.; Sun, Y.; Liu, H.; Wang, X.; Li, Y. Green preparation of high-yield and large-size hydrophilic boron nitride nanosheets by tannic acid-assisted aqueous ball milling for thermal management. *Compos. Part A Appl. Sci. Manuf.* **2023**, *164*, 107266. [CrossRef]

Disclaimer/Publisher's Note: The statements, opinions and data contained in all publications are solely those of the individual author(s) and contributor(s) and not of MDPI and/or the editor(s). MDPI and/or the editor(s) disclaim responsibility for any injury to people or property resulting from any ideas, methods, instructions or products referred to in the content.

polymers

MDPI

Article

Analyzing Temperature Distribution Patterns on the Facing and Backside Surface: Investigating Combustion Performance of Flame-Retardant Particle Boards Using Aluminum Hypophosphite, Intumescent, and Magnesium Hydroxide Flame Retardants

Fangya Pan [1], Hongyu Jia [1], Yuxiang Huang [1], Zhilin Chen [1,2], Shanqing Liang [1] and Peng Jiang [1,*]

1 Research Institute of Wood Industry, Chinese Academy of Forestry, Haidian District, Beijing 100091, China; panfangya@aliyun.com (F.P.); jiahongyu9417@163.com (H.J.); yxhuang@caf.ac.cn (Y.H.); chenzhilin@caf.ac.cn (Z.C.); liangsq@caf.ac.cn (S.L.)

2 Co-Innovation Center of Efficient Processing and Utilization of Forest Resources, Nanjing Forestry University, Nanjing 210037, China

* Correspondence: jiangpeng@caf.ac.cn

Abstract: Particle boards are manufactured through a hot pressing process using wood materials (natural polymer materials) and adhesive, which find common usage in indoor decorative finishing materials. Flame-retardant particleboard, crucial for fire safety in such applications, undergoes performance analysis that includes assessing temperature distribution across its facing surface and temperature increase on the backside surface during facade combustion, yielding critical insights into fire scenario development. In this study, a compact flame spread apparatus is utilized to examine the flame retardancy and combustion behavior of particle boards, with a specific emphasis on the application of cost-effective flame retardants, encompassing aluminum hypophosphite (ALHP), an intumescent flame retardant (IFR) comprising ammonium polyphosphate (APP), melamine (MEL), and Dipentaerythritol (DPE), alongside magnesium hydroxide (MDH), and their associated combustion characteristics. The $D_{300^\circ C}$ values, representing the vertical distance from the ignition point (IP) to $P_{300^\circ C}$ (the temperature point at 300 °C farthest from IP), are measured using a compact temperature distribution measurement platform. For MDH/PB, APP + MEL + DPE/PB, and ALHP/PB samples, the respective $D_{300^\circ C}$ values of 145.79 mm, 117.81 mm, and 118.57 mm indicate reductions of 11.11%, 28.17%, and 27.71%, compared to the untreated sample's value of 164.02 mm. The particle boards treated with ALHP, IFR, and MDH demonstrated distinct flame-retardant mechanisms. MDH/PB relied on the thermal decomposition of MDH to produce MgO and H_2O for flame retardancy, while APP + MEL + DPE/PB achieved flame retardancy through a cross-linked structure with char expansion, polyphosphate, and pyrophosphate during combustion. On the other hand, ALHP/PB attained flame retardancy by reacting with wood materials and adhesives, forming a stable condensed P-N-C structure. This study serves as a performance reference for the production of cost-effective flame-resistant particleboards and offers a practical method for assessing its fire-resistant properties when used as a decorative finishing material on facades in real fire situations.

Keywords: particle board; combustion performance; temperature distribution; flame-retardant mechanism

check for updates

Citation: Pan, F.; Jia, H.; Huang, Y.; Chen, Z.; Liang, S.; Jiang, P. Analyzing Temperature Distribution Patterns on the Facing and Backside Surface: Investigating Combustion Performance of Flame-Retardant Particle Boards Using Aluminum Hypophosphite, Intumescent, and Magnesium Hydroxide Flame Retardants. *Polymers* **2023**, *15*, 4479. https://doi.org/10.3390/polym15234479

Academic Editor: Bob Howell

Received: 13 October 2023
Revised: 16 November 2023
Accepted: 19 November 2023
Published: 21 November 2023

Copyright: © 2023 by the authors. Licensee MDPI, Basel, Switzerland. This article is an open access article distributed under the terms and conditions of the Creative Commons Attribution (CC BY) license (https://creativecommons.org/licenses/by/4.0/).

1. Introduction

Particle board (PB) is a type of engineered wood or non-wood material made from natural polymer materials with wood shavings as its primary raw material. It can be manufactured with or without the use of adhesive by a process involving paving, pre-pressing, and hot pressing to form sturdy panels. The most commonly used particle board typically consists of three layers: two surface layers made of fine wood shavings and a core layer made of coarser shavings. Urea-formaldehyde resin is the prevalent adhesive used in the production of particle board [1]. As a product of resourceful utilization, particle board exhibits remarkable qualities, including high strength, resistance to deformation, a flat surface, and stable physical and mechanical properties. This makes it a versatile material widely employed in furniture manufacturing, construction, packaging, and interior decoration of vehicles and ships, as well as in the production of household appliances, playing a crucial role in addressing the serious shortage of wood resources. In Austria (Kronospan), Canada (Norbord), China (Wanhua), and various other countries worldwide, the particle board industry has experienced rapid growth, with production steadily increasing year after year. Taking China as a prime example, since 2015, the particle board industry has garnered unprecedented attention and experienced substantial growth, primarily fueled by the booming custom furniture sector, indicating a promising developmental trajectory [2]. By the end of 2020, China boasted 348 particle board production lines, capable of producing an annual volume of 36.91 million cubic meters, contributing to the global annual production capacity of particle board, which reached 114.3 million m^3.

In recent years, there has been a surge in fires caused by highly flammable interior decoration materials, significantly affecting people's livelihoods and posing substantial risks to their personal safety and property. To address this issue, relevant laws and regulations impose higher requirements on the combustion performance grade of decoration materials used in various parts of most buildings and locations. Particle board, being both inflammable and flame-retardant, plays a critical role in fire prevention efforts. Initiating flame retardant measures from the source material and applying them during the production of particle board hold significant importance in preventing fires and slowing down their spread at various building sites. By incorporating flame-retardant particle board, we can effectively enhance fire safety and reduce the rapid propagation of fires.

Currently, research on flame-retardant particle board primarily focuses on combustion theory [3], flame retardant mechanisms [4,5], flame retardant agents [6–10], flame retardant treatment processes [4,11,12], and flame retardant property detection methods, among others. Previous studies have explored various types of flame retardants for particle board, including boron-based [7], metal-based [13], phosphonitryl [6], inorganic minerals [12], biomass flame retardants, intumescent flame retardants [8], nano flame retardants [14], and so on. Metal-based flame retardants are widely used in flame-retarding wood materials due to their cost-effectiveness. [15–18]. In particular, a flame-retardant particle board was developed using dried oil palm as the raw material, incorporating aluminum hydroxide and magnesium hydroxide (MDH) as flame retardants, achieving limiting oxygen indices of 28.55% and 27.95%, respectively [19]. The application of intumescent flame retardants was initially adopted for plastic products and later extended to wood materials, yielding improved flame retardancy results. Notably, ammonium polyphosphate (APP) intumescent flame retardants have shown excellent performance when applied to wood-based panels, exhibiting low heat release rates, minimal total heat release, limited smoke emission, a high limiting oxygen index, and overall superior flame retardancy properties [20]. Furthermore, aluminum hypophosphite (ALHP) is frequently employed as a plastic flame retardant [21–25], and when adding 10 wt% ALHP to polylactic acid (PLA) in the UL-94 test, it can achieve a V0 rating [26]. As for testing methods, several approaches are available to assess the flame-retardant performance of particle board, including ignitability, smoke generation, flame propagation, thermal analysis, and more [27–29]. Nevertheless, there remains a requirement for testing methodologies tailored to accurately discern the surface temperature distribution and the corresponding rise in rear temperature for flame-retardant particle board. Therefore, conducting research on the flame-retardant effects and mechanisms of particle board treated with different flame retardants and exploring novel flame-retardant detection methods holds great significance.

The combustion reactions of various parts within artificial boards differ, primarily categorized into the surface gas phase area, and materials progressing from the surface to the interior undergo carbonization, degradation, dehydration, and heating processes, respectively [30]. Thus, this research focuses on enhancing the flame resistance of particle boards by incorporating various flame retardants, achieved by considering the combustion behavior of artificial board materials and the properties of flame retardants in the preparation of a series of flame-retardant particle boards. In this study, samples of flame-retardant particle boards containing ALHP, IFR, and MDH have been prepared, and a combination of a compact temperature distribution measurement platform and a cone calorimeter was employed to evaluate the combustion behavior of flame-retardant particle boards, such as temperature distribution distance, temporal back temperature variations, heat release rate (HRR), and smoke production rate (SPR). Furthermore, a flame-retardant mechanism in the condensed phase was proposed, drawing upon the analysis of char residues following combustion and utilizing scanning electron microscopy (SEM), Raman spectrometry, and X-ray photoelectron spectroscopy (XPS). This study is geared towards offering valuable insights into the cost-effective flame-retardant mechanisms displayed by various flame-retardant particle boards, introducing novel flame-retardant testing approaches, and supplying theoretical backing for the practical implementation of flame retardants in wood-based materials, particularly in real fire scenarios.

2. Materials and Methods

2.1. Materials

In this experiment, wood shavings with a moisture content of 2–3% (measured in percentage) are utilized as the primary raw material. Urea-formaldehyde resin adhesive (UF), along with aluminum hypophosphate (Shandong Taixing New Material Co., Ltd., Jinan, China), DPE (Shanghai Aladdin Biochemical Technology Co., Ltd., Shanghai, China), APP, MEL and MDH (both from Shandong Changsheng flame retardant new material Co., Ltd., Dezhou, China), is incorporated into the experimental setup. The wood shavings and urea-formaldehyde resin adhesive are sourced from Shandong Xingang Enterprise Group Co., Ltd. (Linyi, China) Table 1 provides essential details regarding the adhesives employed in this study.

Table 1. Urea—formaldehyde resin adhesive basic information.

Samples	Solid Content (%)	Viscosity (Pa·s)	pH
Surface adhesive	59.2	39.2	7.7
Core layer adhesive	65.0	150.0	8.0

2.2. Manufacture of Particle Board

In this experiment, ordinary untreated and flame-retardant particle boards with a thickness of 18 cm and a density of 675 kg/m^3 are prepared. The manufacturing process involved weighing, glue mixing, lay-up, and hot pressing, as shown in Figure 1. The shavings consisted of 34% on the surface and 66% in the core. The resin content is 10 wt% on the surface and 10.5 wt% in the core. The compositions of the untreated and flame-retardant particle boards are listed in Table 2. The hot pressing process included prepressing at 3 MPa and room temperature for 20 s, followed by hot pressing at 175 °C and 2.5 MPa for 8 min. The flame retardant is added in the same proportion to both the surface and core layers of the flame-retardant particle boards. The formulations for the untreated and flame-retardant particle boards are listed in Table 3.

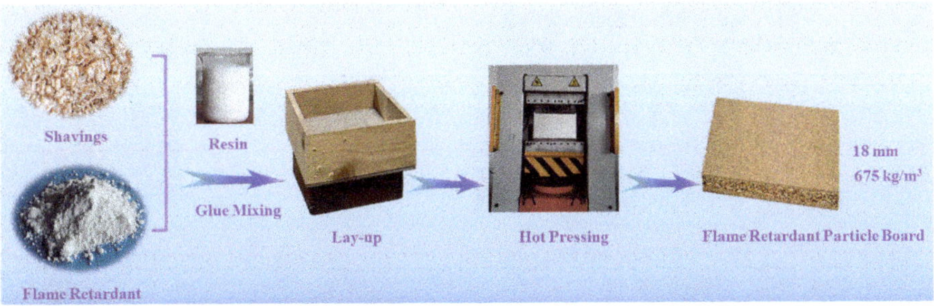

Figure 1. Schematic diagram of particle board manufacturing.

Table 2. The compositions of untreated and flame-retardant particle boards.

Samples	Surface Layer (34 wt%)			Core Layer (66 wt%)		
	Shavings (wt%)	Resin (wt%)	Flame Retardant (wt%)	Shavings (wt%)	Resin (wt%)	Flame Retardant (wt%)
Particle Board	100	10	-	100	10.5	-
MDH/PB	100	10	10	100	10.5	10
APP + MEL + DPE/PB	100	10	10	100	10.5	10
ALHP/PB	100	10	10	100	10.5	10

Table 3. Formulations of untreated and flame-retardant particle boards.

Samples	MDH (wt%)	Intumescent Flame Retardant (IFR) (wt%)			ALHP (wt%)
		APP (wt%)	MEL (wt%)	DPE (wt%)	
Particle Board	-	-	-	-	-
MDH/PB	10	-	-	-	-
APP + MEL + DPE/PB	-	5	2.5	2.5	
ALHP/PB	-	-	-	-	10

2.3. Characterization

Limiting oxygen index (LOI): An oxygen index tester (JF-3, Jiangning Analysis Instrument Company, Nanjing, China) is used for measurement. The specimen size is 120 mm × 10 mm × thickness according to ASTM D2863-17 [31].

Combustion performance: The combustion behavior of the sample is evaluated using a cone calorimeter (Fire Testing Technology Ltd., East Grinstead, UK) in accordance with the ASTM E1354-17 standard [32]. The sample size is 100 mm × 100 mm × thickness. To conduct the experiment, the samples are positioned horizontally and subjected to a heat flux of 50 kW/m^2. They are carefully wrapped in aluminum foil, leaving the upper surface exposed to the heater. The wrapped samples are then placed on a ceramic backing board, maintaining a distance of 35 mm from the cone base.

The distribution of the temperature: The experiment is conducted with a compact temperature distribution measurement platform, as shown in Figure 2, with the sample positioned perpendicular to the horizontal plane. The ignition point (IP) is defined as the intersection of the horizontal axis line (H), situated 3 cm above the lower edge of the sample facing the heat source, and the vertical axis line (V), passing through the center point of the material (at a point 3 cm above the midpoint of the lower edge of the sample facing the heat source).

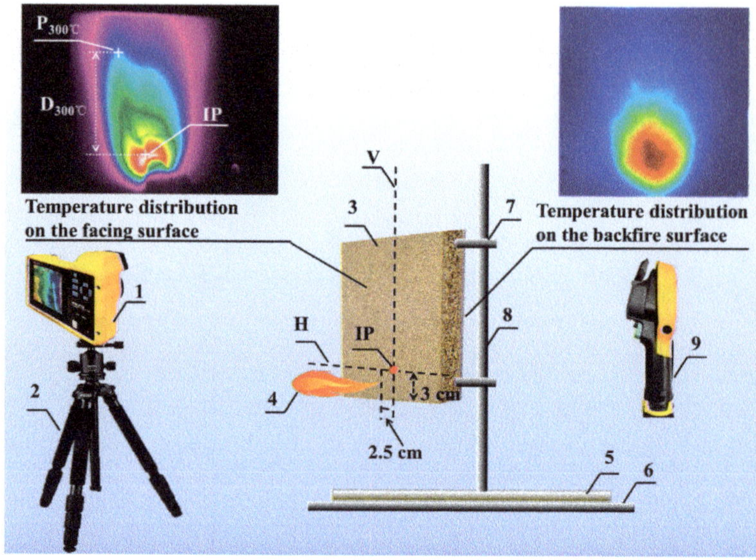

Figure 2. Schematic diagram of a small-scale temperature distribution measurement platform. 1—infrared thermal imager, 2—tripod, 3—sample, 4—spray gun, 5—gypsum board, 6—support pedestal, 7—fixture, 8—support, 9—hand-held temperature detector.

The particle board was continuously lit, and a spray gun was employed, positioned 2.5 cm away from the ignition point. The resulting flame had a length of 10 cm. During the experiment, an infrared thermal imager (TiX580, Fluke, Shanghai, China) is employed to capture video footage of the facing surface, while a handheld temperature detector (Ti100, Fluke, Shanghai, China) is utilized to photograph the backside surface, with the experiment concluding when the maximum temperature of the backside surface reached 275 °C. The sample size in the experiment is 20 cm × 20 cm × 1.8 cm, and to prevent fire hazards arising from the high temperatures during combustion, a piece of gypsum board is placed underneath the support. Smart View 4.4 software is used to obtain the 300 °C temperature point farthest from IP ($P_{300°C}$) on the facing surface of particle board and the highest temperature point on the backside surface. The vertical distance from $P_{300°C}$ to IP ($D_{300°C}$) is calculated, and Origin 8.5 software is employed for data analysis to obtain information such as $D_{300°C}$ and the temporal evolution of backside temperature.

Physical and mechanical performance: The properties of particle board samples, including thickness, density, internal bonding strength, surface bonding strength, modulus of rupture (MOR), and modulus of elasticity (MOE), are determined following the guidelines outlined in the GB/T 17657-2013 standard [33] titled "Test methods of evaluating the properties of wood-based panels and surface-decorated wood-based panels." This standard provides specific procedures and protocols for measuring and evaluating these characteristics of particle boards.

Morphology of char residue: The char residue morphology of untreated and flame-retardant particle boards is captured by a scanning electron microscope (Gemini SEM 300, ZEISS, Oberkochen, Germany) equipped with an energy-dispersive X-ray (EDX) sensor. Scanning electron microscope (SEM) samples are sputtered with gold using a SCD005 Sputter Coater from BAL-TEC(Switzerland) at a current of 40 mA for 180 s.

Raman spectrum of char residue: A Raman spectrometer (LabRAM HR Evolution, HORIBA, Paris, France) is used to study the graphitization degree of char residue. The excitation light source is 532 nm, and the scanning area is 50–4000 cm^{-1}.

X-ray photoelectron spectroscopy (XPS) analysis of char residue: XPS analysis of char residue is conducted using a Thermo ESCALAB 250XI (ThermoFischer, Waltham, MA, USA), with an analysis chamber vacuum level of 4×10^{-9} mbar, an Al Kα ray excitation source (hv = 1486.6 eV), a working voltage of 14.6 kV, a filament current of 13.5 mA, and signal accumulation over 20 cycles. The test pass-energy is 20 eV, the step size is 0.1 eV, and the charge correction is performed with C1s = 284.8 eV as the combined energy standard.

3. Results

3.1. LOI Analysis

The LOI values for both untreated and flame-retardant particle boards have been listed in Table 4. The LOI value for untreated particle board stands at 26.4, indicating a heightened potential for fire hazards. However, when flame retardants ALHP, IFR, and MDH are introduced individually at a 10% mass fraction, LOI values experience significant increases, reaching 35.8, 33.7, and 29.4, respectively. These improvements mark substantial enhancements of 35.61%, 27.65%, and 11.36% in LOI values relative to untreated particle boards. These findings underscore the significant elevation in flame retardancy achieved through the application of ALHP, IFR, and MDH treatments. Table 4 further demonstrates that the most noteworthy LOI values, specifically 35.8 and 33.7, are attained with the inclusion of phosphorus-based ALPH and IFR flame retardants. Significantly, particle board inherently contains oxygen elements present in both the wood and urea-formaldehyde resin adhesive, with the observed enhancement in flame retardancy attributed to their interaction with phosphorus during combustion. Subsequent sections of this study will provide an in-depth exploration of the underlying flame-retardant mechanisms.

Table 4. Limit oxygen index of untreated and flame-retardant particle boards.

Samples	Particle Board	MDH/PB	APP + MEL + DPE/PB	ALHP/PB
LOI/%	26.4 ± 0.1 d	29.4 ± 0.2 c	33.7 ± 0.1 b	35.8 ± 0.3 a

Different letters followed the mean values, which means a significant difference at the 0.05 level ($p < 0.05$).

3.2. Temperature Distribution on the Facing Surface and the Temperature Rise on the Backside Surface

The temperature distribution on the fire-facing surface and the temperature rise on the backside of untreated and flame-retardant particle boards are investigated using a small-scale temperature distribution experimental platform. Figure 3a illustrates the temporal evolution of the vertical distance from $P_{300°C}$ to IP ($D_{300°C}$) for both untreated and flame-retardant particle boards, showing a rapid increase in $D_{300°C}$ after ignition followed by stabilization within distinct ranges, attributed to heat propagation from the ignition source and thermal carbonization of the samples. For untreated particle board, the maximum $D_{300°C}$ was achieved at approximately 164.02 mm (3.17 min after ignition), while the maximum $D_{300°C}$ values and ignition times for MDH/PB, APP + MEL + DPE/PB, and ALHP/PB samples are 145.79 mm (2.40 min), 117.81 mm (1.66 min), and 118.57 mm (15.00 min), respectively. Significantly, the APP + MEL + DPE/PB samples displayed both the lowest peak $D_{300°C}$ and a narrower stable region compared to the other samples, which can be attributed to the effective flame propagation and heat transfer inhibition of the IFR flame retardant. While the peak $D_{300°C}$ of ALHP/PB samples ranked second only to that of IFR-treated samples, it consistently maintained a relatively high level during continuous ignition, indicating that ALHP is less effective than IFR flame retardants in suppressing flame propagation during particle board combustion and highlighting the distinct flame-retardant mechanisms of IFR and ALHP when applied to particle board.

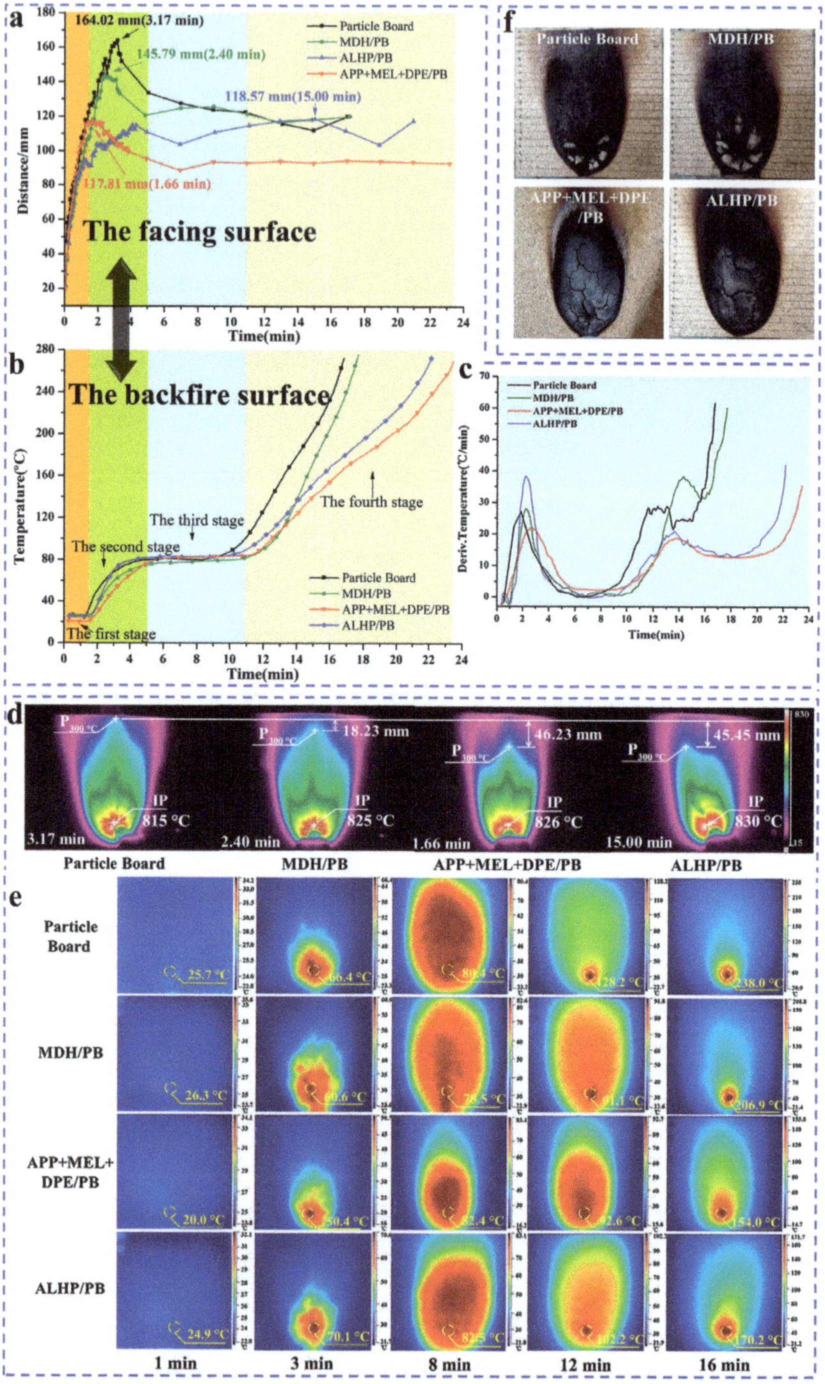

Figure 3. $D_{300^{\circ}C}$—Time curve (**a**); backside surface temperature rise curve (**b**); backside surface temperature rise rate curve (**c**); infrared images of the fire-facing surface (**d**) and backside surface (**e**) of untreated and flame-retardant particle boards; and digital images of char residues (**f**).

Figure 3d displays infrared images depicting the maximum $D_{300°C}$ of the untreated and flame-retardant particle boards. The images indicate the position of the 300 °C temperature point farthest from the IP ($P_{300°C}$) and ignition point (IP), with the 300 °C region represented by a light blue color. Comparing the untreated particle board with MDH/PB, APP + MEL + DPE/PB, and ALHP/PB, the flame-retardant variants demonstrate a reduction in the maximum $D_{300°C}$. Specifically, MDH/PB, APP + MEL + DPE/PB, and ALHP/PB samples reduce the distance by 18.23 mm (11.11% reduction), 46.23 mm (28.17% reduction), and 45.45 mm (27.71% reduction), respectively. The digital photos in Figure 3f depict the char residue following the combustion of the samples. In the cases of untreated particle board and MDH/PB samples, notable ash accumulation is evident at the ignition point (IP), and numerous cracks make the char easily detachable. In contrast, the char residues from the APP + MEL + DPE/PB sample, featuring minimal ash content and narrower cracks compared to other variants, effectively restrained flame spread on the approaching side and minimized heat transfer on the backfiring side, aligning with the trends observed in Section 3.3's backside temperature rise curve. For ALHP/PB samples, wider combustion residue cracks are observed, leading to a delay of up to 300 °C.

The temperature permeability of flame-retardant particle boards in the direction of board thickness is evaluated by monitoring the temperature increase on the backside surface, with the experiment concluding upon reaching a temperature of 275 °C. Figure 3b,c depict the temperature rise and temperature rise rate, respectively, of untreated and flame-retardant particle boards under continuous ignition on the facing surface. According to Figure 3b, the untreated particle board reaches a backside surface temperature of 275 °C in approximately 16.80 min, while MDH/PB, APP + MEL + DPE/PB, and ALHP/PB attain the same temperature levels at 17.65 min, 23.43 min, and 22.13 min, respectively. IFR flame-retardant particle board samples effectively extend the time to reach 275 °C on the backside surface by 6.63 min, a 39.46% increase, which is crucial for fire emergency response.

Figure 3b illustrates four primary stages in the temperature rise over time on the backside surface, with each stage corresponding to distinct characteristics and behaviors that coincide with the rapid increase of $D_{300°C}$ on the fire-facing surface. The second stage occurs near the peak of $D_{300°C}$ on the fire-facing surface. The third stage signifies a relatively stable $D_{300°C}$ on the fire-facing surface, while the fourth stage represents the primary phase of heat penetration into the material. Data regarding the duration of each stage and temperature rise rates, which were obtained from Figure 3c for both untreated and flame-retardant particle board samples, are listed in Table 5. The APP + MEL + DPE/PB sample exhibited the lowest temperature rise rate during the temperature rise on the backside fire surface, primarily in the second and fourth stages. Both the APP + MEL + DPE/PB and ALHP/PB samples exhibit longer backside surface temperature times, primarily due to extended durations in the third stage and the fourth stage compared with the MDH/PB sample. Figure 3e presents infrared images of the backside surface at various characteristic points during the four stages of untreated and flame-retardant particle boards. Examining the time-temperature thermal imaging maps corresponding to the second (3rd minute), third (8th minute), and fourth (12th and 16th minute) stages, it can be observed that there are differences in temperature and thermal imaging range of untreated and flame-retardant particle boards, with the APP + MEL + DPE/PB sample having the lowest temperature and the smallest high-temperature heat distribution range.

Table 5. Key data from the back temperature curve of untreated and flame-retardant particle boards.

Samples	The First Stage (min)	The Second Stage (min)	The Third Stage (min)	The Fourth Stage (min)	Temperature Rise Rate (°C/min)	
					The Second Stage	The Fourth Stage
Particle Board	1.30 ± 0.08 b	3.45 ± 0.23 b	3.57 ± 0.23 c	8.48 ± 0.59 b	27.11 ± 1.89 b	61.64 ± 5.32 a
MDH/PB	1.43 ± 0.09 ab	3.95 ± 0.25 a	5.02 ± 0.32 b	7.25 ± 0.50 b	27.98 ± 2.02 b	60.16 ± 4.21 a
APP + MEL + DPE/PB	1.47 ± 0.08 a	4.15 ± 0.23 a	5.73 ± 0.35 a	12.08 ± 0.80 a	21.94 ± 1.52 c	35.45 ± 2.38 b
ALHP/PB	1.55 ± 0.08 a	2.2 ± 0.13 c	6.25 ± 0.36 a	12.13 ± 0.85 a	38.27 ± 2.87 a	41.96 ± 2.94 b

Different letters followed the mean values, which means a significant difference at the 0.05 level ($p < 0.05$).

3.3. Cone Calorimeter

To more explicitly demonstrate the fire risk and flammability of the particle board samples, cone calorimetry analysis (CONE) was employed. The heat release rate (HRR) and total release (THR) curves of the samples as a function of time are depicted in Figure 4a,b, and the key data received from CONE are listed in Table 6. The flame-retardant particle boards exhibit differences in ignition time (t_{ign}, 21 s) and the time of the second peak heat release rate (t_{pHRR2}, 750 s) compared to the untreated particle board. Specifically, APP + MEL + DPE/PB samples show an extension to 24 s and 865 s for t_{ign} and t_{pHRR2}, respectively, corresponding to delay rates of 14.29% and 15.33%, respectively. APP + MEL + DPE/PB samples exhibit the most significant reduction in peak heat release rate (pHRR) and total heat release at 800 s (THR_{800s}), with percentage decreases of 19.33% and 29.29% compared with the untreated ones, respectively. It is noteworthy that the second peak in this sample is also delayed, which is mainly due to IFR promoting the char formation of particle board materials. APP + MEL + DPE/PB samples exhibit the highest fire performance index (FPI) at 0.151 $m^2 \cdot s/kW$, surpassing the untreated particle board by 41.67%, signifying that APP + MEL + DPE/PB samples present the lowest fire risk among the three types of flame-retardant particle boards.

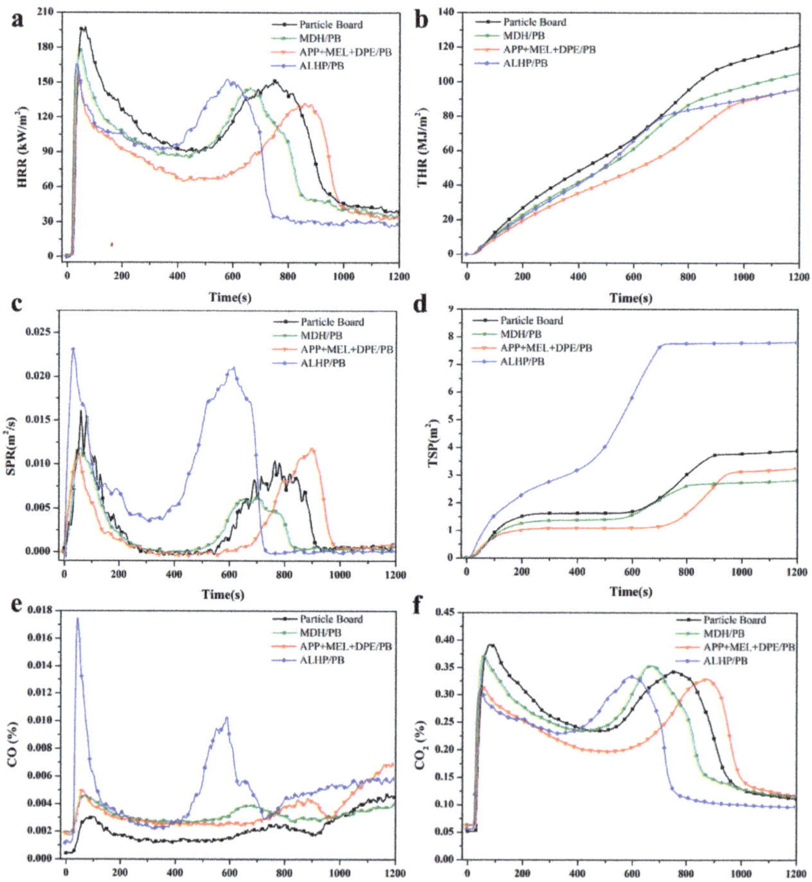

Figure 4. Heat release rates (**a**), total heat residual weight (**b**), smoke production rate (**c**), and total smoke production (**d**), CO concentration (**e**), and CO_2 concentration (**f**) curves for untreated and flame-retardant particle boards.

Table 6. Key data from the cone calorimeter test of untreated and flame-retardant particle boards.

Samples	t_{ign} (s)	t_{pHRR1} (s)	$pHRR_1$ (kW/m^2)	t_{pHRR2} (s)	$pHRR_2$ (kW/m^2)	THR_{800} (MJ/m^2)	Residual Weight (%)	FPI ($m^2 \cdot s/kW$)	pSPR (m^2/s)	TSP (m^2)
Particle Board	21 ± 2 a	65 ± 5 a	197.36 ± 13.06 a	750 ± 31 b	151.31 ± 9.87 a	95.48 ± 10.46 a	22.01 ± 1.54 b	0.106 ± 0.007 bc	0.016 ± 0.001 b	3.89 ± 0.26 b
MDH/PB	21 ± 2 a	50 ± 3 b	177.74 ± 11.74 ab	665 ± 29 c	145.83 ± 9.63 b	86.66 ± 9.42 a	28.19 ± 1.84 a	0.118 ± 0.007 b	0.012 ± 0.001 c	2.82 ± 0.18 c
APP + MEL + DPE/PB	24 ± 2 a	45 ± 3 b	159.21 ± 10.25 b	865 ± 35 a	132.32 ± 8.66 b	67.51 ± 7.24 b	29.59 ± 2.07 a	0.151 ± 0.009 a	0.012 ± 0.001 c	3.26 ± 0.23 c
ALHP/PB	16 ± 1 b	35 ± 3 c	165.74 ± 10.85 b	585 ± 24 d	152.88 ± 10.09 a	83.98 ± 9.15 ab	30.20 ± 2.11 a	0.097 ± 0.006 c	0.023 ± 0.002 a	7.82 ± 0.45 a

Different letters followed the mean values, which means a significant difference at the 0.05 level ($p < 0.05$).

Furthermore, Figure 4c,d present the smoke production rate (SPR) and total smoke production (TSP) curves. From the figure, it is evident that the peak smoke production rate (pSPR) and TSP values of MDH/PB are the lowest, measuring 0.012 m^2/s and 2.82 m^2, respectively, suggesting that magnesium hydroxide is more effective in suppressing smoke during wood combustion. In contrast, ALHP/PB samples exhibit notably higher pSPR and TSP values, measuring 0.025 m^2/s and 7.82 m^2, respectively, which represent approximately 2.08 times and 2.77 times the values observed in the untreated particle board. The variations in smoke production rate (SPR) and total smoke production (TSP) among flame-retardant particle board samples can be attributed to the release of CO and CO_2, a correlation supported by the synchronization of peak CO and CO_2 concentrations with the peak SPR, as depicted in Figure 4e,f. Simultaneously, it is evident that the ALHP/PB samples display the highest pSPR and TSP values, which can be attributed to the emission of CO.

3.4. Physical and Mechanical Performance

The physical and mechanical characteristics of untreated particle board and flame-retardant particle board are detailed in Table 7. It is evident that the incorporation of flame retardants has led to varying degrees of reduction in the physical and mechanical properties of the particle boards. The lower internal and surface bonding strengths of ALHP/PB samples, in comparison to the other two flame-retardant particle boards, may be attributed to the acidic nature of aluminum hypophosphate (pH 3–5), which affects the crosslinking degree of the urea-formaldehyde resin dimer and, in turn, impacts bonding strength. Additionally, ALHP exhibits poor compatibility with the substrate, infiltrating the bonding interface between the adhesive and the unit shavings, thereby diminishing the bonding points. In contrast, the APP + MEL + DPE/PB samples show the highest MOR and MOE due to melamine's presence in the extended flame-retardant system, promoting cross-linking reactions and leading to a stronger three-dimensional network structure, thus enhancing cohesion, rigidity, and resin bonding strength, ultimately boosting the MOR and MOE of the sheet [34]. Remarkably, the considerably higher absorption thickness expansion rate in MDH/PB samples, in comparison to the other two flame-retardant particle boards, may be attributed to the inherent hydrophilic nature of the hydroxyl groups present in magnesium hydroxide.

Table 7. Physical and mechanical data of untreated and flame-retardant particle boards.

Samples	Thickness (mm)	Density ($kg \cdot m^3$)	Internal Bonding Strength (MPa)	Surface Bonding Strength (MPa)	MOR (MPa)	MOE (MPa)	Absorption Thickness Expansion Rate (%)
Particle Board	17.72 ± 1.95 a	669.83 ± 60.21 a	0.89 ± 0.12 a	1.10 ± 0.21 a	15.11 ± 1.43 a	2489.67 ± 290.56 a	13.54 ± 1.47 c
MDH/PB	17.77 ± 2.00 a	657.71 ± 58.13 a	0.57 ± 0.02 b	0.64 ± 0.08 bc	7.39 ± 0.73 b	1757.00 ± 167.14 b	62.22 ± 5.38 a
APP + MEL + DPE/PB	17.70 ± 2.12 a	659.71 ± 58.31 a	0.47 ± 0.19 b	0.67 ± 0.18 b	10.02 ± 3.31 b	1978.00 ± 401.40 bc	22.24 ± 3.24 b
ALHP/PB	17.69 ± 1.94 a	654.61 ± 59.86 a	0.21 ± 0.02 c	0.37 ± 0.10 c	7.47 ± 0.47 b	1342.00 ± 100.44 c	22.47 ± 1.74 b

Different letters followed the mean values, which means a significant difference at the 0.05 level ($p < 0.05$).

3.5. Mechanism Analysis

3.5.1. Morphology of Char Residue

Based on the above analysis, it is interesting to explore the difference in the char residuals between untreated and flame-retardant particle boards. Figure 5 illustrates the digital images, SEM images, and EDX spectra of the combustion residues of untreated and flame-retardant particle boards. As depicted in Figure 5a, the char layer of untreated particle board appears gray after complete wood material combustion, signifying weak char layer strength, with SEM images additionally revealing a loose and collapsed wood fiber microstructure post-combustion. In contrast, the char layer in the MDH/PB samples exhibited a noticeably lighter gray shade, primarily attributed to the lingering magnesium oxide post-combustion and dehydration of MDH. Nonetheless, the char layer structure of the MDH/PB samples is denser compared to the untreated particle board. Following ALHP treatment, the residual char color on the flame-retardant particle boards turned black, with a slight presence of white material on the surface. In contrast to MDH treatment, there is a notable enhancement in the density of the char layer, and the SEM images distinctly displayed the presence of granular material. Concerning the IFR-treated samples, Figure 5c demonstrates a thicker char layer structure, signifying enhanced strength, with SEM images highlighting a significant improvement in densification. According to EDX spectrum analysis, particle boards treated with various flame retardants (MDH, IFR, and ALHP) exhibited varying levels of characteristic elements (Mg, Al, and P/Al elements) remaining after combustion. It is worth emphasizing that the relative atomic percentages of Al and P in the ALHP/PB samples are relatively low, primarily due to the release of P in the form of PH_3 flue gas, thereby enhancing its flame-retardant properties in the gas phase [24]. This finding is consistent with the relatively high pSPR values observed in the ALHP/PB samples (Figure 4c).

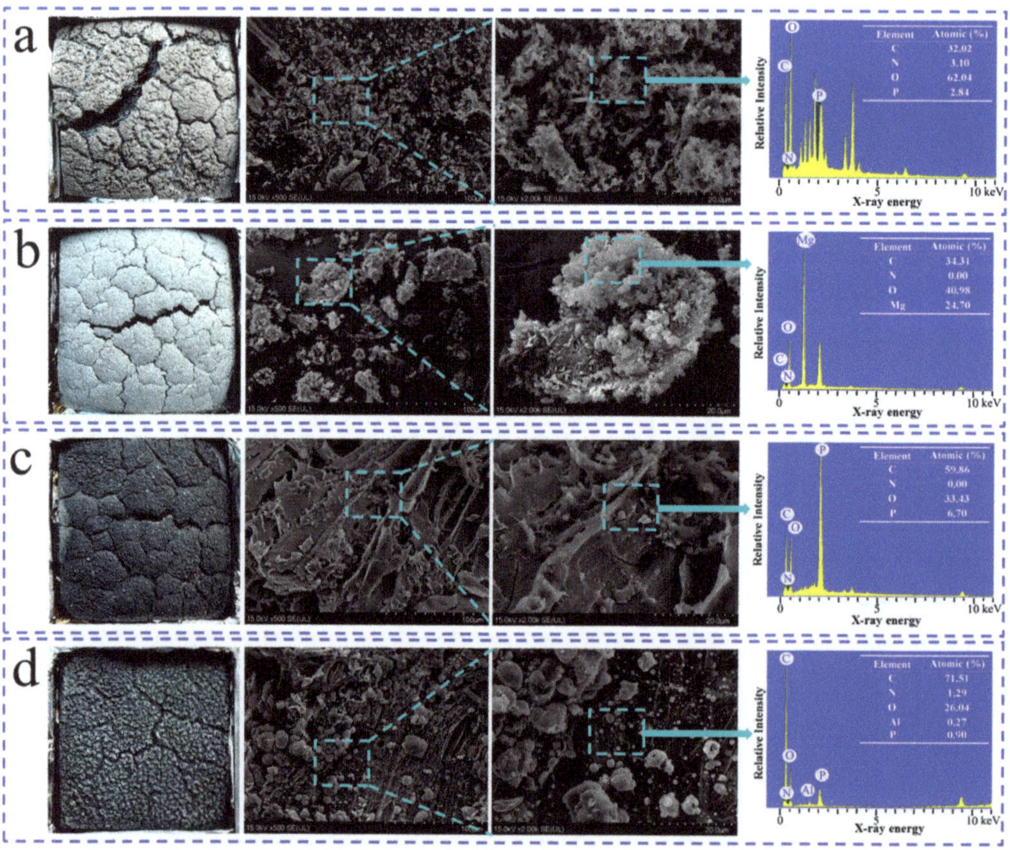

Figure 5. Digital images, SEM images, and EDX spectra of char residues of untreated particle board (**a**), MDH/PB (**b**), APP + MEL + DPE/PB (**c**), and ALHP/PB (**d**).

3.5.2. Raman Spectrum of Char Residue

Raman spectroscopy is a widely employed technique to assess the graphitization degree of carbon materials. In the spectra, the D band corresponds to the vibration of disordered graphite's sp^3 hybrid carbon atoms, while the G band represents the vibration of sp^2 hybrid carbon atoms in a 2D hexagonal lattice [35]. The ID/IG ratio serves as an indicator for evaluating the quality of carbon materials [36]. A lower ID/IG value indicates a higher degree of graphitization and fewer defects. Figure 6 displays the Raman spectra of the combustion residues from untreated and flame-retardant particle boards. The ID/IG values for untreated particle board, MDH/PB, APP + MEL + DPE/PB, and ALHP/PB samples are as follows: 1.16, 1.18, 0.91, and 0.95, respectively. Notably, APP + MEL + DPE/PB exhibit the lowest ID/IG value, signifying a higher degree of graphitization and improved char residue quality compared to ALHP/PB. Additionally, the ID/IG value of MDH/PB is similar to that of the untreated sample, indicating that MDH does not contribute to enhancing the quality of char residue.

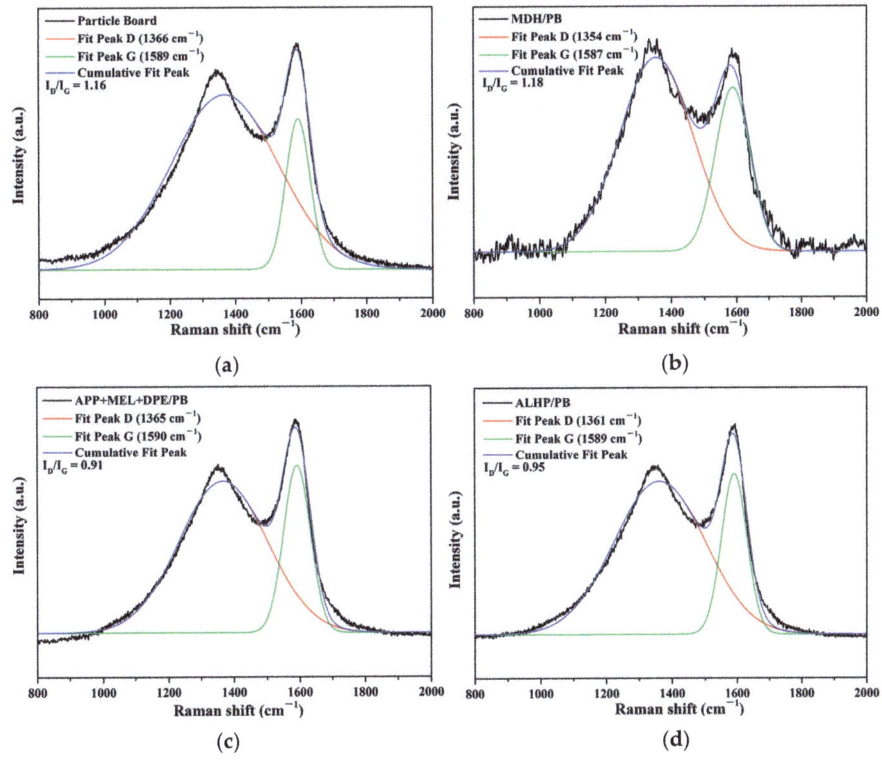

Figure 6. Raman spectra of char residues of untreated particle board (**a**), MDH/PB (**b**), APP + MEL + DPE/PB (**c**), and ALHP/PB (**d**).

3.5.3. XPS Analysis of Char Residue

X-ray photoelectron spectroscopy (XPS) tests are conducted to analyze the chemical structure and composition of the char layer formed by flame-retardant particle boards, and the related results are exhibited in Figure 7. In the full survey XPS spectra of the char residue of untreated and flame-retardant particle boards, characteristic peaks attributed to Mg 1s (1304.9 eV), P 2s (192.25 eV), P 2p (136.31 eV), and Al 2p (76.18 eV) can be clearly observed.

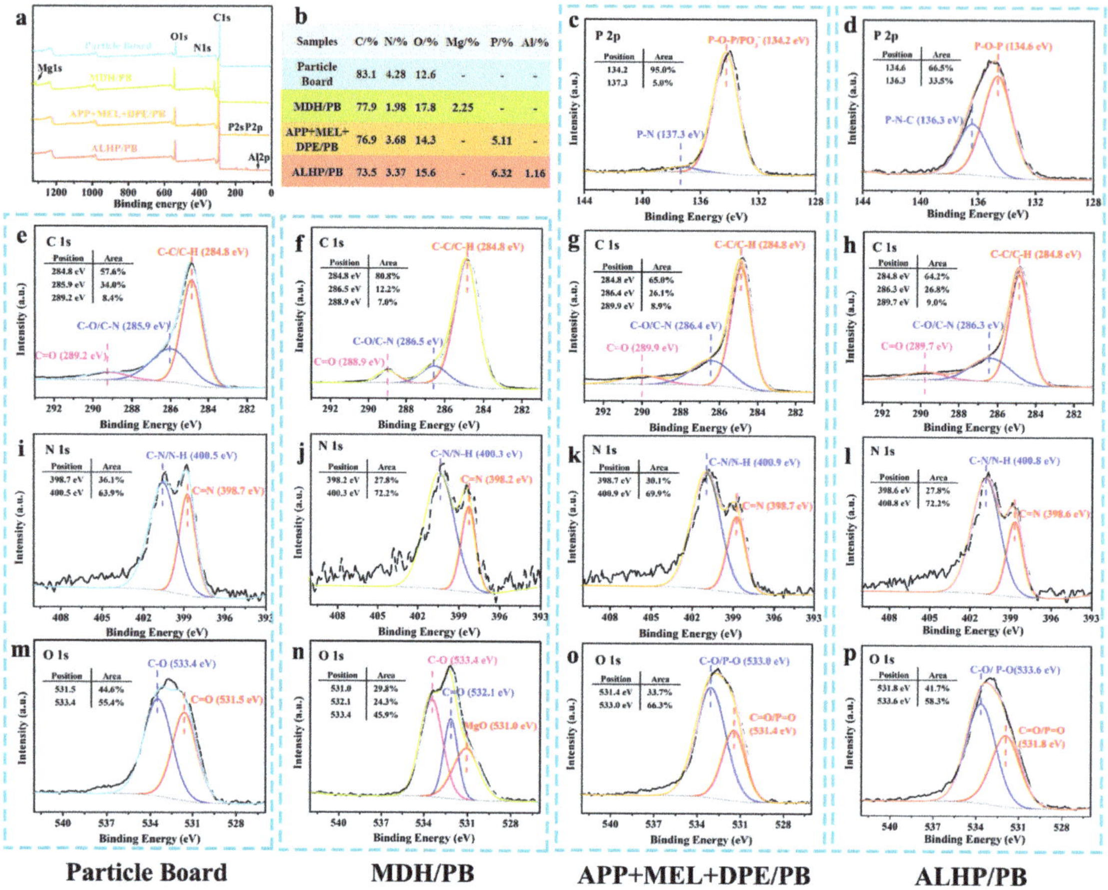

Figure 7. XPS survey spectra, X-ray photoelectron spectroscopy date, and high-resolution XPS spectra of the char residues produced by untreated particle board (**a**,**b**,**e**,**i**,**m**), MDH/PB (**a**,**b**,**f**,**j**,**n**), APP + MEL + DPE/PB (**a**–**c**,**g**,**k**,**o**), and ALHP/PB (**a**,**b**,**d**,**h**,**l**,**p**) after the CONE.

As shown in Figure 7b, compared with the content of various elements of untreated and flame-retardant particle boards, the content of C and N in the Particle Board sample decreased and the content of O increased, suggesting that the addition of MDH, IFR, and ALHP played a role in protecting the matrix during combustion. The C 1s, N 1s, O 1s, and P 2p fine spectra of the char layer are fitted by peak fitting for Particle Board, MDH/PB, APP + MEL + DPE/PB, and ALHP/PB samples. The C 1s spectra of untreated and flame-retardant particle board samples exhibited three fitting peaks: 284.8 eV corresponding to C-C/C-H groups, 285.9–286.5 eV corresponding to C-O and C-N groups in the crosslinked structure, and 288.9–289.9 eV corresponding to completely oxidized C=O groups [37–41]. The peaks of the N 1s high resolution spectrum of untreated and flame-retardant particle board samples at 398.2–398.7 eV and 400.3–400.9 eV are assigned to C=N and pyrrole groups [42–44], respectively, and the content of C=N and pyrrole groups has changed, proving that the addition of flame retardant promotes the char formation of urea-formaldehyde resin adhesive in particle board during combustion. The peaks of the O 1s high resolution spectrum of MDH/PB samples at 532.1 eV, 533.4 eV, and 531.0 eV were assigned to C=O, C-O, and MgO groups, proving the decomposition of $Mg(OH)_2$ into MgO during combustion [45–47]. C=O/P=O and C-O/P-O groups can be found in the O 1s high resolution spectrum of APP + MEL + DPE/PB (531.4 eV and 533.0 eV) and ALHP/PB (531.8 eV and 533.6 eV) samples [38,45]. The P 2p spectra of APP + MEL + DPE/PB and ALHP/PB char residue displayed two fitting peaks. The P 2p binding energy at 134.2 eV for APP + MEL + DPE/PB char residue is attributed to P-O-P/PO^{3-} in cross-linked polyphosphates and pyrophosphates [37,42,45], while 137.3 eV is attributed to P-N [48]. The P 2p binding energy for ALHP/PB char residue is attributed to P-O-P at 134.6 eV and P-N-C at 136.3 eV [49], indicating the formation of a relatively stable P-N-C structure due to the reaction of aluminum hypophosphate with shavings and adhesives.

By analyzing the chemical composition and microstructure of char residue, the possible flame-retardant mechanism of three cost-effective particle boards has been elucidated (as depicted in Figure 8). Flame-retardant particle boards prepared using MDH, IFR, and ALHP all emit CO, CO_2, H_2O, and NH_3 during combustion, exerting flame-retardant effects in the gas phase. Notably, these three flame-retardant particle boards demonstrated distinct combustion mechanisms. $Mg(OH)_2$ exhibits limited reactivity with the substrate during combustion, leading to the formation of MgO, which persists in the ash. In contrast, the IFR compounds form a linked structure with wood and the urea-formaldehyde resin, resulting in the production of polyphosphates, pyrophosphates, phosphorus nitrogen compounds, etc., thereby enhancing flame-retardant performance. ALHP reacts with both shavings and urea-formaldehyde adhesives, establishing a relatively stable P-N-C structure, promoting char formation, and generating gaseous PH_3.

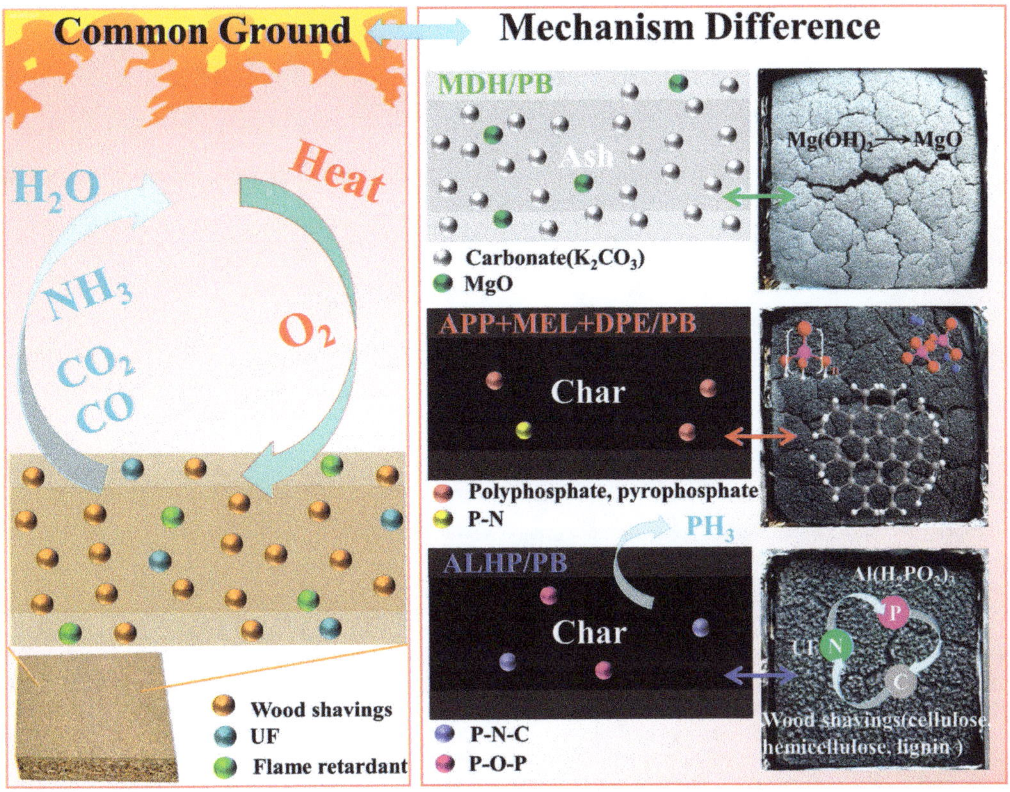

Figure 8. Schematic illustration of a flame-retardant mechanism for flame-retardant particle boards.

4. Conclusions

In this study, we employed infrared imaging devices and other instruments to construct a small-scale temperature distribution experiment platform, enhancing the combustion performance testing methods for artificial board materials. IFR flame-retardant particle board has excellent ability to inhibit temperature diffusion and extend the time it takes for the backside temperature to rise, which is of great significance for emergency rescue work in the event of fire accidents. The self-made temperature distribution experiments revealed that APP + MEL + DPE/PB and ALHP/PB are more effective in reducing the maximum $D_{300°C}$ and extending the back temperature time. Specifically, the maximum temperature diffusion distance is reduced by 46.23 mm (28.17% decrease) and 45.45 mm (27.71% decrease), while the back temperature time is extended by 6.63 min (39.46% extension) and 5.33 min (31.73% extension), respectively. APP + MEL + DPE/PB showed the best performance in reducing pHRR and THR_{800}, with decreases of 19.33% and 29.29%, respectively. SEM, Raman spectra, and XPS analyses revealed that MDH/PBs flame retardant mainly depended on the thermal decomposition of $Mg(OH)_2$ into MgO and H_2O without char formation. In contrast, APP + MEL + DPE/PB expanded into char, forming a cross-linked structure with polyphosphate and pyrophosphate, while ALHP/PB reacted with particles and adhesive to form a stable condensed-phase P-N-C structure.

This study sheds light on flame-retardant mechanisms and combustion behaviors in various flame-resistant particleboards, providing insights for cost-effective production and practical evaluation as decorative building facade materials, especially in real fire scenarios. Moreover, this study's results can guide manufacturers, architects, and builders in selecting appropriate flame-resistant particleboards for different construction projects, ensuring adherence to fire safety regulations, and promoting the development of safer and more fire-resistant structures in real-world scenarios.

Author Contributions: Conceptualization, P.J.; methodology, F.P.; software, F.P., and H.J.; validation, F.P.; formal analysis, Y.H., and S.L.; investigation, F.P.; resources, P.J., and Z.C.; data curation, F.P.; writing—original draft preparation, F.P.; writing—review and editing, P.J., and F.P.; visualization, P.J.; supervision, P.J.; project administration, P.J.; funding acquisition, Z.C. All authors have read and agreed to the published version of the manuscript.

Funding: This research was funded by the National Natural Science Foundation of China (No. 32271788) and the National Key Research and Development Program of China (Grant No. 2022YFD2200702).

Institutional Review Board Statement: Not applicable.

Data Availability Statement: All data included in this study are available upon request by contacting the corresponding author.

Conflicts of Interest: The authors declare no conflict of interest.

References

1. Orelma, H.; Tanaka, A.; Vuoriluoto, M.; Khakalo, A.; Korpela, A. Manufacture of all-wood sawdust-based particle board using ionic liquid-facilitated fusion proces. *Wood Sci. Technol.* **2021**, *55*, 331–349. [CrossRef]
2. Badila, M.; Jocham, C.; Zhang, W.B.; Schmidt, T.; Wuzella, G.; Müller, U.; Kandelbauer, A. Powder coating of veneered particle board surfaces by hot pressing. *Prog. Org. Coat.* **2014**, *77*, 1547–1553. [CrossRef]
3. Tang, Y. Study on the Formaldehyde-Free Flame Retardant Wood Particleboard and Its Adhesive. Master's Thesis, Sichuan University, Chengdu, China, 2007.
4. Wang, K.H.; Wang, S.H.; Meng, D.; Chen, D.; Mu, C.Z.; Li, H.F.; Sun, J.; Gu, X.Y.; Zhang, S. A facile preparation of environmentally-benign and flame-retardant coating on wood by comprising polysilicate and boric acid. *Cellulose* **2021**, *28*, 11551–11566. [CrossRef]
5. Qian, L.J. *Modern Flame Retardant Materials and Technology*; Chemical Industry Press: Beijing, China, 2021; pp. 16–27.
6. Zhang, L.L.; Zhang, W.L.; Peng, Y.; Wang, W.; Cao, J.Z. Thermal behavior and flame retardancy of poplar wood impregnated with furfuryl alcohol catalyzed by boron/phosphorus compound system. *Ind. Crops Prod.* **2022**, *176*, 114361. [CrossRef]
7. Liu, Q.Q.; Chai, Y.B.; Ni, L.; Lyu, W.H. Flame retardant properties and thermal decomposition kinetics of wood treated with boric acid modified silica sol. *Materials* **2020**, *13*, 4478. [CrossRef]

8. Li, L.M.; Chen, Z.L.; Lu, J.H.; Wei, M.; Huang, Y.X.; Jiang, P. Combustion behavior and thermal degradation properties of wood impregnated with intumescent biomass flame retardants: Phytic acid, hydrolyzed collagen, and glycerol. *ACS Omega* **2021**, *6*, 3921–3930. [CrossRef]
9. Liu, D.B.; Qasem, B.; Hou, J.F.; Wang, Z.M.; Cen, J.J.; Arkin, S.; Zhang, X.Y.; Zhang, Y.G.; Yu, Y.M. Recycling construction wastes to fabricate particle boards with admirable flame retardancy, smoke suppression and mechanical performance. *Constr. Build. Mater.* **2023**, *404*, 133265. [CrossRef]
10. Svensson, I.; Butron, A.; Puyadena, M.; González, A.; Irusta, L.; Barrio, A. Bio-Based Phosphate-Containing Polyester for Improvement of Fire Reaction in Wooden Particleboard. *Polymers* **2023**, *15*, 1093. [CrossRef]
11. Kumar, M.; Kumar, M.; Gulati, K.; Bagoria, R.; Arora, S. Thermal, flammability and iso-conversional multiple heating rate kinetic studies of impregnated poplar wood veneers. *J. Indian Acad. Wood Sci.* **2022**, *19*, 30–39. [CrossRef]
12. Guan, P.F.; Li, P.; Wu, Y.Q.; Li, X.G.; Yuan, G.M.; Zuo, Y.F. Comparative study on the properties of inorganic silicate and organic phenolic prepolymer modified poplar wood by vacuum cycle pressurization. *J. Renew. Mater.* **2022**, *10*, 2451–2463. [CrossRef]
13. Zhang, X.T.; Mu, J.; Chu, D.M.; Zhao, Y. Synthesis of N-P fire retardant modified with MH and flame retardance and smoke suppression of poplar impregnated by it. *New Chem. Mater.* **2015**, *43*, 126–129.
14. Zhu, X.D.; Wu, Y.Q.; Tian, C.H.; Yan, Q.; Yao, C.H. Synergistic effect of nanosilica aerogel with phosphorus flame retardants on improving flame retardancy and leaching resistance of wood. *J. Nanomater.* **2014**, *2014*, 7. [CrossRef]
15. Grexa, O.; Lübke, H. Flammability parameters of wood tested on a cone calorimeter. *Polym. Degrad. Stab.* **2001**, *74*, 427–432. [CrossRef]
16. Gürü, M.; Atar, M.; Yildinm, R. Production of polymer matrix composite particleboard from walnut shell and improvement of its requirements. *Mater. Des.* **2008**, *29*, 284–287. [CrossRef]
17. Lee, S.J.; Thole, V. Investigation of modified water glass as adhesive for wood and particleboard: Mechanical, thermal and flame retardant properties. *Eur. J. Wood Wood Prod.* **2018**, *76*, 1427–1434. [CrossRef]
18. Yang, K.; Li, X.G. Preparation of mineral bound particleboards with improved fire retardant and smoke suppression properties based on a mix of inorganic adhesive. *Holzforschung* **2019**, *73*, 599–604. [CrossRef]
19. Baskaran, M.; Hashim, R.; Leong, J.Y.; Ong, Y.N.; Yhaya, M.F.; Sulaiman, O. Flame retardant properties of oil palm trunk particleboard with addition of epoxy resin as a binder and aluminium hydroxide and magnesium hydroxide as additives. *Bull. Mater. Sci.* **2019**, *42*, 138. [CrossRef]
20. Grexa, O.; Poutch, F.; Manikova, D.; Martvonova, H.; Bartekov, A. Intumescence in fire retardancy of lignocellulosic panels. *Polym. Degrad. Stab.* **2003**, *82*, 373–377. [CrossRef]
21. Savas, L.A.; Dogan, M. Flame retardant effect of zinc borate in polyamide 6 containing aluminum hypophosphite. *Polym. Degrad. Stab.* **2019**, *165*, 101–109. [CrossRef]
22. Sun, X.Y.; Xu, Y.; Zhou, R.; Jiang, J.C. Effects of melamine cyanurate and aluminum hypophosphite on the flame retardancy of high-impact polystyrene. *J. Mater. Sci.* **2021**, *56*, 17860–17873. [CrossRef]
23. Liu, J.H.; Wu, H.Z.; Sun, M.M.; Su, S.Q.; Yan, B.; Tang, L.S. Fire-retardant synergy of tris(1-methoxy-2,2,6,6-tetramethyl-4-piperidinyl)phosphite and aluminum hypophosphite/melamine hydrobromide in polypropylene. *J. Vinyl Addit. Technol.* **2021**, *27*, 108–118. [CrossRef]
24. Qiu, Y.Y.; Sun, M.M.; Wu, H.Z.; Tang, M.; Yan, B.; Tang, L.S. Synergistic flame retardancy of bis(1-methoxy-2,2,6,6-tetramethylpiperidin-4-yl)sebacate and aluminium hypophosphite/melamine hydrobromide in PP. *Fire Mater.* **2021**, *45*, 261–272. [CrossRef]
25. Sun, X.Y.; Huang, C.X.; Chen, Z.H.; Zhou, R.; Jiang, J.C. Multi-element synergistic effects to improve the flame retardancy of high impact polystyrene. *Polym. Test.* **2022**, *115*, 107766. [CrossRef]
26. Savas, L.A.; Hacioglu, F.; Hancer, M.; Dogan, M. Flame retardant effect of aluminum hypophosphite in heteroatom-containing polymers. *Polym. Bull.* **2020**, *77*, 291–306. [CrossRef]
27. *ANSI/ASTM E84-2021*; Standard Test Method for Surface Burning Characteristics of Building Materials. ASTM International: West Conshohocken, PA, USA, 2021.
28. *GB/T 17658-2018*; Test of Burning Behavior for Flame Retardant Treated Wood-Method of Test for Fire Propagation. Research Institute of Wood Industry, Chinese Academy of Forestry: Beijing, China, 2018.
29. Li, B.; Wang, J.Q. Utilization of cone calorimeter for the appraisal of the flammability and flame retardancy of polymeric materials. *Polym. Mater. Sci. Eng.* **1998**, *14*, 15–19.
30. Bartlett, A.I.; Hadden, R.M.; Bisby, L.A. A review of factors affecting the burning behaviour of wood for application to tall timber construction. *Fire Technol.* **2019**, *55*, 1–49. [CrossRef]
31. *ASTM D2863-17*; Standard Test Method for Measuring the Minimum Oxygen Concentration to Support Candle-Like Combustion of Plastics (Oxygen Index). ASTM International: West Conshohocken, PA, USA, 2017.
32. *ASTM E1354-17*; Standard Test Method for Heat and Visible Smoke Release Rates for Materials and Products Using an Oxygen Consumption Calorimeter. ASTM International: West Conshohocken, PA, USA, 2017.
33. *GB/T 17657-2013*; Test methods of evaluating the properties of wood-based panels and surface decorated wood-based panels. Standardization Administration of the People's Republic of China: Beijing, China, 2013.
34. Cui, J.Q.; Fen, X.P.; Gu, Y.; Han, S.G.; Cai, Z.W.; Zhu, J.; Dai, X.; Zeng, L. Effect of Feeding Modes of Melamine on Properties of Fiberboard Glued with MUF Adhesive. *J. Northeast. For. Univ.* **2015**, *43*, 134–136.

35. Zhou, C.; Liao, D.J.; Hu, X.P.; Li, W.X.; Xie, C.Q.; Zhang, H.J.; Yang, W.X. Synergistic fire retardant effect between expandable graphite and ferrocene-based non-phosphorus polymer on polypropylene. *Polym. Degrad. Stab.* **2020**, *178*, 109201.
36. Pachfule, P.; Shinde, D.; Majumder, M.; Xu, Q. Fabrication of carbon nanorods and graphene nanoribbons from a metal–organic framework. *Nat. Chem.* **2016**, *8*, 718–724. [CrossRef]
37. Li, W.X.; Zhang, H.J.; Hu, X.P.; Yang, W.X.; Cheng, Z.; Xie, C.Q. Highly efficient replacement of traditional intumescent flame retardants in polypropylene by manganese ions doped melamine phytate nanosheets. *J. Hazard. Mater.* **2020**, *398*, 123001. [CrossRef]
38. Huang, W.J.; He, W.T.; Long, L.J.; Yan, W.; He, M.; Qin, S.H.; Yu, J. Highly efficient flame-retardant glass-fiber-reinforced polyamide 6T system based on a novel DOPO-based derivative: Flame retardancy, thermal decomposition, and pyrolysis behavior. *Polym. Degrad. Stab.* **2018**, *148*, 26–41. [CrossRef]
39. Xu, B.; Wu, X.; Ma, W.; Qian, L.J.; Xin, F.; Qiu, Y. Synthesis and characterization of a novel organic-inorganic hybrid char-forming agent and its flame-retardant application in polypropylene composites. *J. Anal. Appl. Pyrol.* **2018**, *134*, 231–242. [CrossRef]
40. Moulder, J.F.; Stickle, W.F.; Sobol, P.E.; Bomben, K.D. *Handbook of X-ray Photoelectron Spectroscopy*, 2nd ed.; Physical Electronics, Inc.: Chanhassen, MN, USA, 1995.
41. Zhang, N.E.; Zhang, J.; Yan, H.; Guo, X.R.; Sun, Q.; Guo, R.J. A novel organic-inorganic hybrid K-HBPE@APP performing excellent flame retardancy and journal pre-proofsmoke suppression for polypropylene. *J. Hazard. Mater.* **2019**, *373*, 856–865. [CrossRef]
42. Guan, Y.H.; Huang, J.Q.; Yang, J.C.; Shao, Z.B.; Wang, Y.Z. An effective way to flame-retard biocomposite with ethanolamine modified ammonium polyphosphate and its flame retardant mechanisms. *Ind. Eng. Chem. Res.* **2015**, *54*, 3524–3531. [CrossRef]
43. Feng, C.M.; Liang, M.Y.; Jiang, J.L.; Huang, J.G.; Liu, H.B. Flame retardant properties and mechanism of an efficient intumescent flame retardant PLA composites. *Polym. Adv. Technol.* **2016**, *27*, 693–700. [CrossRef]
44. Wang, P.; Xia, L.; Jian, R.K.; Ai, Y.F.; Zheng, X.L.; Chen, G.L.; Wang, J.S. Flame-retarding epoxy resin with an efficient P/N/S-containing flame retardant: Preparation, thermal stability, and flame retardance. *Polym. Degrad. Stab.* **2018**, *149*, 69–77. [CrossRef]
45. Xu, Z.S.; Chu, Z.Y.; Yan, L.; Chen, H.G.; Jia, H.Y.; Tang, W.F. Effect of chicken eggshell on the flame-retardant and smoke suppression properties of an epoxy-based traditional APP-PER-MEL system. *Polym. Compos.* **2019**, *40*, 2712–2723. [CrossRef]
46. Fuggle, J.C. XPS, UPS AND XAES studies of oxygen adsorption on polycrystalline Mg at 100 and 300 K. *Surf. Sci.* **1977**, *69*, 581–608. [CrossRef]
47. Inoue, Y.; Yasumori, I. Catalysis by alkaline earth metal oxides. III. X-ray photoelectron spectroscopic study of catalytically active MgO, CaO, and BaO surfaces. *Bull. Chem. Soc. Jpn.* **1981**, *54*, 1505–1510. [CrossRef]
48. Shul'ga, Y.M.; Bulatov, A.V.; Gould, R.A.T.; Konze, W.V.; Pignolet, L.H. X-ray photoelectron spectroscopy of a series of heterometallic gold-platinum phosphine cluster compounds. *Inorg. Chem.* **1992**, *31*, 4704–4706. [CrossRef]
49. Wang, S.S.; Zhang, L.; Semple, K.; Zhang, M.; Zhang, W.B.; Dai, C.P. Development of biodegradable flame-retardant bamboo charcoal composites, part ii: Thermal degradation, gas phase, and elemental analyses. *Polymers* **2020**, *12*, 2217. [CrossRef]

Disclaimer/Publisher's Note: The statements, opinions and data contained in all publications are solely those of the individual author(s) and contributor(s) and not of MDPI and/or the editor(s). MDPI and/or the editor(s) disclaim responsibility for any injury to people or property resulting from any ideas, methods, instructions or products referred to in the content.

polymers

MDPI

Article

Flame-Retardant Thermoplastic Polyether Ester/Aluminum Butylmethylphosphinate/Phenolphthalein Composites with Enhanced Mechanical Properties and Antidripping

Xue Yang [1], Yan Zhang [2], Jia Chen [3], Liyong Zou [2,3], Xuesong Xing [2], Kangran Zhang [2], Jiyan Liu [2,3,*] and Xueqing Liu [2,3,*]

1 Institute of Noise & Vibration, Naval University of Engineering, Wuhan 430033, China
2 School of Polymer Materials and Engineering, Jianghan University, Wuhan 430056, China
3 Key Laboratory of Optoelectronic Chemical Materials and Devices of Ministry of Education, Jianghan University, Wuhan 430056, China
* Correspondence: liujiyan918@jhun.edu.cn (J.L.); xqliu@jhun.edu.cn (X.L.)

Abstract: Aluminum butylmethylphosphinate AiBMP as a flame retardant and phenolphthalein as a synergistic agent were applied in a thermoplastic polyester elastomer (TPEE)) in the current study. The thermal properties, flame retardancy, crystallization and mechanical properties of TPEE/AiMBP with or without phenolphthalein were investigated using various characterizations, including the limiting oxygen index (LOI), vertical burning test (UL 94), thermogravimetric analysis TG, differential scanning calorimetry, microcombustion calorimeter (MCC), scanning electron microscopy (SEM), and mechanical tests. The results revealed that AiBMP alone is an efficient flame retardant of TPEE. Adding 15 wt.% AiBMP increases the LOI value of TPEE from 20% to 36%. The formula TPEE-15 AiBMP passed the UL 94 V-0 rating with no dripping occurring. The MCC test shows that AiBMP depresses the heat release of TPEE. In comparison with pure TPEE, the heat release rate at peak temperature and the heat release capacity of TPEE-15AiBMP are reduced by 46.1% and 55.5%, respectively. With the phenolphthalein added, the formula TPEE/13AiBMP/2Ph shows a higher char yield at high temperatures (>600 °C), and the char layer is stronger and more condensed than TPEE-15AiBMP.The tensile strength and elongation at break values of TPEE-13AiBMP-2Ph are increased by 29.63% and 4.8% in comparison with TPEE-15AiBMP. The SEM morphology of the fracture surface of the sample shows that phenolphthalein acts as a plasticizer to improve the dispersion of AiBMP within the matrix. The good char charming ability of phenolphthalein itself and improved dispersion of AiBMP make the TPEE composites achieve both satisfying flame retardancy and high mechanical properties.

Keywords: thermoplastic polyether-ester; flame retardancy; metal phosphinate; phenolphthalein; mechanical properties

check for updates

Citation: Yang, X.; Zhang, Y.; Chen, J.; Zou, L.; Xing, X.; Zhang, K.; Liu, J.; Liu, X. Flame-Retardant Thermoplastic Polyether Ester/Aluminum Butylmethylphosphinate/ Phenolphthalein Composites with Enhanced Mechanical Properties and Antidripping. *Polymers* **2024**, *16*, 552. https://doi.org/10.3390/polym16040552

Academic Editor: Marco Zanetti

Received: 15 January 2024
Revised: 14 February 2024
Accepted: 15 February 2024
Published: 18 February 2024

Copyright: © 2024 by the authors. Licensee MDPI, Basel, Switzerland. This article is an open access article distributed under the terms and conditions of the Creative Commons Attribution (CC BY) license (https://creativecommons.org/licenses/by/4.0/).

1. Introduction

Thermoplastic polyether-ester elastomer (TPEE) is produced by combining crystalline (hard) and amorphous (soft) segments to offer properties of a thermal set elastomer with the processability and recycling ability of thermoplastics [1]. They exhibit exceptional toughness, impact resistance, load-bearing capacity, and low-temperature flexibility and are widely used in electronics and electrical appliances, communications, and automotive industries [2,3]. TPEE has inherent flammability and serious dripping during combustion and involves a real hazard to the users of these applications. Therefore, flame retardancy treatment is a requirement [4,5].

Adding flame retardant additives via physical blending is an easy and economical approach to improve the fire resistance of TPEE. Many of the commercially available systems for the flame retardancy of TPEE consist of a halogen-containing additive and a synergistic agent [6]. With the strict limitation on hazardous substances released into the environment during the disposal of electrical and election waste, the industries for these halogen additive-based TPEEs are under pressure to change to flame retardants that are more environmentally friendly and harmless to health [7].

Phosphorus-based flame retardants have become a hot research topic due to their eco-friendly behaviors and efficient flame-retardant properties [8]. In addition, it has been proved that metal components can efficiently enhance the flame retardancy of polymer composites. Metal components bonded with a flame retardant's chemical nature can strengthen the thermal properties and flammability of polymers [9]. Moreover, different compositions of metal components and the chemical nature of flame retardants demonstrated that self-reinforced composite properties can be modified to achieve better properties [10].

The metal salt of phosphinates, with high phosphorus content, good thermal stability, and low affinity to moisture, has been developed as an eco-friendly flame retardant in recent years [11]. Metal phosphinates have advantages such as low water absorption, low current leakage, and high thermal stability. They have received great attention in flame-retardant material for the electrical and electronics (E&E) industry [6,7]. The flame-retardant effectiveness of the metal salt of phosphinates is related to the applied polymer and is also related to its chemical structure [12,13]. Commercial aluminum or zinc diethylphosphinate (AlPi or ZnPi) was an effective flame retardant for polyesters [14–18] and polyamide [19] epoxy resin [20,21]. However, AlPi or ZnPi alone in TPEE cannot achieve satisfying performance on fire retardancy at a dosage below 20 wt%. In addition, the mechanical properties of TPEE are seriously damaged at the loading of the filler to meet the flame retardancy requirement [22].

Balance among various properties such as thermal stability, mechanical properties, and flame retardancy is very important for high-performance flame-retardant TPEE. Tuning the performance of final materials through tailoring the structure of metal salts of the phosphinate has been carried out in recent years. A series of aluminum salts, such as aluminum hypophosphite [23], aluminum isobutylphosphinate [24], aluminum phenylphosphinate [25], aluminum hydroxymethylphosphinate [26], etc., have been comprehensively investigated.

In our previous investigation, aluminum beta-carboxylethylmethylphosphinate, multi-arm aluminum phosphinates, and amide-containing phosphinate salts with varying metal cation and organic groups as flame retardants for epoxy, polybutylene terephthalate and TPEE have been reported [22,27–29]. It was found that the aluminum phosphinate with a longer alkyl group showed better compatibility and fewer negative effects on the mechanical properties of EP and polybutylene terephthalate.

To achieve a required flame-retardant rank while maintaining the mechanical properties at a satisfying stand, adding a synergistic agent to reduce the dosage of the flame retardant is a practical approach. For instance, Wang prepared an aluminum diethyl hypophosphite intercalation-modified montmorillonite nano-size flame retardant (AlPi-MMT) for TPEE. It was found that TPEE with 15 wt% AlPi-MMT exhibited better char formation and flame-retardant properties compared to those incorporating 15 wt% of AlPi or MMT alone [30]. Compared to inorganic compounds, the organic compounds showed better compatibility with TPEE. Some benzene or hetero-ring-rich compounds such as novolac [31], Belta-cyclodextrin [32], and triazine-containing compounds [33] have high-char-forming ability and good compatibility with TPEE. This is because benzene is a key component of residual charring, and benzene-rich compounds can effectively improve the charring performance of composites, thereby enhancing flame retardancy. For instance, Wang synthesized the triazine-based hyper-branched charring agent (CDS) as a char agent for TPEE/aluminum diethlyphosphinate flame-resistant composites. The results showed that the CDS, which could inhibit the melt dropping, improved the fire retardancy and mechanical properties [34]. Wu's group reported a triazine-boron flame retardant (CPB). The CPB can improve the overall residual carbon performance of the material. When only 20% CPB is added, the TPEE composite achieves a high LOI value of 30.2% and passes the UL 94 V-0 rating [35].

In this work, aluminum butylmethylphosphinate (AiBMP), which has a chemical structure different from aluminum diethylphosphinate, was applied in the TPEE. AiBMP is a highly efficient flame retardant for TPEE. With 15 wt% AiBMP added, TPEE passed the UL 94 V-0 ranking with the LOI, improving from 20% to 36%. Phenolphthalein (Ph), a benzene-rich compound, was chosen as the additive combing with AiBMP to enhance the mechanical properties of TPEE. The advantages of phenolphthalein were identified by comparing the properties of TPEE/AiBMP/Ph and TPEE/AiBMP at the fixed filler loading.

2. Experimental Section

2.1. Materials

Thermoplastic elastomer (TPEE, hardness: 55D, melt flow index: 8.6 g/10 min at flowing rate of 2.16 kg at 220 °C, density: 1.17 g/cm^3) was provided by Haiso Plastics Co., Ltd. (Wuhan, China). Aluminum butylmethylphosphinate (AiBMP) with a purity above 98% was provided by Zhenghao Chemical Ltd. (Wuhan, China). Phenolphthalein (Ph) is a chemical reagent that was purchased from Guoyao Chemical Ltd. (Tianjin, China) and used as received.

2.2. Sample Preparation

TPEE pellets were melt-blended with additives using an XK-160 twin–screw internal mixture (Changzhou, Jiangsu, China) at 220 °C for 20 min with a screw speed of 120 rpm. The resulting mixture was then hot-pressed at 225 °C for 5 min under 10 MPa for a sheet of suitable thickness and size for further measurements. The neat TPEE used as a standard was treated in the same way. The composition of the formulations is shown in Table 1.

Table 1. Formula of the flame-retardant TPEE.

Sample	TPEE (wt%)	AilBMP (wt%)	Ph (wt%)	LOI (%)	UL94 Rating	Dripping
TPEE	100	0	0	20	-	yes
TPEE-10AiBMP	90	10	0	29	V-1	yes
TPEE-15AiBMP	85	15	0	36	V-0	no
TPEE-13AiBMP-2Ph	85	13	2	35	V-0	no
TPEE-11AiBMP-4Ph	85	11	4	33	V-1	yes

2.3. Measurements

FTIR was recorded on a Tensor 27 Bruker spectrometer (Bruker, Karlsruhe, German) with KBr powder.

Thermogravimetric analysis (TG) was carried out with a TSDT Q600 spectrometer (TA, New Castle, DE, USA) thermogravimetric analyzer. A sample of about 10 mg was heated in an alumina pan from 30 °C to 700 °C at a linear heating rate of 20 °C/min in the nitrogen atmosphere.

Limiting oxygen index (LOI) measurements were taken using an HC-2-type instrum ent (Jiangning Analytical Instrument Factory, Nanjing, China) in accordance with ASTM D2863-97. The dimensions of the sample were 100 × 6.5 × 3 mm^3, and five samples were carried out in the LOI test.

UL 94 vertical burning tests were conducted on a CZF-3 instrument (Jiangning Analytical Instrument Factory, Nanjing, China). The test was measured according to the vertical burning test standard ASTM D3801. The dimensions of the sample were 100 × 13 × 3 mm^3, and three samples were carried out in the UL 94 test.

The heat release rate (HRR) and total heat release (THR) were measured in an MCC-2 microcombustion calorimeter (MCC) (Govmark Organization Inc., Farmingdale, NY, USA); samples of about 5–7 mg were heated in alumina pans from 40 °C to 700 °C at a heating rate of 1 °C/s. The flow rate of N_2 and O_2 was 80 mL/min and 20 mL/min, respectively.

Differential scanning calorimetry (DSC) measurements were carried out on a TA DSC-Q20 (Waltham, MA, USA) at a heating rate of 10 °C/min in N_2. The flow rate of N_2 was 50 mL/min.

Scanning electron microscopy (SEM) measurements were conducted using a Philips XL-40 instrument with a voltage of 15 kV. The sample was adhibitted on the copper plate.

X-ray diffraction (XRD) measurements were performed with an X-ray diffractometer (X'Pert Powder PANalytical, Almelo, The Netherlands) with CuKa radiation (1.5418 Å) at a scanning rate (2θ) of 5°/min. An X-ray fluorescent spectroscopy (XRF) measurement was conducted with a ZSX Primus II (Rigaku, Tokyo, Japan) XRF spectrometer with a 35 kV Rhanode tube.

The tensile strength and elongation of all specimens were measured by a Reger mechanical instrument (SUNS Company, Shenzhen, China) at a speed of 200 mm min^{-1} according to ASTM D412 standards, and five samples were conducted in the tensile test.

3. Results and Discussion

3.1. Characterization of AiBMP

Figure 1a presents the XRD spectrum of AiBMP. There are peaks at delta 8.1°, 14.0°, 21.5° and 28°, indicating how AiBMP exhibits a crystal structure. The composition of AiBMP was further analyzed with XRF, as shown in Figure 1b. Al is 21.72 wt%, and P is 72.28 wt%; the atomic ratio of P to Al is 3.13:1 (the calculated value is 3:1).

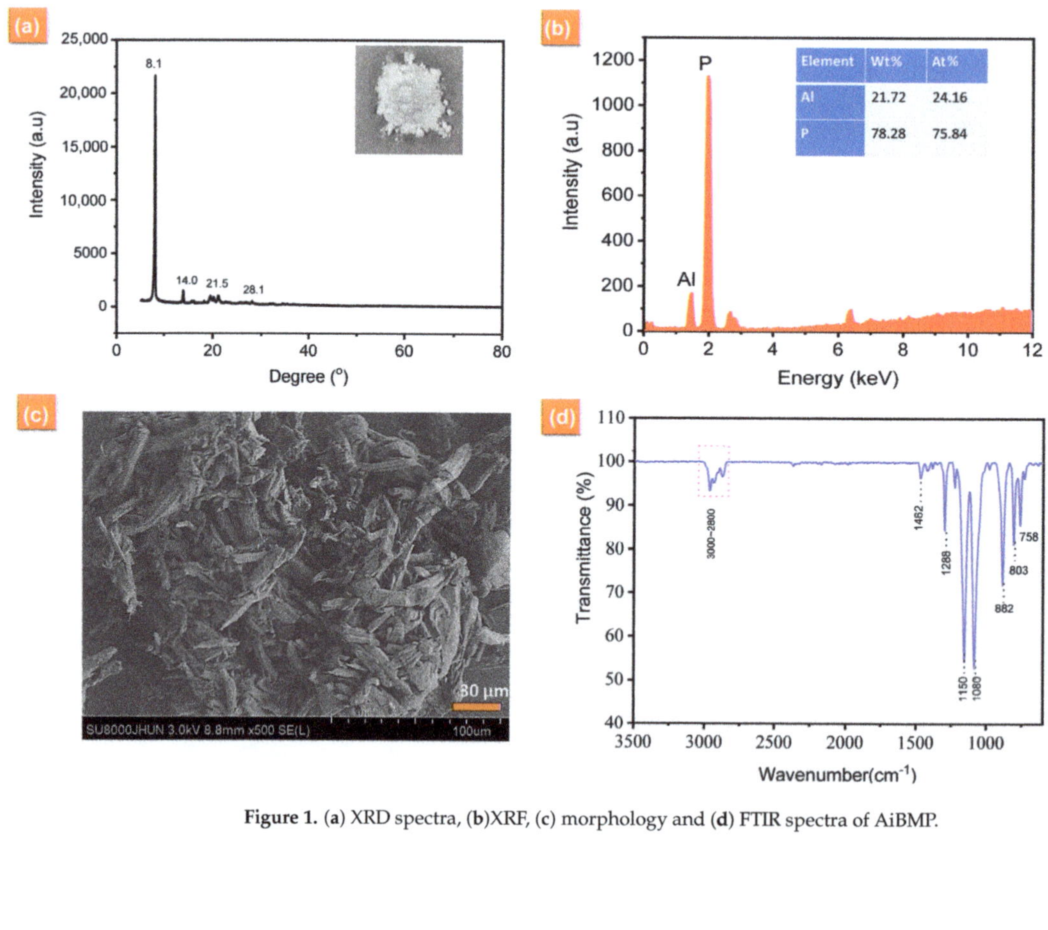

Figure 1. (**a**) XRD spectra, (**b**)XRF, (**c**) morphology and (**d**) FTIR spectra of AiBMP.

The SEM images of as-obtained AiBMP are shown in Figure 1c. AiBMP particles have an oval rod-like morphology with a length of about 50 μm, and rod-like AiBMP is composed of layers. The FTIR spectrum of AiBMP is shown in Figure 1d. The absorption peaks of about 3000–2800 cm^{-1} and 1462 cm^{-1}, 803 cm^{-1}, and 758 cm^{-1} are attributed to CH_3 and CH2. The peak at 1288 cm^{-1} is for P-CH_3. Other absorption peaks are 1150 cm^{-1} (P=O), 1080 cm^{-1} (P-O), and 882 cm^{-1} (Al-O) [23].

3.2. Combustion Characteristics: LOI, UL94, and MCC

The flame retardant properties of TPEE composites are studied using the LOI and UL94 vertical burning test, the results are presented in Table 1. TPEE is easily ignited, releasing smoke and serious flammable dripping The AiBMP is a very efficient flame retardant for TPEE. Adding 10 wt% AiBMP allows the LOI value of TPEE to increase from 20% to 29%. The sample TPEE-10 AiBMP was extinguished within 10 s after the first and second ignition. As the AiBMP content increased to 15 wt%, the LOI of TPEE-15AiBMP reached 36% with the V-0 rating achieved, and no dripping occurred during burning. Adding phenolphthalein has little influence on the fire-resistant properties of TPEE/AiBMP at low dosage (2 wt%). The LOI value of TPEE/TPEE-13AiBMP-2Ph is 35%, and the sample passed the V-0 rating. As the dosage of phenolphthalein increased to 4 wt%, LOI values of the samples TPEE-11AiBMP-4Ph reduced to 33%. The melt dripping was observed after the second ignition.

Figure 2 shows the heat release rate variation in samples with temperatures investigated with MCC. Data such as the peak of the heat release rate (PHRR), the temperature of PHRR (T_{PHRR}), and the total heat release (THR) are presented in Table 2. Adding AiBMP reduces the PHRR and THR of the TPEE greatly. For instance, the PHRR and THR values of TPEE-15AiBMP were reduced by 55.5% and 14.57%, respectively, compared to the TPEE. In addition, the T_{PHRR} of TPEE containing AiBMP shifted slightly to a lower temperature due to the catalytic effect of AiBMP. When the total filler loading was kept at 15 wt%, the AiBMP was partially replaced with phenolphthalein; the PHRR, THR and T_{PHRR} of TPEE-13AiBMP-2Ph was close to that of TPEE-15AiBMP, while the THR and PHRR of TPEE-11AiBMP-4Ph increased to 23.9 kJ/g and 591.9 W/g, respectively. The above results indicate that phenolphthalein shows a negative effect on the heat release of TPEE at a high dosage (4 wt%).

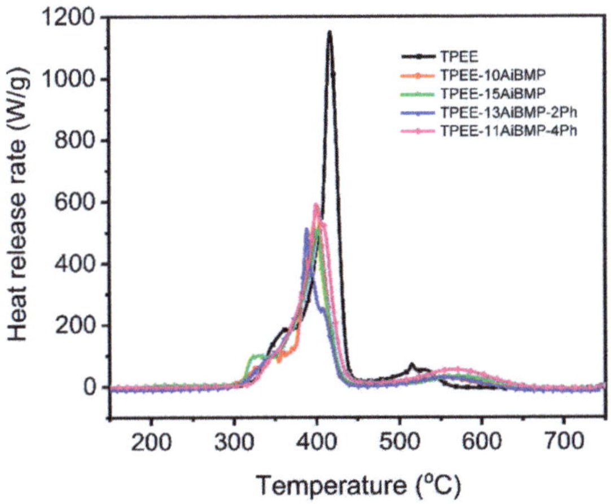

Figure 2. Heat release rate vs. temperature of samples.

Table 2. MCC data of the samples.

Sample	PHRR (W/g)	THR (kJ/g)	T_{PHRR} (°C)
TPEE	1150.4	24.7	417.7
TPEE-10AiBMP	588.1	22.3	402.7
TPEE-15AiBMP	511.9	21.2	403.4
TPEE-13AiBMP-2Ph	513.6	20.9	390.1
TPEE-11AiBMP-4Ph	591.9	23.8	407.4

3.3. Thermal Decomposition Behaviors

Figure 3 presents the TG-DTG curves of TPEE, phenolphthalein, AiBMP, and TPEE composites in N_2. Detailed data, including temperature at 5% weight loss ($T_{5\%}$), the maximum rate degradation temperature (T_{max}), the maximum decomposition rate (DTGmax), and char yields at 700 °C are summarized in Table 3. AiBMP is a highly thermal stable compound with $T_{5\%}$ 418.1 °C, T_{max} 486.1 °C, and residues of about 27.6% at 700 °C. Phenolphthalein decomposes very slowly. The residues of 35.4% are retained at 700 °C, demonstrating that phenolphthalein is a good char-forming agent. TPEE starts to lose weight at 373.2 °C with T_{max} 407.1 °C. It decomposes very fast and loses almost all of its weight before 420 °C.

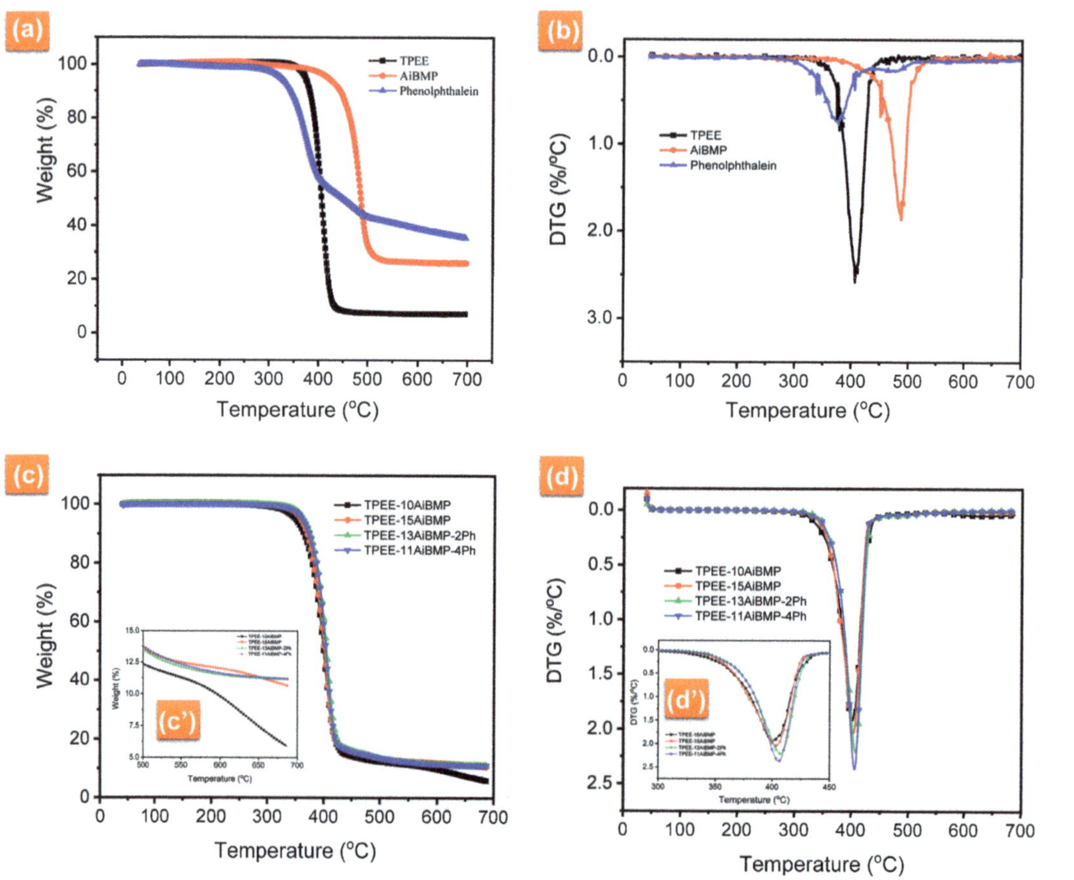

Figure 3. (**a**) TG and (**b**) DTG curves of TPEE, phenolphthalein and AiBMP in N_2. (**c**) TG and (**d**) DTG curves of TPEE composites in N_2. (**c′**) TG curves of TPEE composites between 500–700 °C in N_2, (**d′**) DTG curves of TPEE composites between 300–450 °C in N_2.

Table 3. TG-DTG data of the samples in N_2.

Sample	$T_{5\%}$ (°C)	T_{max} (°C)	DTG_{max} (%/°C)	Residues at 700 °C (%)
TPEE	373.2	407.1	2.58	4.30
AiBMP	418.1	486.1	1.86	26.0
phenolphthalein	323.4	375.8	0.78	35.4
TPEE-10AiBMP	348.7	400.3	1.93	5.91
TPEE-15AiBMP	357.8	402.8	1.90	10.7
TPEE-13AiBMP-2Ph	363.1	407.4	2.21	11.2
TPEE-11AiBMP-4Ph	361.2	405.5	2.37	11.2

With AiBMP added, the $T_{5\%}$ shifts to a lower temperature because of the catalytic effect of aluminum cation. The TPEE contains the ester group. It is well known that the thermal degradation of the polyester is easier when processed by the acid catalyst. AiBMP is a Lewis acid; it can accelerate the degradation of the TPEE in the early stage. The catalytic effect of the metal/organic complex has been reported in the literature [18,19,22,23]. However, the addition of AiBMP reduces the decomposition rate and improves the char yield of the TPEE. For instance, the DTG_{max} of TPEE is 2.58%/°C with a char yield of 4.3% at 700 °C. The DTG_{max} of TPEE-10AiBMP is 1.93%/°C, and the char yield is 5.94% at 700 °C. In the case of TPEE-15AiBMP, the DTG_{max} further reduced to 1.90%, and a char yield as high as 10.65 was achieved. Incorporating phenolphthalein increases the $T_{5\%}$ of the composites. The reason for this is that the phenol group of phenolphthalein as inhibitors decrease the activity of the free radical. Subsequently, the decomposition of TPEE/AiBMP/Ph moves to a higher temperature. In addition, the char residues of TPEE-13AiBMP-2Ph are higher than TPEE-15AiBMP. However, the DTG_{max} of TPEE-11AiBMP-4Ph increases, and the char yield is almost unchanged, with the dosage of phenolphthalein increasing to 4 wt%.

3.4. Residue Characterization

Figure 4 presents the digital and SEM photographs of the residues of TPEE (a), TPEE-15AiBMP (Figure 4b), and TPEE-13AiBMP-2Ph (c) collected by heating samples at 500 °C for 3 min in a muffle oven under a nitrogen atmosphere. The residues of TPEE are loose and composed of tiny ashes. The residues of TPEE-15AiBMP are smoother and condense with cracks and holes on the surface. SEM photo shows holes of about 3–8 um in size. The holes that arise are volatile during combustion. In addition, lots of micro-spheres are seen in the residues of TPEE-15AiBMP under SEM. The micro-spheres are formed by molten TPEE covering AiBMP residues. The micro-spheres are a barrier to hinders TPEE's flow. The surface of TPEE-13AiBMP-2Ph residues is very condensed, smooth, and brilliantly black. A few tiny micro-spheres are found on the surface under SEM. The results from Figure 4 show that AiBMP acts in both the gas and condensed phases. The phenolphthalein acts mainly in the condensed phase by constituting a continuous barrier to prevent or slow down the diffusion of molten TPEE and volatile.

Figure 5 shows the heating and later cooling of the samples recorded using DSC. The crystallization temperature (Tc) and enthalpy (ΔHc) during cooling are listed in Table 4. The Tc of TPEE is 145 °C, and the Tc of TPEE/AiBMP composites moves to around 155 °C. The reason for this is that the AiBMP particles function as a nucleation agent to inhibit super-cooling; later, the chain-folding occurs at a higher temperature.

Table 4. Data from DSC curves of the samples.

Sample	Tc (°C)	ΔHc (J/g)
TPEE	145.0	82.2
TPEE-10AiBMP	154.0	87.7
TPEE-15AiBMP	155.8	50.2
TPEE-13AiBMP-2Ph	155.6	9.1
TPEE-11AiBMP-4Ph	155.1	10.3

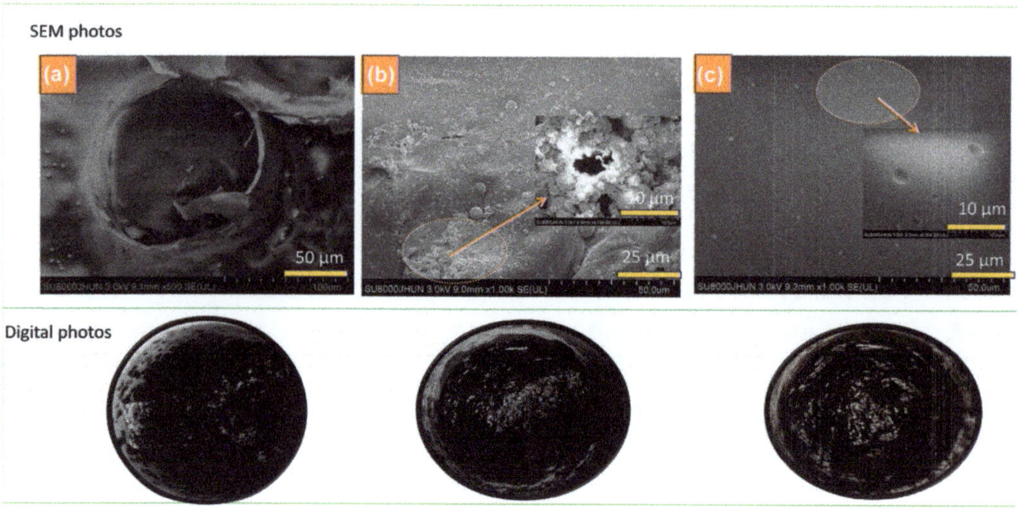

Figure 4. SEM (upper) and digital photos (bottom) of the residues TPEE (**a**), TPEE-15AiBMP (**b**) and TPEE-13AiBMP-2Ph (**c**) obtained by heating samples at 500 °C for 5 min.

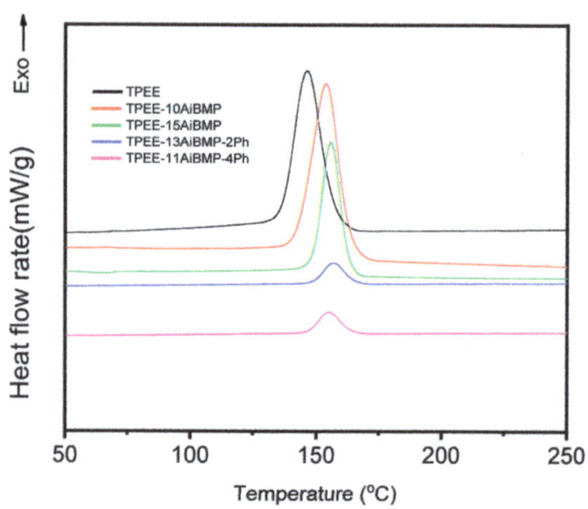

Figure 5. DSC curves of heating and subsequent cooling for samples.

The ΔHc value is the degree of crystallinity. The higher ΔHc value indicates increased crystallinity. The ΔHc value of TPEE is 82.8 J/g. The ΔHc values of TPEE-10AiBMP and PEE-15AiBMP are 87.7 J/g and 50.2 J/g, respectively. The reason for this is that AiBMP particles act as a nucleation agent at low dosage (10 wt%) and promote the crystallization of TPEE, while the AiMBP particles dilute the local concentration of the TPEE at a high dosage (15 wt%) and retard the chain folding. As phenolphthalein was added, the huge benzene group of phenolphthalein, as well as hydrogen bonding between the phenolphthalein and TPEE, constrained the chain movement and results of the degree of decreasing crystallinity. Subsequently, the ΔHc values of TPEE-13AiBMP-2Ph and TPEE-11AiBMP-4Ph reduced to 9.6 J/g and 10.3 J/g.

3.5. Mechanical Properties

The tensile properties of TPEE composites are shown in Figure 6, The tensile strength and elongation at breaking both decrease with increasing the loading of AiMBP and then increase with the addition of phenolphthalein. For instance, the tensile strength and elongation at break of TPEE-15AiBMP were reduced by 53.0% and 42.8%, respectively, compared to the TPEE. Similar results were reported in a previous study. The reason corresponds to the poor compatibility between TPEE and AiBMP, leading to discontinuities at the particle/matrix interface.

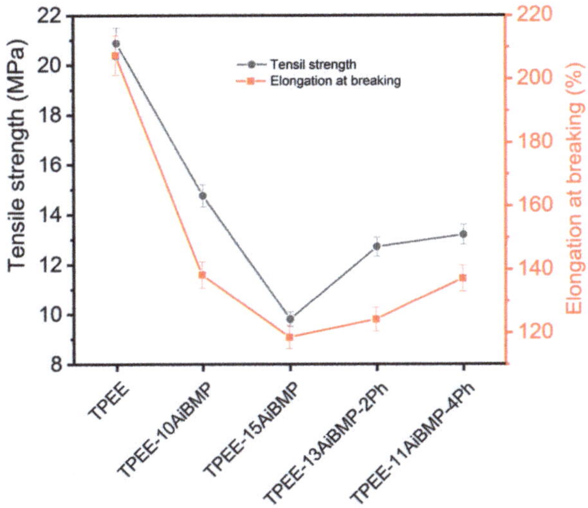

Figure 6. Tensile properties of samples at a crosshead speed of 25 mm/min.

Adding phenolphthalein improved the mechanical properties of composites. For instance, the tensile strength and elongation at break TPEE/13AiBMP/2Ph composites increased by 29.63% and 4.8%, respectively, in comparison to TPEE-15AiBMP. The enhancement in tensile properties resulted from the improved compatibility of phenolphthalein and the decreased mass fraction of AiBMP.

Figure 7 presents the SEM photos of the fractured cross-section of the samples after the tensile test. The AiBMP rods are not found in the matrix because they are wrapped by the matrix. There are lots of white irregular wrinkles on the surface of TPEE-15AiBMP. Many dimples are seen on the surface of TPEE-13AiBMP-2Ph. Generally, breaking often occurs at the interface of the hard filler and the polymer matrix. These wrinkles are caused by the dilute deformation of the polymer near the AiBMP surface. Proportionally, substantial amounts of energy are needed to induce dilute deformation. The dimples are denser and smaller on the surface of TPEE-13AiBMP-2Ph than that of TPEE-15AiBMP, showing how phenolphthalein enhances the interface adhesion between AiBMP and TPEE.

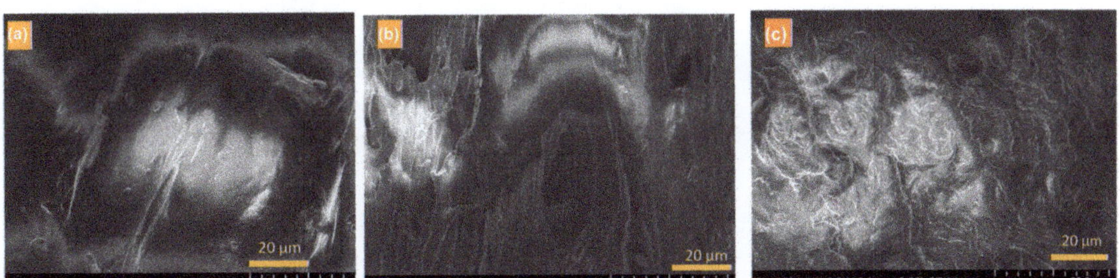

Figure 7. SEM images of fractured surface for TPEE (**a**), TPEE-15AiBMP (**b**) and TPEE-13AiBMP-2Ph (**c**).

4. Conclusions

In this study, a novel aluminum salt of organic phosphinate AiBMP as the main flame retardant and benzine-rich compound phenolphthalein as the synergistic agent were applied as the flame retardants of TPEE. The results found that AiBMP alone is an efficient flame retardant of TPEE. The TPEE containing 15 wt% AiBMP passed the UL 94 V-0 rating and achieved a high LOI value of 30.2%. TG analysis showed how the addition of AiBMP reduced the decomposition rate and improved the char yield of the TPEE. Results from the SEM of residues indicate that AiBMP acts in both gas and condensed phases.

Phenolphthalein exhibits an impressive enhancement in mechanical properties of TPEE. When the total filler loading is kept at 15 wt%, the tensile strength and elongation of the sample TPEE-13AiBMP-2Ph increased by 29.63% and 4.8%, respectively, in comparison to the TPEE-15AiBMP and flame-retardant properties of TPEE. The plasticizing effect of phenolphthalein and the uniform dispersion of AiBMP are the main reasons for the improved tensile strength and high retention of the elongation at the breaking of TPEE-13AiBMP-2Ph.

Moreover, the benzene-rich structure of phenolphthalein endows good char formation and enhances flame retardancy in the solid phase. TG results showed that the char yield of TPEE-13AiBMP-2Ph at 700 °C improved by 0.5% compared to that of TPEE-15AiBMP. SEM morphology indicated that the surface of TPEE-13AiBMP-2Ph residues is more condensed and smoother than that of TPEE-15AiBMP.

With the development and popularization of electric vehicles, corresponding charging facilities with high insulation and safety are required. The AiBMP and phenolphthalein-based flame retardant TPEE have potential applications as a charging cable and pile. In future work, we aim to evaluate the hydrolysis resistance and electrical properties of TPEE/AiBMP/Ph composites under a high-voltage environment.

Author Contributions: X.Y.: data curation, investigation, writing—original draft; Y.Z.: methodology, resources; X.X. and K.Z.: data curation, investigation; J.C. and L.Z.: formal analysis, software, visualization; J.L.: conceptualization, writing—review and editing; X.L.: supervision. All authors have read and agreed to the published version of the manuscript.

Funding: This research was funded by the Program of National Key Research and Development of China [2022YFB3603702], the National key Laboratory on ship vibration and noise [6142204200608], the Fundamental Research Project of Jianghan University [2023ZDCX01], and the Academic Innovation Program for The Outstanding Youth Groups of Universities in Hubei [T201318].

Institutional Review Board Statement: Not applicable.

Data Availability Statement: Data are contained within the article.

Acknowledgments: The authors would like express gratitude for the financial support received from the Program of National Key Research and Development of China [2022YFB3603702]. The authors would like to express gratitude for the financial support received from the National key Laboratory on ship vibration and noise (6142204200608), the Fundamental Research Project of Jianghan University [2023ZDCX01], and the Academic Innovation Program for The Outstanding Youth Groups of Universities in Hubei [T201318].

Conflicts of Interest: The authors declare no conflicts of interest.

References

1. Drobny, J.G. Thermoplastic Polyether Ester Elastomers. In *Handbook of Thermoplastic Elastomers*, 2nd ed.; William Andrew Pub.: Norwich, NY, USA, 2014; pp. 271–286.
2. Liu, C.; Zhang, L.; Mu, L.; Zuo, D.; Wang, Y.; Wu, W. Synergistic Effects between a Triazine-Based Charring Agent and Aluminum Phosphinate on the Intumescent Flame Retardance of Thermoplastic Polyether Ester. *J. Macromol. Sci. Part A* **2019**, *56*, 723–732. [CrossRef]
3. Wang, Z.; Wu, W.; Zhang, W.; Xu, X.; Lin, H.; Wang, W. One-Spot Synthesis of a Benzene-Rich Triazine-Based Hyperbranched Charring Agent and Its Efficient Intumescent Flame-Retardant Performance for Thermoplastic Polyester Elastomer. *Arab. J. Chem.* **2023**, *16*, 104861. [CrossRef]

4. Birley, A.W. *Thermoplastic Elastomers: A Comprehensive Review*; Legge, N.R., Holden, G., Schroeder, H.F., Eds.; Carl Hanser Verlag: Munich, Germany, 1987.
5. Liu, B.; Zhao, H.; Wang, Y. Advanced Flame-Retardant Methods for Polymeric Materials. *Adv. Mater.* **2021**, *34*, e2107905. [CrossRef]
6. Weil, E.D.; Levchik, S.V. 6—Flame Retardants in Commercial Use or Development for Thermoplastic Polyesters. In *Flame Retardants*; Elsevier: Amsterdam, The Netherlands, 2016; pp. 141–160.
7. Weil, E.D.; Levchik, S.V. Further Sources for Flame Retardancy Information (Updated 2014). In *Flame Retardants*, 2nd ed.; Elsevier: Amsterdam, The Netherlands, 2016; pp. 339–348.
8. Abd El-Wahab, H.; Abd El-Fattah, M.; El-Alfy, H.M.Z.; Owda, M.E.; Lin, L.; Hamdy, I. Synthesis, and characterisation of sulphonamide (Schiff base) ligand and its copper metal complex and their efficiency in polyurethane varnish as flame retardant and antimicrobial surface coating additives. *Prog. Org. Coat.* **2020**, *142*, 105577. [CrossRef]
9. Ilyas, R.A.; Sapuan, S.M.; Asyraf, M.R.M.; Dayana, D.A.Z.N.; Amelia, J.J.N.; Rani, M.S.A.; Norrrahim, M.N.F.; Nurazzi, N.M.; Aisyah, H.A.; Sharma, S.; et al. Polymer Composites Filled with Metal Derivatives: A Review of Flame Retardants. *Polymers* **2021**, *13*, 1701. [CrossRef] [PubMed]
10. Pan, Y.T.; Zhang, Z.; Yang, R. The rise of MOFs and their derivatives for flame retardant polymeric materials: A critical review. *Compos. Part B Eng.* **2020**, *199*, 108265. [CrossRef]
11. Wu, W.; Zhong, Y.; Wu, R.; Jiang, C.; Lin, X. Preparation of Flame Retarded Thermoplastic Polyester-Ether Elastomer by Adding Halogen-Free Additives and Novolac. *Gaofenzi Cailiao Kexue Yu Gong Cheng/Polym. Mater. Sci. Eng.* **2015**, *31*, 153–157.
12. Zhong, Y.; Jiang, C.; Ruan, M.; Chen, Y.; Wu, W. Preparation, Thermal, and Flammability of Halogen-Free Flame Retarding Thermoplastic Poly(Ether-Ester) Elastomer/Montmorillonite Nanocomposites. *Polym. Compos.* **2014**, *37*, 700–708. [CrossRef]
13. Wang, A.; Zhang, F.; Xing, L.; Zhu, Y.; Xie, W.; Chen, X.; Cheng, J.; Cheng, Y. Effect of Aluminum Diethylphosphinate and Its Synergist on Flame-Retardant Effects of Epoxy Resin. *J. Therm. Anal. Calorim.* **2022**, *147*, 7277–7287. [CrossRef]
14. Levchik, S.V.; Weil, E.D. Flame Retardancy of Thermoplastic Polyesters—A Review of the Recent Literature. *Polym. Int.* **2005**, *54*, 11–35. [CrossRef]
15. Bauer, H.; Hoerold, S.; Krause, W. Salts of Alkyl Esters of Carboxyethyl (Alkyl)Phosphinic Acid. US2008/0188598A1, 7 August 2008.
16. Bauer, H.; Krause, W.; Sicken, M.; Weferling, N. Use of Salts of Dialkylphosphinic Acids. EP1544206A1, 22 June 2005.
17. Salaün, F.; Lemort, G.; Butstraen, C.; Devaux, E.; Capon, G. Influence of Silica Nanoparticles Combined with Zinc Phosphinate on Flame Retardant Properties of PET. *Polym. Adv. Technol.* **2017**, *28*, 1919–1928. [CrossRef]
18. DoğAn, M.; Erdoğan, S. Mechanical, Thermal, and Fire Retardant Properties of Poly (Ethylene Terephthalate) Fiber Containing Zinc Phosphinate and Organo-Modified Clay. *J. Therm. Anal. Calorim.* **2013**, *112*, 871–876. [CrossRef]
19. Tomiak, F.; Schoeffel, A.; Rathberger, K.; Drummer, D. Expandable Graphite, Aluminum Diethylphospinate and Melamine Polyphosphate as Flame Retarding System in Glass Fiber-Reinforced PA6. *Polymers* **2022**, *14*, 1263. [CrossRef]
20. Liu, X.Q.; Liu, J.Y.; Cai, S.J. Comparative Study of Aluminum Diethylphosphinate and Aluminum Methylethylphosphinate-Filled Epoxy Flame-Retardant Composites. *Polym. Compos.* **2012**, *33*, 918–926. [CrossRef]
21. Chen, W.; Liu, P.; Cheng, Y.; Liu, Y.; Wang, Q.; Duan, W. Flame Retardancy Mechanisms of Melamine Cyanurate in Combination with Aluminum Diethylphosphinate in Epoxy Resin. *J. Appl. Polym. Sci.* **2019**, *136*, 47223. [CrossRef]
22. Zou, L.; Zhou, M.; Liu, J.; Liu, X.; Chen, J.; Hu, Q.; Peng, S. Flame-Retardant Thermoplastic Polyester Based on Multiarm Aluminum Phosphinate for Improving Anti-Dripping. *Thermochim. Acta* **2018**, *664*, 118–127. [CrossRef]
23. Brehme, S.; Schartel, B.; Goebbels, J.; Fischer, O.; Pospiech, D.; Bykov, Y.; Döring, M. Phosphorus Polyester versus Aluminium Phosphinate in Poly (Butylene Terephthalate) (PBT): Flame Retardancy Performance and Mechanisms. *Polym. Degrad. Stab.* **2011**, *96*, 875–884. [CrossRef]
24. Holdsworth, A.F.; Horrocks, A.R.; Kandola, B.K. Novel Metal Complexes as Potential Synergists with Phosphorus Based Flame Retardants in Polyamide 6.6—ScienceDirect. *Polym. Degrad. Stab.* **2020**, *179*, 109220. [CrossRef]
25. Cheng, X.; Wu, J.; Yao, C.; Yang, G. Flame-Retardant Mechanism of Zinc Borate and Magnesium Hydroxide in Aluminum Hypophosphite–Based Combination for TPE-S Composites. *J. Fire Sci.* **2019**, *37*, 073490411985127. [CrossRef]
26. Jian, R.K.; Chen, L.; Chen, S.Y.; Long, J.W.; Wang, Y.Z. A Novel Flame-Retardant Acrylonitrile-Butadiene-Styrene System Based on Aluminum Isobutylphosphinate and Red Phosphorus: Flame Retardance, Thermal Degradation and Pyrolysis Behavior. *Polym. Degrad. Stab.* **2014**, *109*, 184–193. [CrossRef]
27. Cheng, X.; Wu, J.; Yao, C.; Yang, G. Aluminum Hypophosphite and Aluminum Phenylphosphinate: A Comprehensive Comparison of Chemical Interaction during Pyrolysis in Flame-Retarded Glass-Fiber-Reinforced Polyamide 6. *J. Fire Ences* **2019**, *37*, 073490411983620. [CrossRef]
28. Lin, G.P.; Chen, L.; Wang, X.L.; Jian, R.K.; Zhao, B.; Wang, Y.Z. Aluminum Hydroxymethylphosphinate and Melamine Pyrophosphate: Synergistic Flame Retardance and Smoke Suppression for Glass Fiber Reinforced Polyamide 6. *Ind. Eng. Chem. Res.* **2013**, *52*, 15613–15620. [CrossRef]
29. Liu, X.; Liu, J.; Chen, J.; Cai, S.; Hu, C. Novel Flame-Retardant Epoxy Composites Containing Aluminium β-Carboxylethylmethylp hosphinate. *Polym. Eng. Sci.* **2015**, *55*, 657–663. [CrossRef]

30. Wang, Z.; Wu, W.; Zhang, W.; Shen, H.; Feng, Y.; Li, J. Preparation of aluminum diethyl hypophosphite intercalation-modified montmorillonite AlPi-MMT and its effect on the flame retardancy and smoke suppression of thermoplastic polyester elastomer. *J. Appl. Polym. Sci.* **2023**, *141*, 55103. [CrossRef]
31. Jana, S.C. *Proceedings of PPS-30: The 30th International Conference of the Polymer Processing Society—Conference Papers*; AIP Publishing: Melville, NY, USA, 2015.
32. Zou, L.; Liu, J.; Liu, X.; Wang, X.; Chen, J. Synthesis and Performance of Star-Shaped Aluminum Phosphinate Flame Retardant. *J. Therm. Anal. Calorim.* **2016**, *124*, 1399–1409. [CrossRef]
33. Zhong, Y.; Wu, W.; Wu, R.; Luo, Q.; Wang, Z. The Flame Retarding Mechanism of the Novolac as Char Agent with the Fire Retardant Containing Phosphorous–Nitrogen in Thermoplastic Poly (Ether Ester) Elastomer System. *Polym. Degrad. Stab.* **2014**, *105*, 166–177. [CrossRef]
34. Zhang, L.; Wu, W.; Li, J.H.; Wang, Z.; Wang, L.; Chen, S. New Insight into the Preparation of Flame-Retardant Thermoplastic Polyether Ester Utilizing β -Cyclodextrin as a Charring Agent. *High Perform. Polym.* **2017**, *29*, 422–430. [CrossRef]
35. Wang, Z.; Wu, W.; Liu, Z.; Shen, H.; Feng, Y. Study on novel boron-containing triazine flame retarded thermoplastic polyester elastomer composites and the flame-retardant mechanism. *React. Funct. Polym.* **2023**, *190*, 105621. [CrossRef]

Disclaimer/Publisher's Note: The statements, opinions and data contained in all publications are solely those of the individual author(s) and contributor(s) and not of MDPI and/or the editor(s). MDPI and/or the editor(s) disclaim responsibility for any injury to people or property resulting from any ideas, methods, instructions or products referred to in the content.

polymers

MDPI

Article

Development of a Zr-Based Metal-Organic Framework (UiO-66) for a Cooperative Flame Retardant in the PC/ABS

Shaojun Chen [1], Zerui Chen [1], Weifeng Bi [1], Wei Du [1], Ling Lin [1], Dasong Hu [1] and Haitao Zhuo [2,*]

[1] College of Materials Science and Engineering, Shenzhen University, Shenzhen 518053, China; chensj@szu.edu.cn (S.C.); czr1079975344@163.com (Z.C.); a773122527@163.com (W.B.); 15368563584@163.com (W.D.); 2023200132@email.szu.edu.cn (L.L.); 2021200135@email.szu.edu.cn (D.H.)

[2] College of Chemistry and Environment Engineering, Shenzhen University, Shenzhen 518053, China

* Correspondence: haitaozhuo@szu.edu.cn

Abstract: Polycarbonate/acrylonitrile butadiene styrene (PC/ABS) blends are widely used as engineering plastic alloys; however, they have a low fire safety level. To improve the flame-retardant property of PC/ABS, a zirconium-based metal-organic framework material (UiO-66) was synthesized with zirconium chloride and terephthalic acid and used as a flame-retardant cooperative agent. Its flame-retardant performance and mode of action in the PC/ABS blends were carefully investigated. The results showed that UiO-66 had good thermal stability and delayed the pyrolysis of the materials, thus significantly enhancing the efficiency of intumescent flame retardants. By compounding 7.0 wt% hexaphenyloxy-cyclotri-phosphazene (HPCTP) with 3.0 wt% UiO-66, the PC/ABS blends reached a limiting oxygen index value of 27.0% and V0 rating in the UL-94 test, showing significantly improved resistance to combustion dripping. In addition, UiO-66 enhanced the smoke and heat suppression characteristics of the intumescent flame-retardant materials. Finally, the flame-retardant mode of action in the blends was indicative of UiO-66 having a cooperative effect on the flame-retardant performance of PC/ABS/HPCTP materials. This work provides good ideas for further development of the flame-retardant ABS/PC.

Keywords: metal-organic framework; flame retardance; cooperative effect; PC/ABS

check for updates

Citation: Chen, S.; Chen, Z.; Bi, W.; Du, W.; Lin, L.; Hu, D.; Zhuo, H. Development of a Zr-Based Metal-Organic Framework (UiO-66) for a Cooperative Flame Retardant in the PC/ABS. *Polymers* **2024**, *16*, 2083. https://doi.org/10.3390/polym16142083

Academic Editor: Bob Howell

Received: 25 June 2024
Revised: 18 July 2024
Accepted: 19 July 2024
Published: 21 July 2024

Copyright: © 2024 by the authors. Licensee MDPI, Basel, Switzerland. This article is an open access article distributed under the terms and conditions of the Creative Commons Attribution (CC BY) license (https://creativecommons.org/licenses/by/4.0/).

1. Introduction

Polycarbonate (PC)/acrylonitrile butadiene styrene (ABS) is a typical polymer blend made of PC and ABS resin. PC confers strength and a high heat-deflection temperature, whereas ABS possesses an easy processing ability and is cheap [1,2]. Thus, PC/ABS has become one of the most widely used plastics in the electrical and electronics industries [3]. However, it is easily ignited in the air. To expand its flame retardancy application in automotive parts, 5G communication equipment, and household appliances, it is, therefore, necessary to improve the flame-retardant properties of PC/ABS. Various kinds of flame retardants have been used in the PC/ABS. For example, phosphazene compounds are known for releasing less smoke and having a lower toxicity than other phosphorus compounds when burning, making them environmentally friendly flame retardants [4]. Particularly, hexaphenyloxy-cyclotri-phosphazene (HPCTP) provides flame-retardant properties through condensed and partial gas-phase mechanisms. For example, the phosphonitrile-triazine bis-alkyl flame retardant (A3) with an aniline end-group could form a more complete and dense char layer when PC burnt and released PO· to act as a flame retardant in the gas phase [5]. It was also reported that the aromatic polyimide (API) charging agent could fill in the toughness gap of HPCTP and then enhance the flame-retardant properties of PC, acting as a co-efficient flame retardant to improve notched impact properties [6]. During combustion, HPCTP decomposes into phosphoric acid, polyphosphoric acid, and N_2 non-flammable gases. This promotes the dehydration of PC/ABS into a carbonaceous char that expands to form an intumescent char layer, thus inhibiting heat transfer and oxygen access [7]. However, many previous studies have shown that HPCTP has a relatively low flame-retardant efficiency, requiring the addition of 15 parts of cooperative agents to achieve a UL94 V0 rating [4,8]. Therefore, there is an urgent need to develop a new flame-retardant cooperative agent to modify the intumescent flame-retardant properties of HPCTP.

Metal-organic frameworks (MOFs) are a class of porous crystalline materials formed via the coordination between metal ions or clusters and organic ligands. Compared to traditional inorganic porous materials, MOFs have distinct advantages such as high porosity, tunable pore sizes and topologies, organic–inorganic hybrid properties, and the natural coexistence of metallic and heteroatoms. Thus, they have a wide range of applications, such as for smart sensors [9], gas storage [10], CO_2 fixation [11], and are commonly referred to as "versatile materials". For example, carboxyl-substituted porphyrin derivatives were successfully synthesized via MOFs materials (UiO-66 and UiO-67) [12], and they contributed to the development of porphyrin-based functional materials. In recent years, MOFs have gained increasing attention as emerging flame retardants. Owing to their natural porous structure, tunable and modifiable characteristics, and rich metallic components, MOFs are beneficial for enhancing the fire safety performance of polymers. In 2008, researchers reported the successful and reproducible synthesis of a new zirconium-based MOF named UiO-66, which comprises zirconium clusters and organic ligands with the molecular formula $[Zr_6O_4(OH)_4L_6]_n$, where L represents linear dicarboxylic acid ligand [13]. However, scholars have found broad diffraction peaks in XRD patterns, indicating low crystallinity [14,15]. It implied that the synthesis of UiO-66 could be regulated with acetic acid to improve its regularity and particle size [16], thereby significantly enhancing the success rate of synthesis. In 2019, zirconium-based MOFs (Zr-BDC) were used to improve the fire safety performance of PC, increasing the time to ignition (TTI) and reducing fire hazards [17]. The addition of 4 wt% Zr-BDC allowed PC to achieve a UL-94 V0 rating, conferring PC with a better char-forming ability (catalytic charring) and smoke suppression property during pyrolysis, which are necessary for personnel evacuation and rescue measures. However, there are only a few reports on their application in PC/ABS blends.

In this work, a flame retardant with zirconium-based metal-organic frameworks (MOFs) was synthesized to enhance the flame-retardant properties of PC/ABS. First, MOFs (UiO-66) was synthesized by $Zr_6O_4(OH)_4$ clusters and 1,4-benzenedicarboxylate ligands. Then, a small amount of UiO-66 combined with HPCTP was melt-blended with PC/ABS to produce the intumescent flame-retardant material, termed UiO-66@HPCTP@PC/ABS. Finally, the structures of the UiO-66 samples was characterized, and their flame-retardant properties and flame-retardant mode of action in the blends were studied carefully. This work presents a highly efficient flame-retardant cooperative agent for the development of high-performance flame-retardant plastics.

2. Experimental Section

2.1. Materials

PC (injection grade, QiMei Company, Taiwan), ABS (injection grade, QiMei Company, Taiwan, China), zirconium chloride (98% $ZrCl_4$, Shanghai Aladdin Biochemical Company, Shanghai, China), terephthalic acid (99% TPA, Shandong Yusuo Chemical Technology Co., Ltd., Heze, China), hexaphenyloxy-cyclotri-phosphazene (99% HPCTP, Jinan Sino New Material Technology Co., Ltd., Heze, China), N,N-dimethylformamide (99.5% DMF, Shanghai Aladdin Biochemical Company, Shanghai, China), ethanol (99% EtOH, Shanghai Aladdin Biochemical Company, Shanghai, China), and acetic acid (99.8% EtOH, Shanghai Aladdin Biochemical Company, Shanghai, China) were used without further purification for the experiments.

2.2. Synthesis of UiO-66

The synthetic route of UiO-66 is presented in Figure 1, and the 3D/2D structure is shown in Figure S1. First, 1.98 g $ZrCl_4$ and 1.76 g TPA were dissolved in 150 mL anhydrous DMF, followed by sonication for 20 min. Then, 4 mL of acetic acid was added, followed by sonication for 10 min. The solution was transferred to a blue bottle and placed into a 120 °C electric blast drying chamber. After solvent removal for 36 h, the powder was cooled down to 25 °C, centrifuged at 6000 rpm for 15 min to filter the white precipitate, and washed with DMF and ethanol three times. Finally, UiO-66 was obtained after drying in an 85 °C constant-temperature vacuum drying oven for 36 h.

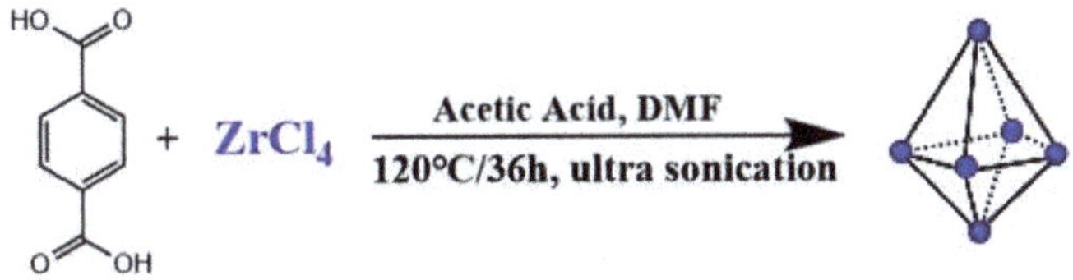

Figure 1. Synthetic route of UiO-66.

2.3. Synthesis of UiO-66@HPCTP@PC/ABS

The PC/ABS flame-retardant materials (UiO-66@HPCTP@PC/ABS) were prepared by melt blending and a twin-screw extrusion. The components and ratios of samples are shown in Table 1. First, PC and ABS resins were weighed and mixed (70:30) and dried in a 70 °C electric heating blast drying oven for 4 h. Then, specified amounts of HPCTP and UiO-66 were added to PC and ABS, according to Table 1. After all components were mixed evenly, the materials were added to a twin-screw extruder. The temperature of each zone in the twin-screw extruder was set at 210, 220, 230, 240, 250, 250, 250, 245, and 245 °C, and the screw speed was set to 15~20 rpm. After melt blending, pellets were prepared in the pelleting machine and further dried in an electric blast drying box at 70 °C for 4 h. The test specimens were injected according to the GB/T 17037.1-1997 standard or GB/T 9341-2008 [18,19]. The injection temperature was 230 °C, and the injection pressure was 90 MPa.

Table 1. Composition of flame-retardant PC/ABS blends.

Sample	PC/ABS (wt%)	HPCTP (wt%)	UiO-66 (wt%)
PC/ABS	70:30	0	0
HP-3.5	70:30	3.5	0
HP-7.0	70:30	7.0	0
HP-10.5	70:30	10.5	0
U-1.5	70:30	0	1.5
U-3.0	70:30	0	3
U-4.5	70:30	0	4.5
HPU-1	70:30	7.0	1
HPU-2	70:30	7.0	2
HPU-3	70:30	7.0	3

2.4. Characterizations

FT-IR spectroscopy was conducted on a Nicolet 6700 infrared spectrometer (Thermo Fisher Scientific, Waltham, MA, USA), with a scanning speed of 4 cm^{-1} and scanning range of 4000–500 cm^{-1}. The attenuated total reflection (ATR) mode and the KBr pressurization method was used to measure the samples.

X-ray photoelectron spectroscopy (XPS) was conducted on a Kratos Axis Ultra DLD XPS instrument from Shimadzu Corporation. The powder was cut into a 5 mm × 4 mm sample. The chemical states of C1s, Zr 3d, P 2p, and O1s, among others, were determined by scanning the sample surface.

SEM was conducted using a scanning electron microscope (NGB 4-DXS-10AC, Nanjing Grand Technology Co., Ltd., Nanjing, China). The sample was broken in liquid nitrogen and was adhered to the powder sample.

Differential scanning calorimetry (DSC) curves were recorded using a DSC 25 cell (TA Instrument, New Castle, DE, USA) with a nitrogen protection of 40 mL/min. Approximately 5–10 mg was dried in an 85 °C electric thermostatic drying oven for 2 h and pressed into an aluminum crucible tablet.

The TG and DTG curves were obtained using the TGA 55 (TA Instrument, New Castle, DE, USA) at a flow rate of 50 mL/min in a nitrogen atmosphere. The heating rate was 10 °C/min, and the temperature range varied from room temperature to 800 °C.

A vertical burning test was performed to measure flame-retardant performance. Each sample was analyzed three times, and the average of the three readings was used to determine its corresponding flame-retardant level.

The limiting oxygen index (LOI) was determined using a JF-3 oxygen index instrument (Shenzhen, China) according to standard GB/T 2406.2-2009 [20], with specimen dimensions of 150 mm × 4 mm × 10 mm.

The CCT was conducted using a British FTT cone calorimeter, according to ISO5660-1 and ASTM D7309 standards [21,22], with an irradiance power of 50 kW/m^2 and sample dimensions of 100 mm × 100 mm × 3 mm.

The fire growth index (FGI) was used to evaluate the fire safety performance of materials. FGI was calculated as the ratio of PHRR to HRR, with a lower value indicating that the material takes less time to reach a state of intense combustion, thus presenting a lower fire hazard.

Raman spectroscopy was performed using a Laser Confocal Microscope Raman Spectrometer (INVIA, RENISHAW, London, England). The excitation source was a 514.5 nm argon ion laser. The experiments were performed at 25 °C and tested at a scanning range of 500~2500 cm^{-1}.

TG-IR was performed on an FT-IR spectrometer (SPECTRUM TWO, PerkinElmer, MA, USA) equipped with a TGA analyzer (TGA 4000, PerkinElmer, MA, USA) through a heat transfer line, allowing for in situ characterization of decomposition products with a spectral resolution of 4 cm^{-1} and a scanning interval of 2.23 s.

XRD spectroscopy was performed on an X-ray diffractometer (Miniflex600, Rigaku, Tokyo, Japan), with a scanning speed of 10°/min and a scanning range of 5° to 70°.

3. Results and Discussion

3.1. Preparation of UiO-66

Infrared spectroscopy was used to characterize the synthesized UiO-66 and the selected flame retardant of HPCTP. As shown in the FT-IR spectra in Figure 2a, the asymmetric stretching vibration of Zr-(OC) appears at 669 cm^{-1}. The absorption peaks at 1017 cm^{-1} and 1507 cm^{-1} confirm the coordination of terephthalic acid ligands with Zr. The peaks near 1405 cm^{-1} and 1560 cm^{-1} belong to the symmetric and asymmetric vibration peaks of O-C-O in terephthalic acid ligands, respectively. The absorption at the range of 3200~3600 cm^{-1} features the characteristic peak of an O-H stretching vibration, indicating the successful synthesis of UiO-66 [23]. In the FT-IR spectrum of HPCTP, the characteristic peak for the stretching vibration of unsaturated =C-H in the benzene ring is observed at 3060 cm^{-1}. The peaks at 1589 cm^{-1} and 1487 cm^{-1} belong to the characteristic peaks of the benzene ring's skeletal vibration, indicating the presence of a benzene ring structure. The peaks at 1270 cm^{-1} and 1181 cm^{-1} belong to the P=N stretching vibration of cyclic triphosphazene, and the characteristic absorption peak at 954 cm^{-1} represents P-O-C. These results are consistent with previous reports [24].

In the XRD spectra, the main diffraction peaks of UiO-66 are essentially consistent with those of simulated UiO-66. There are distinct sharp peaks at 7.38°, 8.52°, 12.06°, 14.15°, and 14.78°, corresponding to the (111), (200), (220), (311), and (222) crystal planes, respectively. These results confirm the successful synthesis of UiO-66 [25]. Moreover, HPCTP has a high degree of crystallinity and good molecular symmetry, with many sharp diffraction peaks in its spectrum. The positions of peaks at 7.2°, 9.4°, 10.7°, 17.5°, and 20.4° match those previously reports [7].

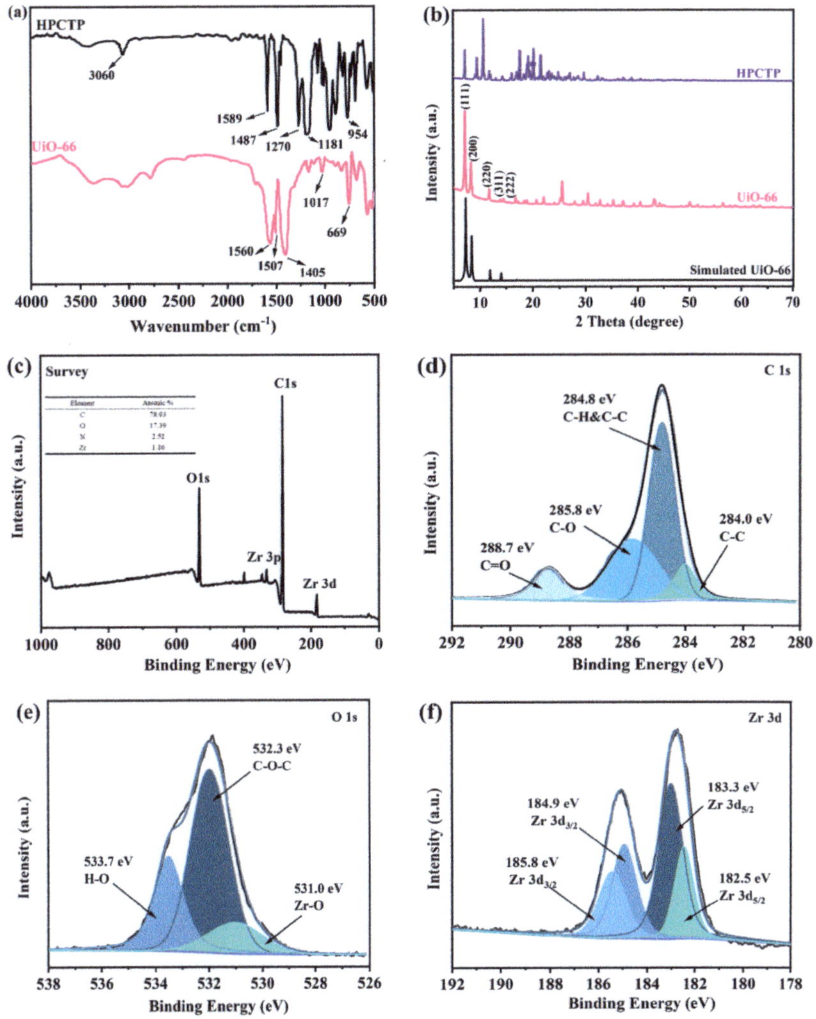

Figure 2. Structure analysis: (**a**) FT—IR spectra of UiO-66 and HPCTP, (**b**) XRD patterns and simulations of UiO-66 and HPCTP, (**c**) XPS survey spectra of UiO-66, (**d**) XPS spectra of C 1s, (**e**) XPS spectra of O 1s, and (**f**) XPS spectra of Zr 3d.

XPS was performed to detect the types of elements and bonding in UiO-66. From the XPS survey spectra, C 1s, O 1s, and Zr 3d XPS bands were detected. The C 1s peaks were distributed across four groups of peaks at 284.0 eV, 284.8 eV, 285.8 eV, and 288.7 eV. The signals at 284.0 eV and 284.8 eV corresponded to C-C bonds, while those at 285.8 eV and 288.7 eV corresponded to C-O and C=O or the connection between the benzene ring and carboxyl groups. In addition, the O 1s spectra can be fitted to three peaks. The peak near 532.3 eV indicates that C-O-C at 533.7 eV is attributed to H_2O, and that at a peak at 531.0 eV corresponds to Zr-O bonds. The Zr 3d has four peaks: the signals at 184.9 eV and 185.8 eV belong to Zr 3d3/2, and those at 183.3 eV and 182.5 eV correspond to Zr 3d5/2. Therefore, these results confirm the successful synthesis of UiO-66.

SEM images show that the UiO-66 nanocrystals have dispersed octahedral morphology with a uniform crystal size (Figure 3). The average size of the nanocrystals is approximately 350 nm. EDS shows that the elements zirconium and oxygen are evenly distributed throughout the materials (Figure S1). TGA is performed to analyze the thermal stability of HPCTP and UiO-66. UiO-66 has a slight weight loss peak near 144.8 °C, which is attributed to the desorption of DMF and ethanol. At the temperature range of 200~500 °C, there is a minor weight loss of UiO-66 that may be attributed to the removal of hydroxyl groups from the zirconium clusters to form water. Above 500 °C, the MOF structure begins to collapse and pyrolyze, with the maximum thermal weight loss rate temperature (T_{MAX2}) reaching approximately 548.9 °C, indicating that the obtained UiO-66 has high thermal stability with a residual mass rate of 33.92 wt% at 800 °C. Additionally, the pure HPCTP has only one T_{MAX} (392.3 °C), and its mass rapidly decreases as the temperature is increased, with an approximate residual rate of 0 wt% at 400 °C (Table S1). These findings imply that HPCTP by itself does not possess good thermal stability and needs to be mixed with UiO-66 to achieve good flame retardancy.

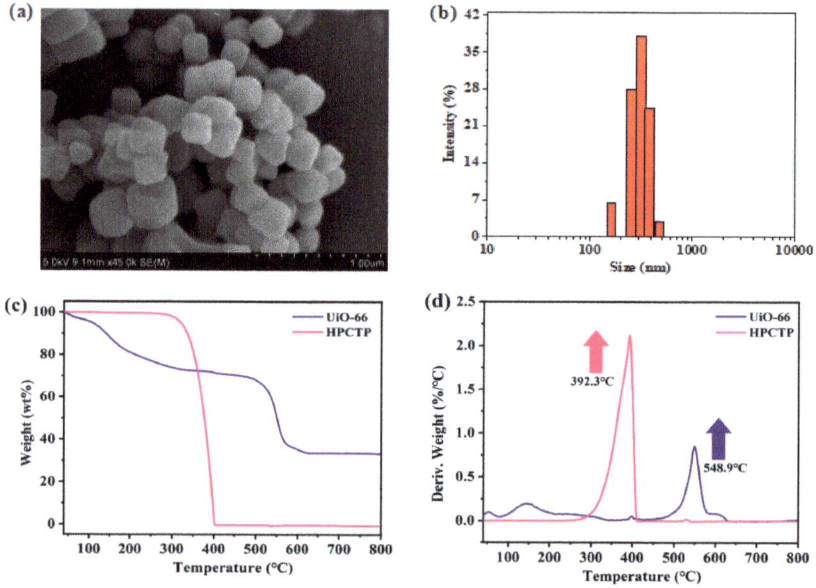

Figure 3. Morphology ((**a**) SEM, (**b**) particle size distribution) of UiO-66 and thermal-stability ((**c**) TG curves, (**d**) DTG curves) of UiO-66 and HPCTP.

3.2. Preparation of UiO-66@HPCTP@PC/ABS

UiO-66 and HPCTP were both used to modify the flame retardancy of PC/ABS. FT-IR spectroscopy was performed to characterize the structure of PC/ABS and the flame-retardant samples, e.g., HP-7, U-3, and HPU-3. The stretching vibration peaks of aromatic and aliphatic C-H in the PC/ABS were detected at 3200~3000 cm^{-1} and 3000~2800 cm^{-1}, respectively. The peaks at 2237 cm^{-1} and 911 cm^{-1} correspond to the nitrile C≡N of ABS and the C-H bond connected to PB, respectively. Those peaks at 1772 cm^{-1}, 1495 cm^{-1}, and 1200 cm^{-1} belong to the carbonyl stretching vibration, aromatic ring stretching vibration, and C-O stretching vibration in PC, respectively, representing the main groups of PC/ABS. After adding HPCTP, a weak absorption peak near 949 cm^{-1} was detected in HP-7, corresponding to P-O-C stretching. The U-3 sample showed a new absorption peak at 1,405 cm^{-1}, which was attributed to the symmetric vibration peak of O-C-O from the terephthalic acid ligands of UiO-66. The HPU-3 sample displayed both P-O-C and O-C-O characteristic peaks. These results indicate that the UiO-66@HPCTP@PC/ABS blends were successfully prepared.

Figure 4b shows a broad and diffuse PC/ABS diffraction peak. With an increase in the amount of UiO-66, characteristic sharp peaks at 7.38 ° and 8.52 ° appeared, corresponding to the (111) and (200) planes of UiO-66, respectively. Thus, UiO-66 was successfully introduced to PC/ABS.

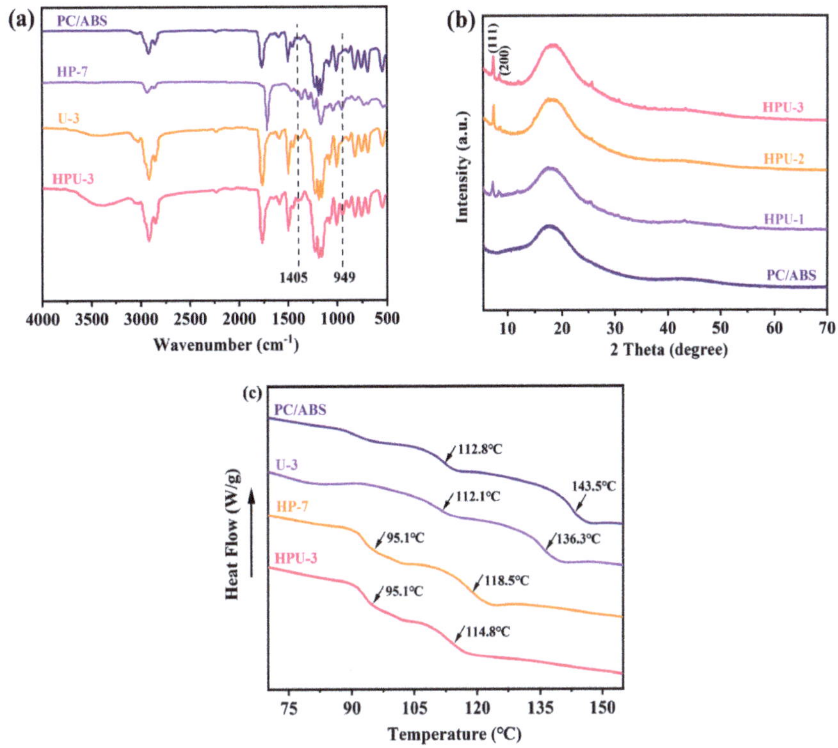

Figure 4. (**a**) FT-IR spectra of PC/ABS, HPc7, U-3, and HPU-3 samples. (**b**) XRD spectra of PC/ABS and HPU-n series. (**c**) DSC curves of PC/ABS, HP-7, U-3, and HPU-3 samples.

DSC was performed to characterize the phase structure and transition behaviors of UiO-66 and its blends. As shown in Figure 4c, the incorporation of UiO-66 led to a slight reduction in the Tg of the PC and SAN phases in ABS, indicating that UiO-66 particles might be incorporated to the molecular chains of resins to serve as lubricants, which increases free volume and enhances segmental mobility. Moreover, HPCTP significantly reduces the Tg of the aforementioned two phases. The benzene ring structure in HPCTP interacts with the SAN and PC phases through π-π stacking to form weak hydrogen bonds [26], plasticizing the PC/ABS system containing benzene ring structures. Therefore, with the addition of HPCTP and UiO-66, the flame-retardant material, PC/ABS, is endowed with better processing performance.

Figure 5 presents the TGA and DTG curves of the HPCTP intumescent, UiO-66, and the UiO-66 cooperative intumescent flame-retardant series. Pure PC/ABS has a two-step degradation process—matrix thermal degradation (410~500 °C) and deep cracking, such as chain cyclization and crosslinking (540~630 °C). However, after introducing UiO-66, the main degradation process of UiO-66 occurs between the two degradation stages of

PC/ABS, implying that UiO-66 may play a role throughout the thermal degradation process of the material.

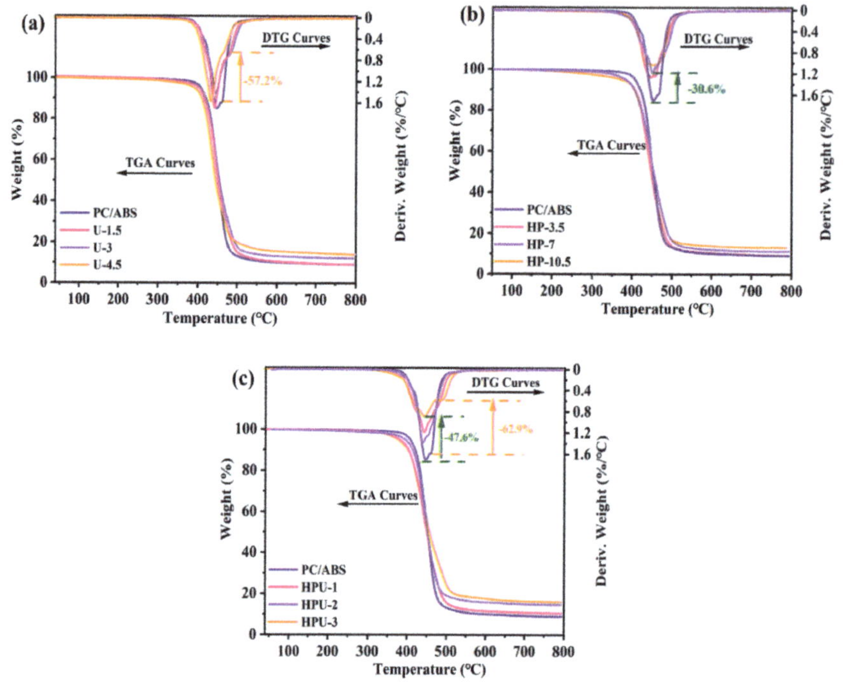

Figure 5. TGA and DTG curves of the PC/ABS blends: (**a**) UiO-66 flame-retardant series sample (U), (**b**) intumescent flame-retardant system sample (HP), and (**c**) UiO-66 cooperative intumescent flame-retardant system sample (HPU).

Figure 5a shows that the U-3 sample's DTG shoulder peak was reduced by 57.2% at 490 °C and that char yield was significantly increased at 500 °C and 800 °C (Table S2), suggesting that UiO-66 had a significant delay effect on the thermal decomposition of PC/ABS. Additionally, samples U-1, 2, and 3 tended to degrade slightly at a low temperature range, possibly due to minor PC/ABS chain scission in the early stages of TGA. As shown in Figure 5b, compared to the pure PC/ABS, sample HP-7 has a maximum weight loss rate of 30.6%. This implies that the intumescent flame-retardant material promotes the dehydration of combustion products into char to form an expanded char layer, thereby improving the thermal stability. Figure 5c displays the TGA and DTG curves of UiO-66 cooperative intumescent flame-retardant materials. The HPU-3 sample shows a 47.6% decrease in the maximum thermal weight loss rate and a 62.9% reduction in the DTG shoulder peak. The temperature of the DTG shoulder peak increases from 463.2 °C to 493 °C. The char yield at 800 °C increases from 9.0 wt% to 16.4 wt% (Table S2). These results imply that UiO-66 has a significant cooperative flame-retardant effect and causes a noticeable delay in the thermal decomposition process. The increase in char yield suggests that zirconium and organic ligands may participate in the carbonization process during thermal degradation.

3.3. *Performances of Flame Retardant*

Vertical burning and LOI tests are two common methods used to characterize the flame retardancy of polymers. Table 2 summarizes the results of vertical burning tests for materials containing different amounts of UiO-66 and HPCTP. Samples HP-3.5, HP-7, and HP-10.5 exhibited dripping phenomena during combustion, but only HP-10.5 containing

10.5 wt% HPCTP achieved a UL-94 V-2 rating. However, as engineering plastics, they still did not meet the requirements for fire safety standards.

Table 2. LOI and vertical burning test results of PC/ABS blends.

Sample Name	HPCTP (wt%)	UiO−66 (wt%)	LOI (%)	UL-94	t_1/s	t_2/s	Dripping
PC/ABS	0	0	21.2	NR	>60	>60	YES
HP-3.5	3.5	0	22.7	NR	>60	>60	YES
HP-7.0	7.0	0	24.0	NR	50.2	>60	YES
HP-10.5	10.5	0	25.3	V-2	18.4	26.7	YES
U-1.5	0	1.5	23.2	NR	>60	>60	YES
U-3.0	0	3	23.7	NR	50.8	>60	YES
U-4.5	0	4.5	24.1	NR	39.6	45.1	NO
HPU-1	7.0	1	23.8	NR	39.2	>60	YES
HPU-2	7.0	2	25.2	V-2	17.2	28.8	YES
HPU-3	7.0	3	27.0	V-0	9.4	17.2	NO

The pure PC/ABS has an LOI value of 21.2% and shows noticeable melt dripping that can ignite cotton during combustion. After adding UiO-66, particularly in sample U-4.5, this melt dripping phenomenon was effectively suppressed, and its LOI value increased to 24.1%. Since the preparation of UiO-66 is rather costly and intricate owing to complex procedures, it is rational to combine it with intumescent flame-retardant systems. From Table 1, by compounding 7.0 wt% HPCTP with 1, 2, and 3 wt%, UiO-66, HPU-1, HPU-2, and HPU-3 had increased LOI values of 23.8%, 25.2%, and 27.0%, respectively. Among them, only HPU-3 reached a V-0 rating, showing significantly improved resistance to combustion dripping. Therefore, UiO-66 alone cannot significantly enhance the efficiency of intumescent flame retardants in HPCTP.

At fire scenes, heat radiation or toxic smoke produced during combustion leads to many casualties [27]. Currently, the CCT is widely used to simulate the fire behavior of polymeric materials in real fire scenarios [28] and has two main parameters: heat release rate (HRR) and total heat release (TRR). As shown in Figure 6, compared to those of PC/ABS, the peak HRR (PHRR) and TRR of sample HP-7 decrease by 15.1% and 6.3%, respectively, while those of sample HPU-3 decrease by 32.2% and 33.1% after introducing UiO-66 (Table S3). This confirms that UiO-66 enhances the suppression of heat release in the intumescent flame-retardant system. In the initial stage of combustion, HPU-3 releases heat more slowly, as the MOF does not decompose in the early stages of pyrolysis, acting as a physical flame retardant. Furthermore, MOFs can provide smoke suppression effects [29]. As shown in Figure 6c,d, compared to that of the pure PC/ABS and sample HP-7, the smoke production rate (SPR) of HPU-3 significantly decreases as the peak SPR (TSPR) and total smoke release (TSR) are reduced by 39.3% and 49.4%, respectively (Table S3). From Figure 3d, the UiO-66 framework structure begins to collapse at approximately 450 °C, requiring higher temperatures for complete collapse. Although the internal temperature of polymeric materials can reach 700 °C or even higher during combustion, the framework structure remains relatively intact in the early stages of pyrolysis. When smoke is produced, most pyrolysis gases need to overcome the tunneling effect, traveling through the labyrinthine interior of the crystals, thereby delaying the smoke release in the early stages of pyrolysis. Additionally, UiO-66, which contains metals, metal oxides, and ligands with benzene ring structures, participates in the matrix carbonization reaction, making the char more compact, thus enhancing the smoke and heat suppression characteristics of the intumescent flame-retardant materials.

Figure 7 illustrates the changes in CO_2 and O_2 contents during CCT. Upon the addition of UiO-66, the O_2 peak gradually increases, whereas that of CO_2 decreases. This result indicates that the flame-retardant system reduces the intensity of PC/ABS combustion by inhibiting the contact between the matrix and O_2, thereby diminishing its reaction with O_2. Figure 7c shows that the FGI value of HP-7 containing only HPCTP is 1.24, whereas those of

HPU-2 and HPU-3 containing both UiO-66 and HPCTP decrease to 1.15 and 0.98, showing 7.3% and 21.0% reductions, respectively. This result confirms that UiO-66 significantly improves fire safety performance.

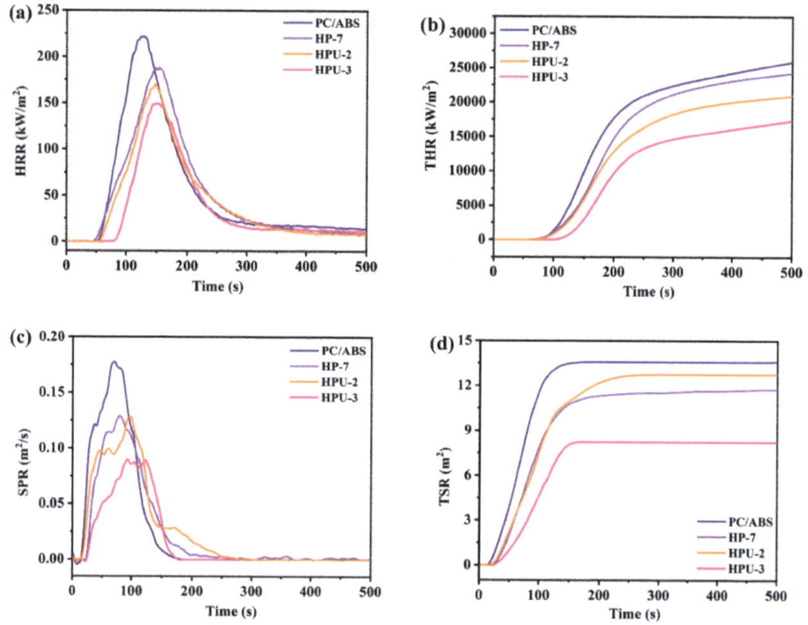

Figure 6. (**a**) Heat release rate (HPR), (**b**) total heat release (THR), (**c**) smoke production rate (SPR), and (**d**) total smoke production (TSR) curves of the PC/ABS blends.

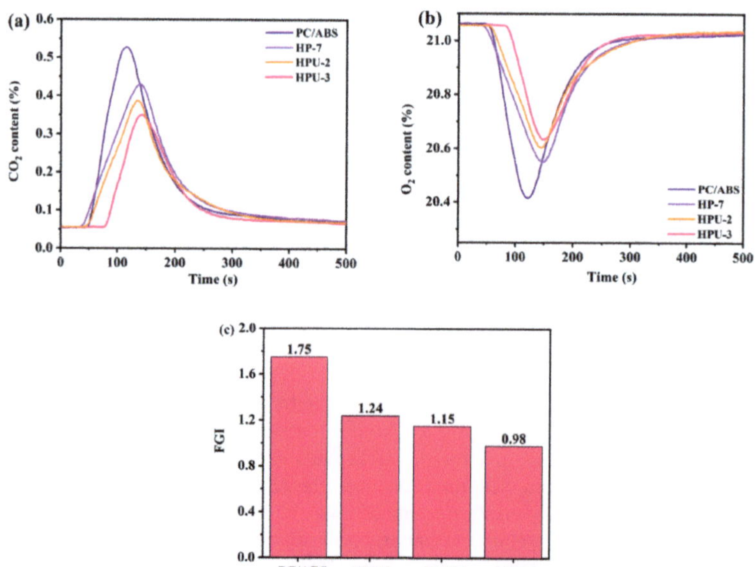

Figure 7. Changes in CO_2 (**a**) and O_2 contents (**b**) in relation to the time and (**c**) fire growth index (FGI) of PC/ABS blends.

3.4. Flame-Retardant Mode of Action

The flame-retardant mode of action was investigated by analyzing the composition of pyrolysis gases and residues from vertical burning tests. TG-IR was used to identify the composition of pyrolysis gases from PC/ABS at different temperatures. As shown in Figure 8, the absorption peaks at 2349 cm^{-1} and 667 cm^{-1} represent CO_2 stretching and bending vibrations, respectively. The region between 3750~3735 cm^{-1} corresponds to phenol and its derivatives, while that between 3720~3600 cm^{-1} corresponds to water vapor. The peak at 1780 cm^{-1} represents ester oligomers, and that at 1500 cm^{-1} is associated with aromatic compounds. Minor ether features are detected between 1285~1120 cm^{-1}. Figure 8b shows the infrared spectrum measured at T_{MAX}, where 7 wt% HPCTP has reduced the volatilization concentration of CO_2 from pure PC/ABS. Characteristic peaks for phosphorus-containing organics are not found, which may be attributed to their release in amounts that are too small to be detectable. At T_{MAX}, the absorption peaks for phenolic derivatives, ester compounds, and other hydrocarbons, such as aromatics, in HPU-2 and HPU-3 are significantly lower than those in HP-7 and pure PC/ABS. This result suggests that UiO-66 affects the thermal decomposition process of this material as a flame retardant in the solid phase. This might have resulted from the formation of a new compound with a more cross-linked and higher-molecular-weight network structure, reducing the release of organic volatiles.

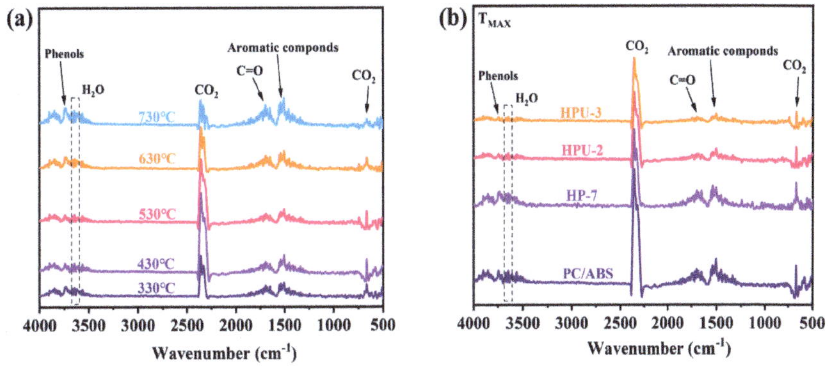

(a) (b)

Figure 8. TG-IR spectra: (**a**) PC/ABS at different temperatures; (**b**) PC/ABS, HP-7, HPU-1, HPU-2 and HPU-3 at maximum weight loss temperature (T_{MAX}).

The char residue formed during combustion affects the release of heat and smoke. The carbonization process in burning is often closely related to the amount of residual char, macro-morphology, and micro-structure. Therefore, it is necessary to observe the morphology of charred residues. As shown in Figure 9a, pure PC/ABS has the lowest char formation height, with more products dispersing into the atmosphere as smoke during combustion. In contrast, HPU-3 has a higher expansion degree, continuity, and integrity. Figure 9b presents the corresponding SEM images of the char layers for each sample. The char layer of pure PC/ABS shows many pores and low density owing to the extensive escape of various aromatic derivatives, CO_2, H_2O, and phenolic compounds during pyrolysis. The penetration of O_2 through the loose and porous char layer further exacerbates the burning process on the substrate surface. Although the density of the char layer of HP-7 increases, many pores remains present. Sample HPU-3 has a reduced pore size and a dense and smooth surface char layer. Hence, it is confirmed that UiO-66 enhances the char-forming process of the intumescent flame-retardant material during combustion.

Figure 9. Micromorphology and SEM images of carbon residue after vertical combustion test: (**a1**,**b1**) PC/ABS, (**a2**,**b2**) HP-7, (**a3**,**b3**) HPU-2, and (**a4**,**b4**) HPU-3.

Raman spectroscopy was used to characterize the degree of graphitization of the char layer. The D peak near 1,350 cm^{-1} and G peak near 1,600 cm^{-1} correspond to amorphous and graphitic carbons, respectively. A higher degree of graphitization results in better barrier effects against heat and oxygen. The integral ratio of the D peak to the G peak (ID/IG) was used to assess the degree of graphitization. As shown in Figure 10, HPU-3 has the lowest ID/IG value, whereas pure PC/ABS has the largest ID/IG value. After UiO-66 is added to the sample HPU-2, its ID/IG value becomes higher than that of HP-7. These results suggest that HPU-3 has the highest degree of graphitization in the char layer, and the UiO-66 enhances char formation during combustion.

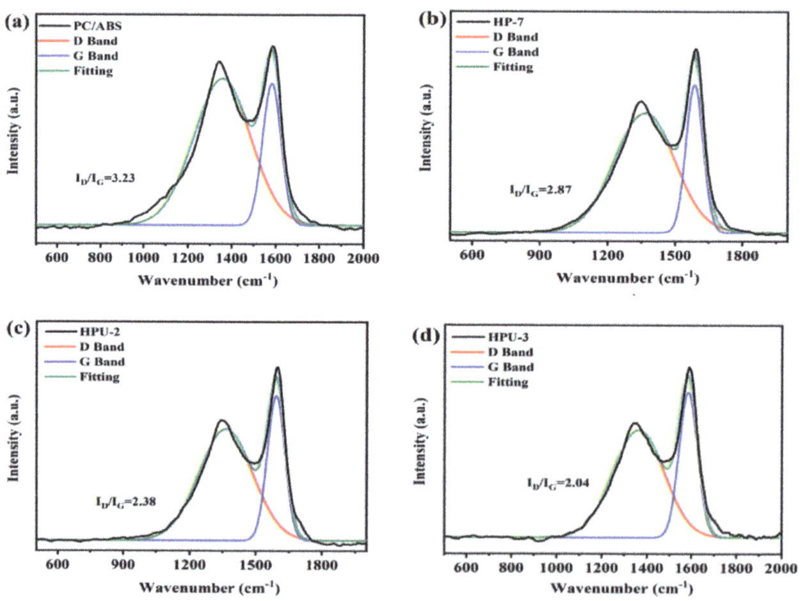

Figure 10. Raman spectra of PC/ABS (**a**), HP-7 (**b**), HPU-2 (**c**), and HPU-3 (**d**) carbon residues.

XRD was used to analyze the crystalline structure of the pyrolysis products of UiO-66. The prepared UiO-66 was placed in a muffle furnace at 600 °C for 0.5 h and collected for testing. As shown in Figure 11, the XRD pattern shows the characteristic peaks of the thermal barrier material, zirconium oxide. When the material burns, zirconium oxide can slow down the heat transfer rate and prevent the combustion products from spreading from the PC/ABS matrix to the burning area, to some extent. Therefore, after the material underwent intense combustion, although a large amount of the framework structure of UiO-66 was destroyed, its pyrolysis product—zirconium oxide—could still participate in smoke suppression.

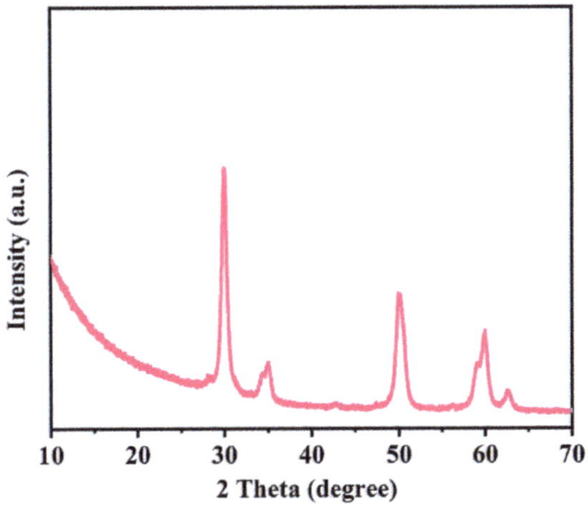

Figure 11. XRD patterns of UiO-66 after thermal degradation.

Based on the above analysis, Figure 12 proposes the flame-retardant mode of action for UiO-66 in the PC/ABS blends. Generally, under constant heating from an external heat source, thermal degradation occurs at the first stage. Oxygen at the surface of the degrading polymer greatly promotes the thermo-oxidative process. The degrading polymer thus continuously produces volatile combustibles, which catch fire when they reach a certain concentration and temperature. A portion of the heat released by combustion further feeds the degrading polymer. In this work, by melt blending, UiO-66 and HPCTP can be dispersed homogenously into the PC/ABS blends. During the burning process, the incorporation of UiO-66 contributes to the formation of a continuous char layer, which acts as a protective shield for the underlying materials. One reason for this is that the metal oxides produced by UiO-66 during combustion expose metal sites that can positively participate in and catalyze the dehydrogenation or cross-linking reactions of PC/ABS pyrolysis into char. Additionally, UiO-66 is conducive to form a dense, stable graphitized char layer since the terephthalic acid ligand can enhance the degree of graphitization of the char layer[12]. As discussed above, this char layer can prevent heat radiation and reduce the release of volatile combustibles. Moreover, the porosity of UiO-66 remains stable in the early and middle stages of combustion, and it requires PC/ABS pyrolysis volatiles to overcome the tunneling effect. In the later stage of pyrolysis, UiO-66 partially decomposes into zirconium oxide, which can slow down the heat transfer rate as a thermal barrier material, and it prevents combustion products escaping from the PC/ABS matrix to the combustion zone. Therefore, in the PC/ABS/HPCTP blends, UiO-66 shows a cooperative effect on the flame-retardant performance of intumescent flame-retardant materials.

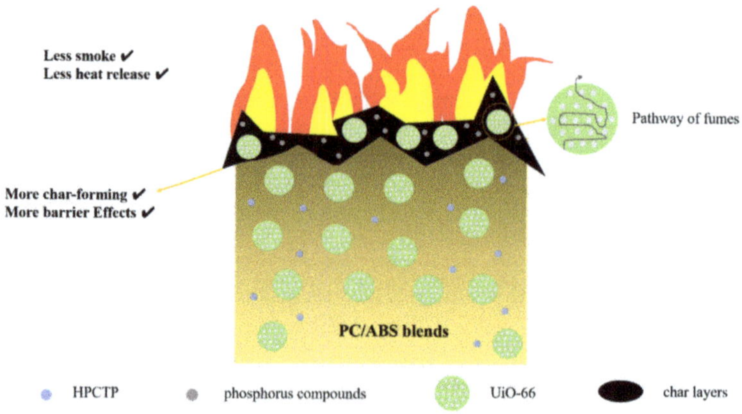

Figure 12. Illustration for PC/ABS blends' flame-retardant mode of action.

4. Conclusions

In this work, a zirconium-based metal-organic framework material (UiO-66) was synthesized to cooperatively enhance the flame retardancy of PC/ABS intumescent materials constructed with HPCTP via melt blending and two-screw extrusion. The successful preparation of UiO-66 was confirmed using FT-IR, XRD, XPS, and SEM analyses. Results indicate that UiO-66 has good thermal stability and a delaying effect on the pyrolysis of the intumescent flame-retardant materials. UiO-66 can significantly enhance the efficiency of HPCTP's intumescent flame retardants. The flame-retardant PC/ABS blends with both UiO-66 and HPCTP show good flame retardancy. By compounding 7.0 wt% HPCTP with 3 wt% UiO-66, the blend shows LOI values of 27.0%, reaches a V-0 rating, and significantly improves resistance to combustion dripping. The CCT showed that UiO-66 suppresses heat release and smoke release in these materials. Finally, the flame-retardant mode of action in the blends is proposed, and UiO-66 has a cooperative effect on the flame-retardant performance of PC/ABS/HPCTP intumescent flame-retardant materials. This work provides a basis for the further development of flame-retardant PC/ABS.

Supplementary Materials: The following supporting information can be downloaded at: https://www.mdpi.com/article/10.3390/polym16142083/s1, Figure S1: Element distribution of UiO-66; Figure S2: 3D and 2D structures of UiO-66; Figure S3: tensile strain at break and tensile strength of PC/ABS and HPU-1, 2, 3; Table S1:Thermogravimetric data of UiO-66 and HPCTP; Table S2: Thermogravimetric data of PC/ABS composites; Table S3: PC/ABS composite cone calorimetry test data.

Author Contributions: Methodology, W.B. and S.C.; software, W.D., D.H. and L.L.; formal analysis, Z.C. and S.C.; investigation, W.B. and Z.C.; data curation, W.B., Z.C., L.L., D.H. and W.B.; writing—original draft preparation, Z.C.; writing—review and editing, S.C. and H.Z.; supervision, S.C. and H.Z.; project administration, H.Z. All authors have read and agreed to the published version of the manuscript.

Funding: This research was supported by the Basic and Applied Basic Research Foundation of Guangdong province (Grant No. 2024A1515010738; 2022A1515011985), the Shenzhen Science and Innovation Commission (Grant No. 20231120143631001), the Heyuan Science and Technology Innovation Strategy (Grant No. 221028141607508), and the Shenzhen University Cross-research program (Grant No. 2023YG028).

Institutional Review Board Statement: Not applicable.

Data Availability Statement: Data are contained within this article.

Conflicts of Interest: The authors declare that this research was conducted in the absence of any commercial or financial relationships that could be construed as potential conflicts of interest.

References

1. Truc, N.T.T.; Lee, B.-K. Selective separation of ABS/PC containing BFRs from ABSs mixture of WEEE by developing hydrophilicity with ZnO coating under microwave treatment. *J. Hazard. Mater.* **2017**, *329*, 84–91. [CrossRef]
2. Andrzejewski, J.; Mohanty, A.K.; Misra, M. Development of hybrid composites reinforced with biocarbon/carbon fiber system. The comparative study for PC, ABS and PC/ABS based materials. *Compos. Part B Eng.* **2020**, *200*, 108319. [CrossRef]
3. Wang, J.; Wang, H.; Yue, D. Optimization of surface treatment using sodium hypochlorite facilitates coseparation of ABS and PC from WEEE plastics by flotation. *Environ. Sci. Technol.* **2019**, *53*, 2086–2094. [CrossRef] [PubMed]
4. Yan, C.; Yan, P.; Xu, H.; Liu, D.; Chen, G.; Cai, G.; Zhu, Y. Preparation of continuous glass fiber/polyamide 6 composites containing hexaphenoxycyclotriphosphazene: Mechanical properties, thermal stability, and flame retardancy. *Polym. Compos.* **2022**, *43*, 1022–1037. [CrossRef]
5. Chen, Y.; Wu, X.; Qian, L. Flame-retardant behavior and protective layer effect of phosphazene-triazine bi-group flame retardant on polycarbonate. *J. Appl. Polym. Sci.* **2020**, *137*, 49523. [CrossRef]
6. Feng, H.; Qian, L.; Lu, L. Synergistic effect of polyimide charring agent and hexaphenoxycyclotriphosphazene on improving fire safety of polycarbonate: High graphitization to strengthen the char layer. *Polym. Adv. Technol.* **2021**, *32*, 1135–1149. [CrossRef]
7. Seraji, S.M.; Gan, H.; Swan, S.R.; Varley, R.J. Phosphazene as an effective flame retardant for rapid curing epoxy resins. *React. Funct. Polym.* **2021**, *164*, 104910. [CrossRef]
8. Hoang, D.; Kim, J. Synthesis and applications of biscyclic phosphorus flame retardants. *Polym. Degrad. Stab.* **2008**, *93*, 36–42. [CrossRef]
9. Olorunyomi, J.F.; Sadiq, M.M.; Batten, M.; Konstas, K.; Chen, D.; Doherty, C.M.; Caruso, R.A. Advancing metal-organic frameworks toward smart sensing: Enhanced fluorescence by a photonic metal-organic framework for organic vapor sensing. *Adv. Opt. Mater.* **2020**, *8*, 2000961. [CrossRef]
10. Elsaidi, S.K.; Ongari, D.; Mohamed, M.H.; Xu, W.; Motkuri, R.K.; Haranczyk, M.; Thallapally, P.K. Metal organic frameworks for xenon storage applications. *ACS Mater. Lett.* **2020**, *2*, 233–238. [CrossRef]
11. Hao, Y.-C.; Chen, L.-W.; Li, J.; Guo, Y.; Su, X.; Shu, M.; Zhang, Q.; Gao, W.-Y.; Li, S.; Yu, Z.-L. Metal-organic framework membranes with single-atomic centers for photocatalytic CO2 and O2 reduction. *Nat. Commun.* **2021**, *12*, 2682. [CrossRef]
12. Birin, K.; Abdulaeva, I.; Polivanovskaya, D.; Sinel'shchikova, A.; Demina, L.; Baranchikov, A.; Gorbunova, Y.G.; Tsivadze, A.Y. Immobilization of Heterocycle-Appended Porphyrins on UiO-66 and UiO-67 MOFs. *Russ. J. Inorg. Chem.* **2021**, *66*, 193–201. [CrossRef]
13. Cavka, J.H.; Jakobsen, S.; Olsbye, U.; Guillou, N.; Lamberti, C.; Bordiga, S.; Lillerud, K.P. A new zirconium inorganic building brick forming metal organic frameworks with exceptional stability. *J. Am. Chem. Soc.* **2008**, *130*, 13850–13851. [CrossRef] [PubMed]
14. Viana, A.M.; Ribeiro, S.O.; Castro, B.d.; Balula, S.S.; Cunha-Silva, L. Influence of UiO-66 (Zr) preparation strategies in its catalytic efficiency for desulfurization process. *Materials* **2019**, *12*, 3009. [CrossRef] [PubMed]
15. DeStefano, M.R.; Islamoglu, T.; Garibay, S.J.; Hupp, J.T.; Farha, O.K. Room-temperature synthesis of UiO-66 and thermal modulation of densities of defect sites. *Chem. Mater.* **2017**, *29*, 1357–1361. [CrossRef]
16. Schaate, A.; Roy, P.; Godt, A.; Lippke, J.; Waltz, F.; Wiebcke, M.; Behrens, P. Modulated synthesis of Zr-based metal–organic frameworks: From nano to single crystals. *Chem. A Eur. J.* **2011**, *17*, 6643–6651. [CrossRef] [PubMed]
17. Sai, T.; Ran, S.; Guo, Z.; Fang, Z. A Zr-based metal organic frameworks towards improving fire safety and thermal stability of polycarbonate. *Compos. Part B Eng.* **2019**, *176*, 107198. [CrossRef]
18. *GB/T 17037.1-1997*; Preparation of Injection Molding Samples for Thermoplastic Materials Part 1. China Standard Press: Beijing, China, 1997.
19. *GB/T 9341-2008*; Determination of Plastic Bending Performance. China Standard Press: Beijing, China, 2008.
20. *GB/T 2406.2-2009*; Determination of Combustion Behavior of Plastics by Oxygen Index Method. China Standard Press: Beijing, China, 2009.
21. *ISO5660-1*; Reaction to Fire Tests-Heat Release, Smoke Production and Mass Loss Rate-Part 1: Heat Release Rate (Cone Calorimeter Method. ISO: Geneva, Switzerland, 2015.
22. *ASTM D7309*; Standard Test Method for Determining Flammability Characteristics of Plastics and Other Solid Materials Using Microscale Combustion Calorimetry. ASTM International: West Conshohocken, PA, USA, 2007.
23. Stawowy, M.; Ciesielski, R.; Maniecki, T.; Matus, K.; Łużny, R.; Trawczynski, J.; Silvestre-Albero, J.; Łamacz, A. CO_2 hydrogenation to methanol over Ce and Zr containing UiO-66 and Cu/UiO-66. *Catalysts* **2019**, *10*, 39. [CrossRef]
24. Ahmadijokani, F.; Mohammadkhani, R.; Ahmadipouya, S.; Shokrgozar, A.; Rezakazemi, M.; Molavi, H.; Aminabhavi, T.M.; Arjmand, M. Superior chemical stability of UiO-66 metal-organic frameworks (MOFs) for selective dye adsorption. *Chem. Eng. J.* **2020**, *399*, 125346. [CrossRef]
25. Wang, D.; Xin, Y.; Li, X.; Wang, F.; Wang, Y.; Zhang, W.; Zheng, Y.; Yao, D.; Yang, Z.; Lei, X. A universal approach to turn UiO-66 into type 1 porous liquids via post-synthetic modification with corona-canopy species for CO_2 capture. *Chem. Eng. J.* **2021**, *416*, 127625. [CrossRef]
26. Rungswang, W.; Kotaki, M.; Shimojima, T.; Kimura, G.; Sakurai, S.; Chirachanchai, S. Directing thermoplastic elastomer microdomain parallel to fiber axis: A model case of SEBS with benzoxazine through π–π stacking. *Macromolecules* **2011**, *44*, 9276–9285. [CrossRef]

27. Fang, F.; Song, P.; Ran, S.; Guo, Z.; Wang, H.; Fang, Z. A facile way to prepare phosphorus-nitrogen-functionalized graphene oxide for enhancing the flame retardancy of epoxy resin. *Compos. Commun.* **2018**, *10*, 97–102. [CrossRef]
28. Fang, F.; Ran, S.; Fang, Z.; Song, P.; Wang, H. Improved flame resistance and thermo-mechanical properties of epoxy resin nanocomposites from functionalized graphene oxide via self-assembly in water. *Compos. Part B Eng.* **2019**, *165*, 406–416. [CrossRef]
29. Hou, Y.; Hu, W.; Zhou, X.; Gui, Z.; Hu, Y. Vertically aligned nickel 2-methylimidazole metal–organic framework fabricated from graphene oxides for enhancing fire safety of polystyrene. *Ind. Eng. Chem. Res.* **2017**, *56*, 8778–8786. [CrossRef]

Disclaimer/Publisher's Note: The statements, opinions and data contained in all publications are solely those of the individual author(s) and contributor(s) and not of MDPI and/or the editor(s). MDPI and/or the editor(s) disclaim responsibility for any injury to people or property resulting from any ideas, methods, instructions or products referred to in the content.

polymers

MDPI

Article

Development of a UiO-66 Based Waterborne Flame-Retardant Coating for PC/ABS Material

Shaojun Chen [1], Youhan Zeng [1], Weifeng Bi [1], Haitao Zhuo [2,*] and Haiqiang Zhong [3,*]

1 College of Materials Science and Engineering, Shenzhen University, Shenzhen 518053, China; chensj@szu.edu.cn (S.C.); 2110343072@email.szu.edu.cn (Y.Z.); a773122527@163.com (W.B.)
2 College of Chemistry and Environment Engineering, Shenzhen University, Shenzhen 518053, China
3 Guangdong Provincial Enterprise Key Laboratory of Smart Automotive Display, Varitronix (Heyuan) Display Technology Limited, Heyuan 517000, China
* Correspondence: haitaozhuo@163.com (H.Z.); zhongkennyhq@boevx.com (H.Z.)

Abstract: The flame-retardancy of polymeric materials has garnered great interest. Most of the flame retardants used in copolymers are functionalized additives, which can deteriorate the intrinsic properties of these materials. As a new type of flame retardant, functionalized metal–organic frameworks (MOFs) can be used in surface coatings of polymers. To reduce the flammability, a mixture of phytic acid, multi-wall carbon nanotubes, zirconium-based MOFs, and UiO-66 was coated on a PC/ABS substrate. The structure of the UiO-66-based flame retardant was established by FT-IR, XRD, XPS, and SEM. The flammable properties of coated PC/ABS materials were assessed by LOI, a vertical combustion test, TGA, CCT, and Raman spectroscopy. The presence of a UiO-66-based coating on the PC/ABS surface resulted in a good flame-retardant performance. Heat release and smoke generation were significantly reduced. Importantly, the structure and mechanical properties of PC/ABS were less impacted by the presence of the flame-retardant coating. Hence, this work presents a new strategy for the development of high-performance PC/ABC materials with both excellent flame-retardancy and good mechanical properties.

Keywords: MOFs; flame retardant polymer; PC/ABS; coating

check for updates

Citation: Chen, S.; Zeng, Y.; Bi, W.; Zhuo, H.; Zhong, H. Development of a UiO-66 Based Waterborne Flame-Retardant Coating for PC/ABS Material. *Polymers* **2024**, *16*, 275. https://doi.org/10.3390/polym16020275

Academic Editors: Fei Xiao, Fubin Luo and Ke Sun

Received: 28 November 2023
Revised: 13 January 2024
Accepted: 14 January 2024
Published: 19 January 2024

Copyright: © 2024 by the authors. Licensee MDPI, Basel, Switzerland. This article is an open access article distributed under the terms and conditions of the Creative Commons Attribution (CC BY) license (https://creativecommons.org/licenses/by/4.0/).

1. Introduction

The fire risk posed by polymeric materials has received considerable attention. The hydrocarbon skeleton and organic composition of these materials make them highly flammable. Heat and toxic gases generated during combustion are a great threat to human beings and the environment [1,2]. Many studies have been conducted to prevent the combustion of polymer materials and thereby reduce the harmful effects of combustion [3–7]. The simplest method is to blend specific flame retardants with the polymer matrix. An intumescent flame-retardant system is commonly used, which obstructs the transfer of heat and oxygen by forming an expansive carbon layer on the polymer surface during thermal degradation of the polymer matrix [8–10]. Another flame-retardant system that is used is based on catalytic principles, wherein specific catalysts promote a reduction in volatile fragments and toxic gases during the degradation of the polymer matrix. Although the use of these flame-retardant systems and others have been effective in reducing the flammability of polymer, there is a huge demand for flame retardants having a high efficiency.

A flame-retardant coating on the surface of a substrate represents an important method for controlling flammability. Compared with the method of adding flame retardants, the intrinsic properties of the polymer are not compromised in the surface coating method. This method also has the advantage of easy processing and can be used for a variety of materials [11]. In recent years, metal organic frameworks (MOFs), which are organic–inorganic hybrid materials comprising metal ions and organic ligands, have been found to have flame-retardant properties. At high temperatures, MOFs decompose to form catalytic species that promote oxidation and coking [12–14]. In particular, a zirconium organic framework, UiO-66, has been found to be a promising material for use in retardant systems. It has a high specific surface area (1000 m^2/g) [15] and excellent thermal, water, and acid stability, which is conducive to the post-synthesis modification of UiO-66. However, applying MOFs alone cannot provide sufficient flame-retardancy.

A mixture of materials containing UiO-66 can provide excellent flame-retardancy to the system [16]. As a green biomass material, phytic acid (PA) has a high phosphorus content. It can readily chelate with MOFs to provide a composition with better flame-retardancy. In addition, multi-walled carbon nanotubes (MWCNT) have a high carbon content and excellent thermal properties [17–19]. These materials are commonly used as components of high-efficiency green flame retardants. Polycarbonate/acrylonitrile-butadiene-styrene copolymer (PC/ABS) materials are widely used in many products, due to their excellent impact strength. However, the presence of a flame-retardant additive greatly influences the mechanical properties of the blend. In an effort to obtain an excellent flame-retardant PC/ABS material, an aqueous flame-retardant system for coating of a PC/ABS surface was prepared using UiO-66, PA, and MWCNT. The as-prepared UiO-66@PA@MWCNT was carefully characterized by Fourier transform infrared (FT-IR) spectroscopy, X-ray diffractometry (XRD), X-ray photoelectron spectroscopy (XPS), and scanning electron microscopy (SEM). The effectiveness of the UiO-66 flame-retardant coating for PC/ABS was evaluated by the limiting oxygen index (LOI), vertical combustion test, thermogravimetric analysis (TGA), cone calorimeter test (CCT), and Raman spectroscopy.

2. Experiments

2.1. Materials

Polycarbonate (PC, injection grade, QiMei Company, Taiwan), acrylonitrile-butadiene-styrene copolymer (ABS, injection grade, QiMei Company, Taiwan), zirconium chloride ($ZrCl_4$, 98%, Shanghai Aladdin Biochemical Company, Shanghai, China), terephthalic acid (99% TPA, Shandong Yusuo Chemical Technology Co., Ltd., Shandong, China), carboxylated multi-walled carbon nanotubes (95% MWCNT-COOH, Suzhou Tianke Trading Co., Ltd., Suzhou, China), N,N-dimethylformamide (99.5% DMF, Shanghai Aladdin Biochemical Company, Shanghai, China), ethanol (99% EtOH, Shanghai Aladdin Biochemical Company, Shanghai, China), and SiO_2 (99.5%, Jiangsu Tianxing New Materials Co., Ltd., Changshu, China) were used for the experiments.

2.2. Synthesis of UiO-66

$ZrCl_4$ (1.98 g) and TPA (1.76 g) were weighed and dissolved in 150 mL of DMF. After ultrasonication for 20 min, 4 mL acetic acid was added and the solution was stirred continuously for 10 min. Thereafter, the reaction mixture was transferred to a blue bottle and placed in a constant-temperature vacuum-drying oven at 120 °C. After 36 h, the bottle was taken out and cooled to room temperature naturally. After centrifugation at 6000 r/min for 15 min, the white precipitate was filtered and washed three times with DMF and ethanol. Finally, the white product was dried in a vacuum oven at 85 °C for 36 h, and UiO-66 was collected for further use.

2.3. Synthesis of MWCNT@UiO-66

$ZrCl_4$ (1.98 g) and MWCNT-COOH (0.20 g) were weighed and dissolved in a blue bottle containing 150 mL of anhydrous DMF. After dispersion by ultrasonication for 20 min, TPA (1.76 g) and 4 mL of acetic acid was added and ultrasonicated for another 20 min. After 36 h of reaction, the blue bottle was taken out and cooled to room temperature naturally. After centrifugation at high speed of 6000 r/min for 15 min, the black precipitate obtained was filtered out and further centrifuged three times each with DMF and ethanol. The product was placed in a constant-temperature vacuum-drying oven at 85 °C for 36 h. Finally, the MWCNT@UiO-66 was collected and stored.

2.4. Synthesis of PA@MWCNT@UiO-66

MWCNT@UiO-66 (1.0 g) was weighed and dispersed in a blue bottle containing 150 mL DMF. Then, PA (10.68 mL) was added to the above MWCNT@UiO-66 suspension, followed by ultrasonic dispersion for 20 min. Thereafter, the mixture was continuously stirred for 24 h at room temperature using a magnetic agitator. The black precipitate obtained was filtered and centrifugally washed with DMF and ethanol 3 times. The final PA@MWCNT@UiO-66 product was then dried in a vacuum-drying oven at 85 °C for 24 h. The synthetic route is presented in Figure 1.

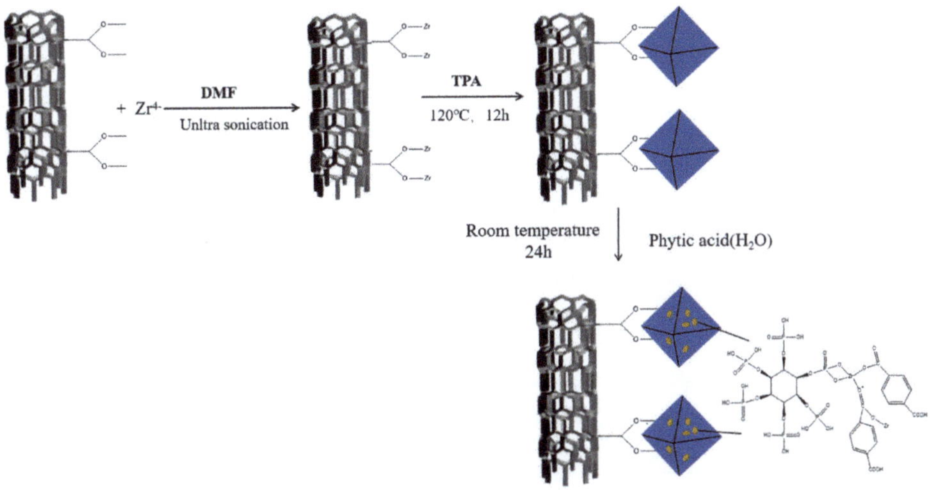

Figure 1. Route for the synthesis of PA@MWCNT@UiO-66.

2.5. Preparation of Aqueous Flame-Retardant Coatings

For comparison, different MOFs-based flame retardants were evenly dispersed in water-based acrylic emulsion by adding an appropriate amount of water and a small amount of SiO_2 according to Table 1. SiO_2 was added as a filler to improve the hardness of coatings and reduce the viscosity. The aqueous emulsions were prepared in a centrifugal tube, by stirring evenly at room temperature using a vortex oscillator, and then dispersed by ultrasonication for 20 min.

Table 1. Composition of aqueous flame retardants.

Sample	Solid Content (wt%)	Water-Based Acrylic Adhesive (g)	UiO-66 (g)	MWCNT @UiO-66 (g)	PA@MWCNT @UiO-66 (g)	SiO_2 (g)	H_2O (g)
WA	41.9	2	0	0	0	0	0.34
WAU	41.9	2	1	0	0	0.2	2
WAUM	41.9	2	0	1	0	0.2	2
WAUPM	41.9	2	0	0	1	0.2	2

2.6. Preparation of Flame Resistant PC/ABS

The dried PC/ABS particles were transferred to a polytetrafluoroethylene (PTFE) mold in a plate vulcanization machine, with the temperature set at 240 °C. After heating for 6 min, hot pressing at 18 MPa for 4 min, and cold pressing at room temperature for 2 min, the PC/ABS films were cut into three types of specimens having dimensions of 125 mm × 13 mm × 3 mm, 150 mm × 4 mm × 10 mm, and 100 mm × 100 mm × 3 mm. Some control specimens were reserved for the vertical combustion test, oxygen index test, and cone calorimetry test, whereas the other specimens were sprayed with the aqueous flame retardants, and were placed in a 60 °C electric blast-drying oven for 4 h. The steps for spraying and drying were repeated until the coating thickness reached the standard specification (±10%), as shown in Table 2. The thickness of the coating was determined by measuring the spline thickness before and after coating, using a Vernier caliper.

Table 2. Thickness of flame-retardant coating on PC/ABS.

Sample *	WAU	WAUM	WAUPM	Coating Thickness (μm)
Untreated	×	×	×	0
WAU-3	√	×	×	750
WAUM-3	×	√	×	750
WAUPM-1	×	×	√	250
WAUPM-2	×	×	√	500
WAUPM-3	×	×	√	750

* × represents not using this sample for coating and √ represents using this sample for coating.

2.7. Characterizations

Fourier transform infrared spectra (FT-IR) were acquired on a Nicolet 6700 infrared spectrometer in reflectance mode. The spectral range was 4000–500 cm^{-1} at a resolution of 4 cm^{-1}. The powdered samples were generally measured by the KBr pellet method.

An AXIS Ultra XPS instrument (Shimadzu Company of Japan, Kyoto, Japan) was used for testing the chemical states of elements. Its cathode was made of a lanthanum–aluminum–molybdenum alloy, and the analysis was conducted with a power input of 400 W. The background was a non-rotating background compensation type, and the peak function was the Lorentz–Gaussian function. The powder to be tested was adhered to the tin foil with viscose, folded and pressed, then unfolded and cut into specimen of 5 mm × 4 mm. Chemical states, such as C1s, Zr3d, P2p, and O1s, were determined by scanning the surface of the sample.

The morphologies and elemental distributions of the samples were studied by scanning electron microscopy (NGB4-DXS-10AC, Nanjing Grand Technology Co., Ltd., Nanjing, China). The section of the block sample to be observed was made brittle in liquid nitrogen, and the powder sample and block sample were spray coated with gold for 75 s, after which they were attached to the sample table with conductive adhesive for observation.

TGA and DTG analyses of the samples were conducted using a TGA 55 instrument in a nitrogen atmosphere at a 50 mL/min flow rate. Except when specified, the heating rate was 10 °C/min, and the temperature range was from room temperature to 800 °C.

An X-ray diffractometer (SmartLab, Boston, MA, USA) was used to characterize the material phases in order to analyze their internal structure and morphology. The wavelength was 1.5406 A and the scanning speed was 10°/min in the scanning range of 5–70°.

Vertical combustion tests were performed on uncoated and coated samples according to the GB/T2408-2008 combustion test standard [20]. Each sample was tested thrice and the average value of three readings was used to determine its corresponding flame-retardant level.

The limiting oxygen index (LOI) was expressed according to the percentage of oxygen in the volume. A JF-3 oxygen index instrument was used for the LOI test. The specimen dimensions were 150 mm × 4 mm × 10 mm, according to the standard GB/T2406.2-2009 test [21].

The British FTT cone calorimeter was used for the cone calorimetry test (CCT), according to the ISO5660-1 and ASTM D7309 test standard [22]. The irradiation power was 50 kW/m^2, sample dimensions were 100 mm × 100 mm × 3 mm, and three parallel tests were carried out for each sample.

The structure, morphology, and graphitization degree of the carbon layer were characterized by FT-IR, SEM, and Raman spectroscopy. The Raman spectra of the residual carbon residue after combustion by CCT were acquired using an excitation wavelength of 514.5 nm. Based on the above results, the mode of action of flame-retardant materials was clarified.

3. Results and Discussion

3.1. Structural Analysis of Flame-Retardant UiO-66@PA@MWCNT

First, the zirconium organic framework, UiO-66, was synthesized using $ZrCl_4$ and TPA. The obtained UiO-66 was then functionalized with MWCNT and PA to obtain MOFs, labeled as UiO-66@PA@MWCNT. The structure was carefully characterized by FT-IR, XRD, XPS, and TG-DTG during the process. FT-IR spectroscopy was used to determine the molecular structure of UiO-66@PA@MWCNT. In Figure 2a, strong absorption peaks appeared at 1560 cm^{-1} and 1395 cm^{-1} in the FT-IR spectrum of UiO-66. They corresponded to the O-C-O asymmetrical and symmetrical stretching vibrations of TPA, which served as ligands for MOFs. The absorption peaks at 1507 cm^{-1} corresponded to the C=C of the benzene ring, while the peak at 669 cm^{-1} was consistent with the asymmetric stretching vibrations of Zr-(OC). In MWCNT-COOH, the peak for C=O stretching vibrations appeared at 1720 cm^{-1}. The small peak at 1655 cm^{-1} was due to the O-H bending vibrations. This confirmed that the carboxyl group was successfully grafted onto MWCNT-COOH, suggesting a good compatibility with PA and UiO-66. The FT-IR spectrum of PA also showed some typical peaks of phosphate groups, including P=O (1130 cm^{-1}), p-O-C (1060 cm^{-1}), and P-O (1012 cm^{-1}). The high phosphorus content can provide a synergistic flame-retardant effect in UiO-66@PA@MWCNT. After MWCNT-COOH, PA, and UiO-66 were reacted, several absorption peaks of PA shifted to higher wave numbers in UiO-66@PA@MWCNT. These results indicated that there was a complex formed between the P-O bond and metal zirconium from UiO-66, verifying the successful synthesis of UiO-66@PA@MWCNT. The XRD patterns in Figure 2b show that the diffraction peaks of UiO-66 were basically consistent with that of the standard profile. After modification with PA and MWCNT, the peak pattern of UiO-66 remained almost unchanged. This indicated that the crystal structure of UiO-66 remained intact during the preparation of functionalized UiO-66.

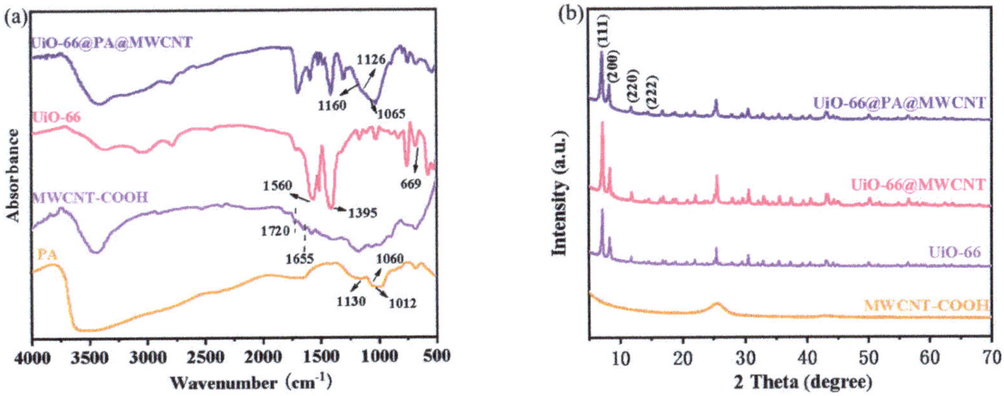

Figure 2. FT-IR spectra (**a**) and XRD patterns (**b**) of UiO-66@PA@MWCNT, UiO-66@MWCNT, MWCNT−COOH, and UiO-66.

XPS was used to detect the elements and bonding nature in UiO-66@PA@MWCNT. Firstly, the overall spectrum in Figure 3a showed that the obtained powders were rich in C, O, Zr, and P. Secondly, the energy spectrum of C1s is presented in Figure 3b. After deconvolution of the peaks, it was found that the peaks at 284.1 eV and 284.6 eV corresponded to the neutral bond and sp2-hexagonal network structure of UiO-66@PA@MWCNT, respectively.

The peak at 286.4 eV was due to the carbon atom of the C-O bond from UiO-66 and the C-P bond from PA. The peak for C=O at 288.5 eV was consistent with the carboxyl group of the ligand. Thirdly, the energy spectrum of O1s in Figure 3c can be deconvoluted into four peaks. Among them, the peak at 530.1 eV corresponded to the P-O bond of PO_4^{3-} and HPO_4^{2-} of PA, while the peak at 531.2 eV was consistent with the Zr-O bond. The peaks at 532.1 eV and 533.3 eV corresponded to C-O-C and P-O-C and C-OH and P-OH, respectively. Fourthly, Figure 3d also shows that the Zr3d spectrum consisted of two peaks, which were further deconvoluted into four peaks. The peaks at 185.1 eV and 185.9 eV corresponded to Zr3d3/2, while the peaks at 182.7 eV and 183.5 eV corresponded to Zr3d5/2. Finally, the P2p spectrum in Figure 3e was composed of three peaks at 133.1 eV (P=O), 133.9 eV (P-OH), and 134.8 eV (P-O-ZR and HPO_4^{2-}). Additionally, TGA and DTG curves (see Supporting Information Figure S1) showed that the carbon residue percentage at 800 °C of UiO-66@PA@MWCNT was 61.3 wt%, which was much higher than that of pure UiO-66. It implied that PA and MWCNT promoted carbon formation in UiO-66@PA@MWCNT. The SEM image of UiO-66 showed a dispersed and smooth octahedral nanocrystal structure (see Supporting Information Figure S2). The reason is that the carboxyl groups on the surface of MWCNT provided sites for the growth of UiO-66, which reduced the stacking of carbon tubes during the in situ growth of UiO-66 along the carbon tubes. The steric hindrance of the carbon tube limited the crystal size of UiO-66, and the grain refinement guaranteed the uniformity of film coating on the substrate by water-based flame-retardant coatings. Hence, these results confirmed the synthesis of functionalized MOFs UiO-66@PA@MWCNT.

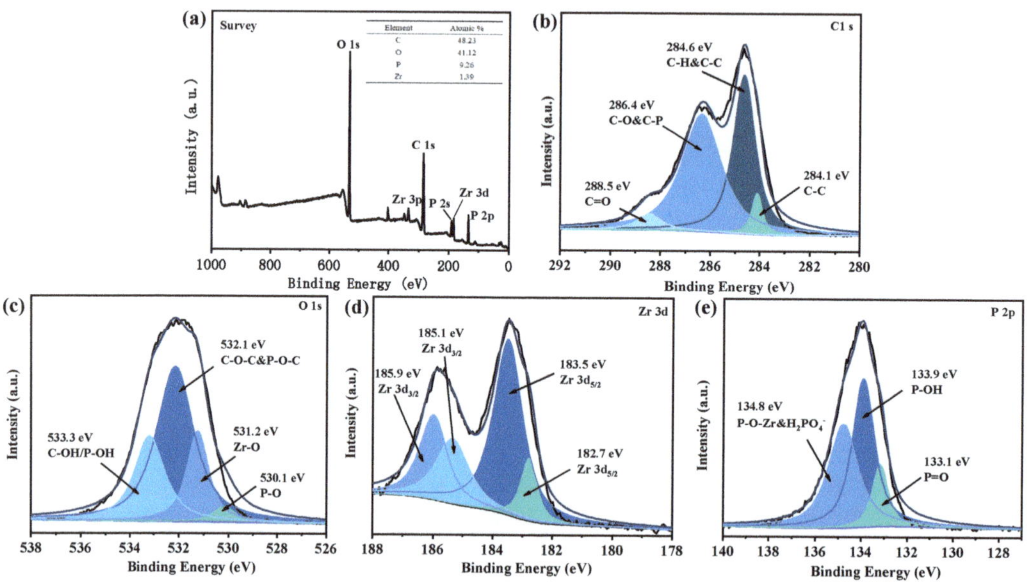

Figure 3. XPS spectra of UiO-66@PA@MWCNT: (**a**)survey scan; (**b**) C1s; (**c**) O1s; (**d**) Zr3d; (**e**) P_{2p}.

3.2. Structure Analysis of Flame Resistant PC/ABS Coating

To improve surface adhesivity, UiO-66@PA@MWCNT was dispersed in a water-based acrylic emulsion. The MOFs-based acrylic emulsion was characterized carefully before use. For characterization, films of the MOFs-based acrylic resin were prepared by drying the emulsion in the oven at 85 °C. Figure 4 presents the FT-IR spectra, XRD patterns, and TG-DTG curves of the films. In the FT-IR spectra, the peaks at 1170 cm^{-1}, 1450 cm^{-1}, 1730 cm^{-1}, and 2926 cm^{-1} for the WAUPM sample were attributed to C-O-C stretching vibrations, -OH bending vibrations, C=O stretching vibrations, and methylene stretching vibrations of the acrylic resin, respectively. Additionally, the vibrations for P=O (1160 cm^{-1}),

P-O-C (1126 cm^{-1}), P-O (1065 cm^{-1}), Si-O-Si (1092 cm^{-1}), and O-C-O symmetric vibrations (1405 cm^{-1}) and asymmetric vibrations (1560 cm^{-1}) of the terylene ligand were also observed for the WAUPM sample. Comparison with the FT-IR spectra of raw materials, viz. WA, PA@MWCNT@UiO-66, and SiO_2, showed no deviation of peaks for the WAUPM sample. This implied that the flame retardant did not react with the acrylic resin, which was added as an adhesive.

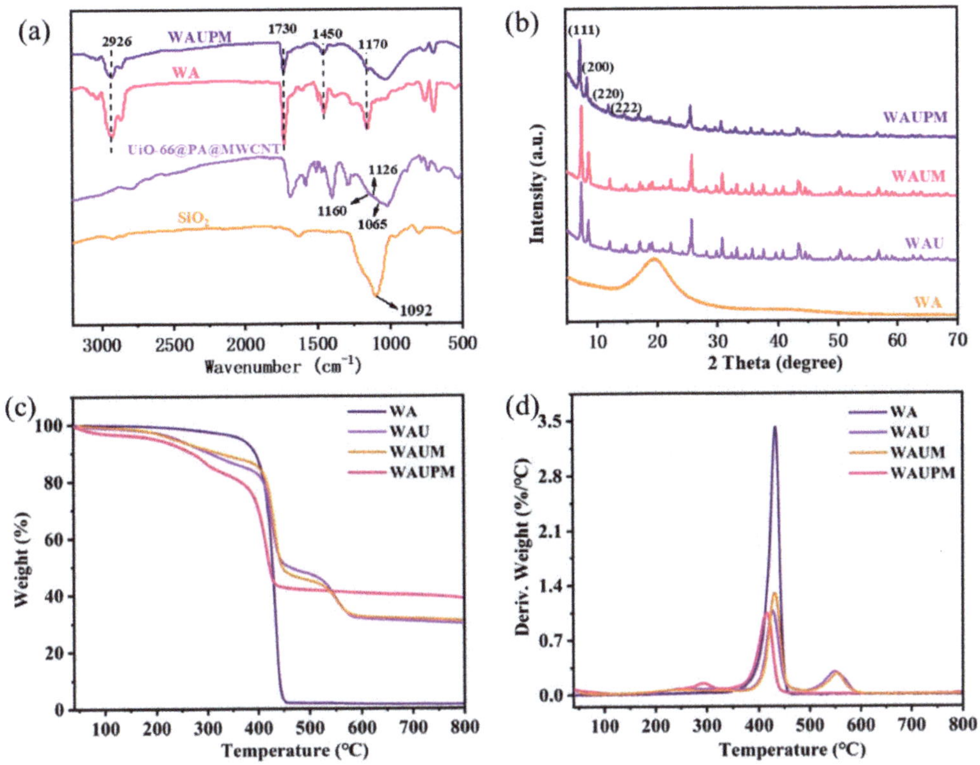

Figure 4. FT-IR spectra (**a**) and XRD patterns (**b**) of WAUPM, WA, UiO-66@PA@MWCNT and TGA (**c**) and DTG (**d**) curves of WAUPM, WAUM, WAU, and WA.

Furthermore, the XRD spectra showed diffraction peaks consistent with those of UiO-66 for the samples WAM, WAUM, and WAUPM [23]. This indicated that the crystal structure of UiO-66 did not change after functionalization. The TGA curve of WA indicated a good thermal stability, since the weight remained constant until around 443.1 °C. After the addition of UiO-66 or MWCNT@UiO-66, the initial degradation temperature decreased, and a secondary weight loss stage occurred at 550 °C, due to the collapse of the framework structure. WAU and WAUM showed more than 30.9 wt% carbon residue below 800 °C. After further modification with PA, the thermal stability decreased in the initial stage, but the residual carbon rate increased to 38.8 wt% below 800 °C for WAUPM. This result indicated that PA further promoted the char formation process during pyrolysis and also provided a rich carbon source to flame retardants.

Different from the additive flame retardants, the PC/ABS flame retardants with different coating thicknesses were prepared by spraying the MOFs-based acrylic emulsions on the surface of PC/ABS. The steps of spraying and drying were repeated until the coating thickness reached the required standard (±10%). The micro-structures of the coated PC/ABS substrates were studied by SEM. As shown in Figure 5, the untreated PC/ABS showed a slightly but uniformly fluctuating surface, while the PC/ABS coated with a MOFs-based flame retardant showed many granular structures on the substrate, presumed to be UiO-66 crystals. It was also found that the substrate became smoother when the thickness of the flame-retardant coating was increased. As seen from Figure 5c,d, the PC/ABS substrates were coated completely above a 750 μm thickness. There were less prominent granular structures on the WAUM-3 surface. WAUPM-3 showed a choppy surface morphology, but with fewer bright MOFs and smoother surfaces. In the magnified SEM images, smaller SiO_2 nano-particles and larger MOFs were observed, as shown in Figure 5e. Compared with WAUM-3, WAUPM-3 showed an improved surface smoothness. It implied that PA improved the compatibility of MWCNT@UiO-66 with acrylic resin. It also improved the wrapping and maximized the interfacial bonding between the adhesive layer and the PC/ABS substrate. Thus, PA@MWCNT@UiO-66 can be expected to provide a good flame-retardancy to PC/ABS.

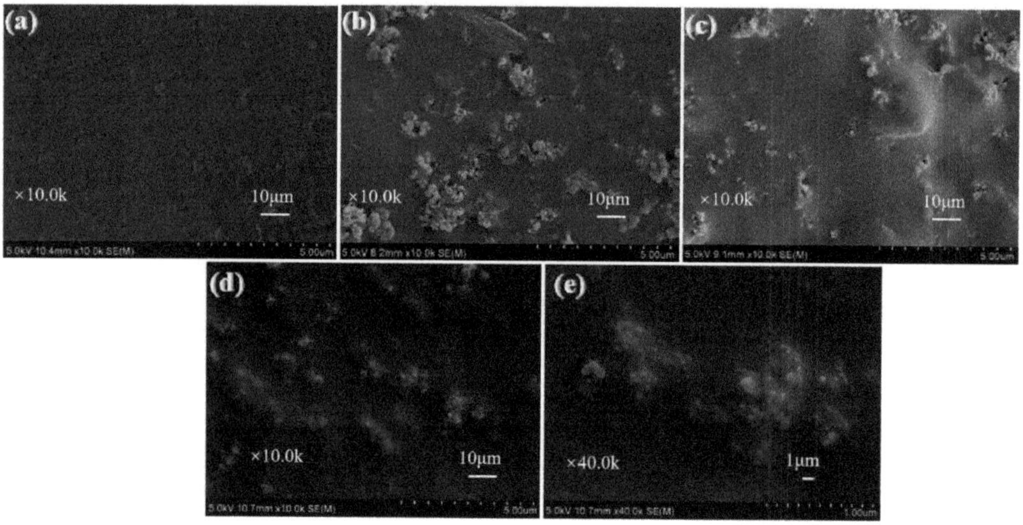

Figure 5. SEM images of untreated PC/ABS (**a**), WAU-3 (**b**), WAUM-3 (**c**), and WAUPM-3 (**d**,**e**).

3.3. *Performances of Flame Retardant*

Samples with different thicknesses of coating (250 μm, 500 μm, and 750 μm) were dried to test the flame-retardant performance, and the results are presented in Figure 6a and Supplement Material Table S2. As the coating thickness increased, the LOI values of WAUPM-1, -2, and -3 samples gradually increased (22.9%, 24.8%, and 27.5%). WAUPM-3 achieved the UL-94 V-0 grade, while WAUM-3 with the same thickness achieved a V-1 grade. The decrease in flame-retardancy was due to the absence of phosphorus source from PA. The flame-retardancy of the WAU-3 sample was poor and it failed to pass the V-2 grade. This was because this coating easily fell off during the vertical combustion test, reflecting weak adhesion, which was consistent with SEM analysis. Overall, WAUPM-3 achieved ideal flame-retardancy through the coordination of phosphate–zirconium–carbon flame-retardant elements. As shown in Table S2, the total of the two combustion times (t_1 + t_2) was about 12.5 s, and the presence of PA was reduced (t_1 + t_2) by 26.2 s in the sample WAUPM-3. Therefore, PA played a significant role in flame-retardancy in the combustion process of PC/ABS.

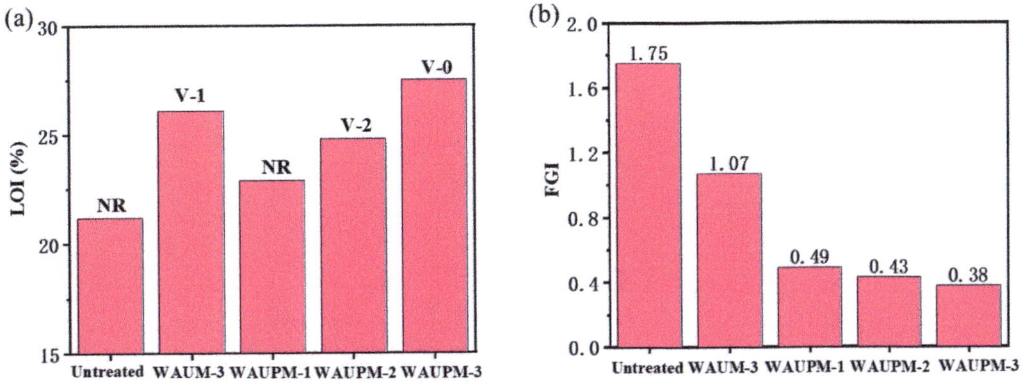

Figure 6. Limiting oxygen index and vertical burning grade (**a**) and fire growth index (FGI) (**b**) of PC/ABS samples coated with a waterborne flame-retardant coating.

In addition, the fire growth index (FGI) was introduced to evaluate the impact of the functionalized MOF coating system on the fire safety performance of PC/ABS. The smaller the value, the less time was required for the material to reach the intense burning state, and the lower the fire risk. As shown in Figure 6b, the FGI value of the untreated PC/ABS was 1.75, while it dropped to 1.07 after coating with MWCNT@UiO-6. The FGI was only 0.38 after the introduction of PA. Therefore, the fire safety performance of PC/ABS was significantly improved by coating with PA@MWCNT@UiO-66.

Smoke suppression performance is an important factor for the evaluation of the fire resistance of materials. CCT was used to simulate the real fire environment and evaluate the combustion behavior of materials in terms of heat, smoke, CO_2 release, etc. Figure 7 presents the heat release rate (HPR), total heat release (THR), smoke production rate (SPR), total smoke production (TSR), and curves for CO_2 and O_2 content of the samples. Figure 7a showed that the HRR value of the WAUM-3 sample was 153.7 kW/m^2, which decreased by 26.2% compared with the untreated sample. Figure 7b showed that the TRR value decreased by 26.9% compared with the untreated sample, which implied that the MWCNT@UiO-66 coating could effectively inhibit the heat release of PC/ABS. After the introduction of PA, the HRR and TRR values decreased by 57.5% and 33.6%, respectively, compared with the untreated sample (Figure 7a,b). This result indicated that PA could further inhibit the heat release of PC/ABS. In addition, as the thickness of the PA@MWCNT@UiO-66 coating increased from 250 μm to 750 μm, the peak smoke production rate decreased significantly and smoke release had a delayed effect. WAUM-3 and WAUPM-3 emitted similar amounts of smoke (11.44 m^2 and 12.03 m^2), but the WUPM-1, -2, and -3 samples extended the smoke release time and increased the escape time. This implied that MOFs played an important role, due to their high specific surface area and highly ordered porous structure. In the early stage of pyrolysis, the organic flammable volatiles released by PC/ABS passed through a complicated path. PA, as a phosphate, easily decomposes in a degraded polymer matrix to generate phosphoric acid. Phosphoric acid promotes surface cation crosslinking and carbonization, acting as an insulation barrier in the carbonization layer on the polymer surface to suppress thermal feedback from the combustion zone and act as a flame retardant [24]. Therefore, the coating with MOFs resulted in delayed smoke release. As the combustion intensified, PA produced meta-phosphate, polyphosphate, and other compounds, which could undergo dehydration with the hydroxyl group from PC/ABS pyrolysis and promote the formation of a carbon layer. Thus, the cross-linking of PC/ABS with the pyrolysis products of PA@MWCNT@UiO-66 promoted the carbonization process and improved the flame-retardancy. Figure 7e,f showed the changes in CO_2 and O_2 contents during the cone calorimetry test. Compared with the untreated sample, the consumption of O_2 and generation of CO_2 were reduced and delayed significantly in

WAUM-3. Particularly, the time was delayed by more than two times upon the addition of PA in the case of WAUPM. As the coating thickness increased from 250 μm to 750 μm, the overall O_2 consumption and CO_2 generation of the material continued to decrease. This result further showed that the thermal oxygen stability of flame-retardant materials increased after the coating of an MOFs-based flame retardant onto PC/ABS materials. It also indicated that the MOFs-based flame-retardant system reduced the reactivity with oxygen by inhibiting the contact between the matrix and O_2. Thus, the MOFs-based flame-retardant coating inhibited the violent combustion of PC/ABS.

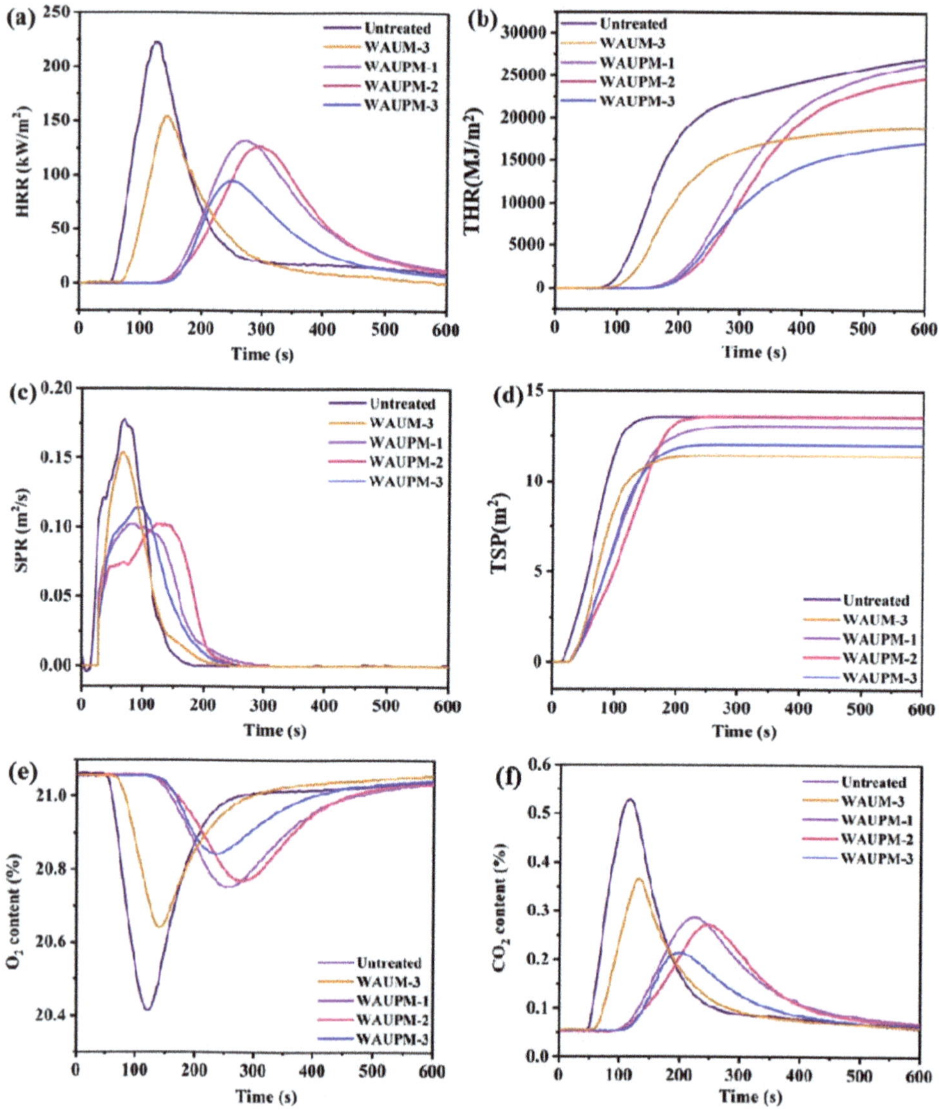

Figure 7. Curves for heat release rate (HPR) (**a**), total heat release (THR) (**b**), smoke production rate (SPR) (**c**), total smoke production (TSR) (**d**), CO_2 (**e**), and O_2 (**f**) contents of untreated, WAUM-3, WAUPM-1, WAUPM-2, and WAUPM-3 samples.

3.4. Mode of Action of Flame Retardant

In order to understand the mode of action of MOFs-based flame-retardant coatings, the carbon residues were carefully characterized by FT-IR spectroscopy, SEM, and Raman spectroscopy. Figure 8 presents the FT-IR spectra of carbon slags after the cone calorimetry test. WAUM-3 showed weak absorption peaks at 806 cm^{-1} and 752 cm^{-1}, ascribed to the out-of-plane deformation vibrations of the benzene ring =C-H, and para-substituted and mono-substituted benzene ring, respectively. The peaks at 1622 cm^{-1} and 471 cm^{-1} corresponded to the aromatic structure and Zr-O stretching vibrations, respectively. This result indicated that zirconia was produced after the combustion of WAUM-3 and that the aromatic structures of UiO-66 promoted the formation of carbon layers. Furthermore, the WAUPM coated samples showed P-O-C stretching vibrations at 980 cm^{-1} and peaks at 1187 cm^{-1} and 1092 cm^{-1} for P=O absorption and Si-O-Si absorption, respectively. This suggested that PA was decomposed during the combustion process to phosphoric acid, meta-phosphate, and other compounds. These compounds were esterified with the pyrolysis products of PC/ABS containing hydroxyl groups. Moreover, the presence of nano silica and carbon nanotubes promoted the formation of dense and continuous coke layers. Hence, as the coating thickness increased, the intensities of both P-O-C and Zr-O peaks in the carbon slag increased, as shown in Figure 8 for WAUPM-1, WAUPM-2, and WAUPM-3. Zirconia produced by the decomposition of zirconium organic framework also served as an efficient thermal barrier when mixed with the carbon layer. Thus, the product improved the thermal stability and thermal shielding property of the carbon layer.

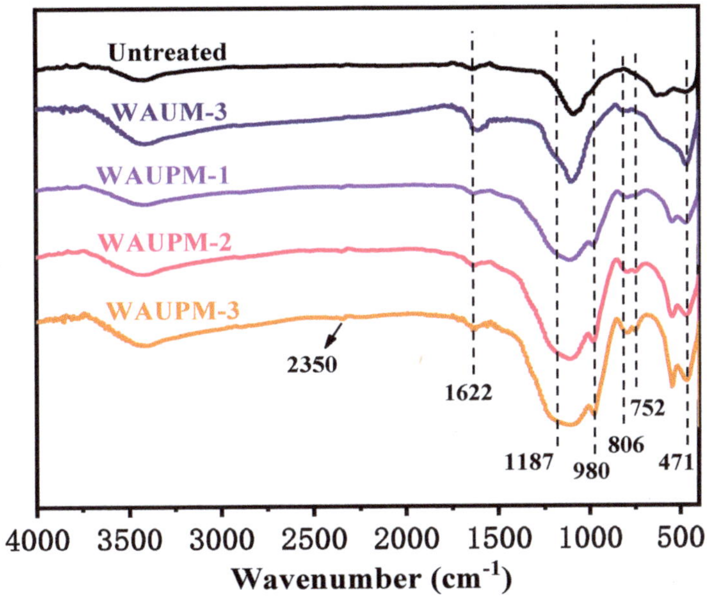

Figure 8. FT-IR spectra of carbon residues after the cone calorimetry test.

The residual amounts and the macro- and micro-morphologies of the carbon layers after polymer combustion also reflected the carbonization process. The morphology of the carbon slag in the condensation stage was investigated carefully. From a macro perspective, Figure 9a–d showed that the untreated PC/ABS generated the least amount of carbon, because most of the elements were released into the atmosphere in the form of smoke during combustion. In contrast, the amounts of carbon residues were much higher for WAUM-3 and WAUPM-3. The expanded size of carbon increased as the coating thickness increased for WAUPM-1, WAUPM-2, and WAUPM-3. The morphology of the carbon layer was further studied by SEM. The untreated PC/ABS showed many pores and the carbon layer was relatively loose (Figure 9f). The quality of the carbon layer improved in WAUM-3, but still showed a porous structure with traces of incomplete combustion (Figure 9f). However, the carbon layer of WAUPM-3 showed a smoother surface (Figure 9j). This implied that the carbon layer acted as a physical barrier against heat and inhibited the transfer of oxygen and materials between the condensed and gas phases. Thus, the MOFs-based flame-retardant coating effectively inhibited the thermal degradation of the underlying PC/ABS.

Figure 9. Macroscopic images and SEM images of carbon residues after vertical combustion test: (**a**,**f**) Untreated; (**b**,**g**) WAUM-3; (**c**,**h**) WAUPM-1; (**d**,**i**) WAUPM-2; and (**e**,**j**) WAUM-3.

Raman spectroscopy was employed to further characterize the degree of graphitization. Figure 10 presents the Raman spectra of samples. Peak D and peak G at about 1350 cm^{-1} and 1600 cm^{-1} corresponded to the peaks of amorphous carbon and graphitized carbon, respectively. The higher the degree of graphitization, the better the effect of heat insulation and oxygen isolation. As shown in Figure 10, as the thickness of the coating increased in WAUPM-1, WAUPM-2, and WAUPM-3, the I_D/I_G value decreased. Finally, the carbon residue of WAUPM-3 had the lowest I_D/I_G value. Compared with untreated PC/ABS and WAUM-3, the I_D/I_G value was also much smaller for WAUPM-3. These results indicated that WAUPM-3 had the highest degree of graphitization after carbon combustion, which displayed a good barrier effect and effectively inhibited the thermal degradation of the underlying PC/ABS. This observation was consistent with the SEM results.

Finally, based on the above analysis, the possible mode of action of the flame-retardant WAUPM-3 is described in Figure 11. When an external heat source or flame was applied, the MOFs-based coating acted as a thermal barrier due to the better thermal stability of PA@MWCNT@UiO-66. This prevented the flame from coming into direct contact with the PC/ABS substrate. In the second stage, when PC/ABS combustion was caused by cracking of the coating due to thermal expansion, the porous structure of MOFs delayed the partial release of smoke. This was because the porous structure of MOFs played a role of catalytic carbonization, forming dense coke that covered the surface of the material. This layer acted as an insulation barrier to inhibit heat transfer from the combustion zone and decrease the rate of formation of volatile fuel fragments. Moreover, MWCNT played another role of promoting the construction of a carbon cross-linking network. In general, the coatings comprised of a phosphorus-carbon-zirconium flame-retardant system endowed PC/ABS with excellent flame-retardancy.

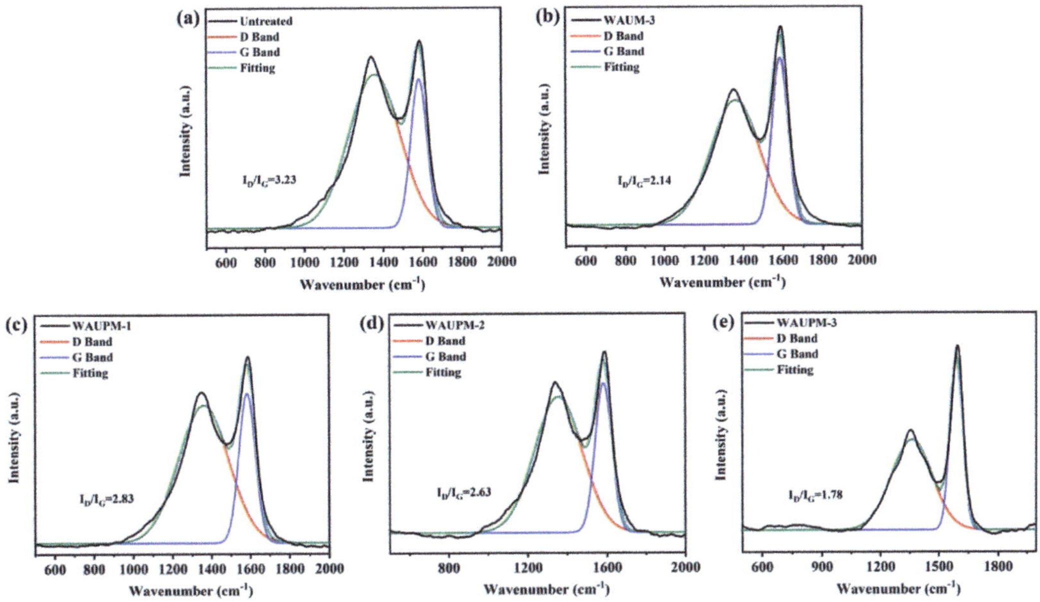

Figure 10. Raman spectra of carbon residues from untreated PC/ABS (**a**), WAUM-3 (**b**), WAUPM-1 (**c**), WAUPM-2 (**d**) WAUM-3, and (**e**) coated PC/ABS.

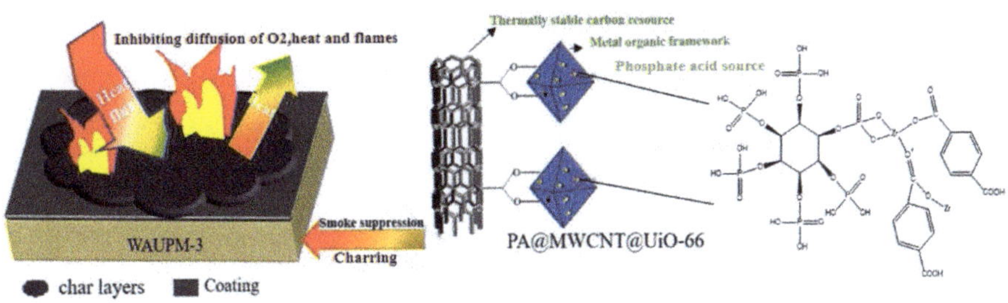

Figure 11. Illustration of the mode of action of fire-retardancy of the MOFs-based flame-retardant coating.

4. Conclusions

A new kind of MOFs-based flame retardant was prepared by loading phytic acid (PA) and multi-walled carbon nanotubes (MWCNT) onto UiO-66 by a solvothermal method. It was further applied as a flame-retardant surface coating onto PC/ABS. The results showed that PA improved the compatibility between UiO-66 and the acrylic resin and also helped to maximize the interfacial bonding between the adhesive layer and PC/ABS substrate and improve the wrapping property. When the thickness of the coating was 750 μm, the thermal barrier effect was significant and the LOI value of the material was 27.5%. The cone calorimetry test showed that the heat release and smoke release of the material had significant hysteresis and inhibition effects. Raman spectral analysis showed that a high-quality carbon layer was constructed during the combustion process, which effectively shielded the heat and oxygen transfer to the internal matrix. The phosphorus, carbon, and zirconium elements of PA, MWCNT, and UiO-66 contained in the coating showed flame-retardancy in the condensed phase. The hexagonal structure and thermal stability of MWCNT promoted the formation of a cross-linked network in the carbon layer. UiO-66 was involved in char formation and the tunnel effect during the early and middle stages of pyrolysis to provide the flame-retardant effect. In addition, the flame-retardant coating did not affect the mechanical properties of PC/ABS as in the case of physical filling, and the brushing was easy to repeat.

Supplementary Materials: The following supporting information can be downloaded at: https://www.mdpi.com/article/10.3390/polym16020275/s1, Figure S1: TGA (a) and DTG (b) curves of UiO-66@PA@MWCNT, UiO-66 and MWCNT-COOH; Figure S2: SEM images of (a) UiO-66; (b) MWCNT-COOH; (c) UiO-66@PA@MWCNT, and elemental distribution (d1–d4) of UiO-66@PA@MWCNT; Table S1: TGA data of samples for WAUPM, WAUM, WAU and WA; Table S2: Results of flame retardant coated PC/ABS for LOI and vertical burning test; Table S3: Results of flame retardant coated PC/ABS for Cone Calorimetry Test.

Author Contributions: Methodology, W.B. and H.Z. (Haitao Zhuo); Software, Y.Z.; Formal analysis, S.C.; Investigation, S.C., W.B. and H.Z. (Haitao Zhuo); Data curation, Y.Z., W.B. and H.Z. (Haiqiang Zhong); Writing—original draft, Y.Z.; Writing—review & editing, H.Z. (Haitao Zhuo); Supervision, S.C.; Project administration, H.Z. (Haitao Zhuo) and H.Z. (Haiqiang Zhong); Funding acquisition, H.Z. (Haiqiang Zhong). All authors have read and agreed to the published version of the manuscript.

Funding: This research was supported by the Guangdong Basic and Applied Basic Research Foundation (Grant No. 2022A1515011985), Shenzhen Science and Innovation Commission (20231120143631001), the Key Project of Department of Education of Guangdong Province (Grant No. 2020ZDZX2040), and the Heyuan Science and Technology Innovation Strategy (Grant No. 221028141607508).

Institutional Review Board Statement: Not applicable.

Data Availability Statement: Data are contained within the article.

Acknowledgments: This research was supported by the Guangdong Basic and Applied Basic Research Foundation (Grant No. 2022A1515011985), Shenzhen Science and Innovation Commission (20231120143631001), the Key Project of Department of Education of Guangdong Province (Grant No. 2020ZDZX2040), the Heyuan Science and Technology Innovation Strategy (Grant No. 221028141607508), and the Shenzhen University Cross-research program (no. 2023YG028).

Conflicts of Interest: Author Haiqiang Zhong was employed by the company Varitronix (Heyuan) Display Technology Limited. The remaining authors declare that the research was conducted in the absence of any commercial or financial relationships that could be construed as a potential conflict of interest.

References

1. Araby, S.; Philips, B.; Meng, Q.S.; Ma, J.; Laoui, T.; Wang, C.H. Recent advances in carbon-based nanomaterials for flame retardant polymers and composites. *Compos. Part B Eng.* **2021**, *212*, 29. [CrossRef]
2. Morgan, A.B. The Future of Flame Retardant Polymers—Unmet Needs and Likely New Approaches. *Polym. Rev.* **2019**, *59*, 25–54. [CrossRef]
3. Chiang, W.Y.; Hu, H.C.H. Phosphate-containing flame-retardant polymers with good compatibility to polypropylene. II. Effect of the flame-retardant polymers on polypropylene. *J. Appl. Polym. Sci.* **2001**, *82*, 2399–2403. [CrossRef]
4. Hirschler, M.M. Flame retardants and heat release: Review of data on individual polymers. *Fire Mater.* **2015**, *39*, 232–258. [CrossRef]
5. Huang, G.B.; Huo, S.Q.; Xu, X.D.; Chen, W.; Jin, Y.X.; Li, R.R.; Song, P.A.; Wang, H. Realizing simultaneous improvements in mechanical strength, flame retardancy and smoke suppression of ABS nanocomposites from multifunctional graphene. *Compos. Part B Eng.* **2019**, *177*, 10. [CrossRef]
6. Kim, C.; Lee, J.; Yoon, H.; Suhr, J. High Flame Retardancy and High-strength of Polymer Composites with Synergistically Reinforced MOSw and EG. *Compos. Res.* **2022**, *35*, 359–364. [CrossRef]
7. Zhao, S.K.; Chen, X.Y.; Zhou, Y.; Zhao, B.; Hu, Q.; Chen, S.T.; Pan, K. Molecular design of reactive flame retardant for preparing biobased flame retardant polyamide 56. *Polym. Degrad. Stab.* **2023**, *207*, 12. [CrossRef]
8. Feng, Y.Z.; Hu, J.; Xue, Y.; He, C.G.; Zhou, X.P.; Xie, X.L.; Ye, Y.S.; Mai, Y.W. Simultaneous improvement in the flame resistance and thermal conductivity of epoxy/Al2O3 composites by incorporating polymeric flame retardant-functionalized graphene. *J. Mater. Chem. A* **2017**, *5*, 13544–13556. [CrossRef]
9. Geoffroy, L.; Samyn, F.; Jimenez, M.; Bourbigot, S. Intumescent Polymer Metal Laminates for Fire Protection. *Polymers* **2018**, *10*, 995. [CrossRef] [PubMed]
10. Wu, J.N.; Chen, L.; Fu, T.; Zhao, H.B.; Guo, D.M.; Wang, X.L.; Wang, Y.Z. New application for aromatic Schiff base: High efficient flame-retardant and anti-dripping action for polyesters. *Chem. Eng. J.* **2018**, *336*, 622–632. [CrossRef]

11. Lu, H.D.; Song, L.; Hu, Y.A. A review on flame retardant technology in China. Part II: Flame retardant polymeric nanocomposites and coatings. *Polym. Adv. Technol.* **2011**, *22*, 379–394. [CrossRef]
12. Andrzejewski, J.; Mohanty, A.K.; Misra, M. Development of hybrid composites reinforced with biocarbon/carbon fiber system. The comparative study for PC, ABS and PC/ABS based materials. *Compos. Part B Eng.* **2020**, *200*, 13. [CrossRef]
13. Liu, Y.C.; Xu, W.Z.; Chen, R.; Cheng, C.M.; Hu, Y.Z. Effect of different zeolitic imidazolate frameworks nanoparticle-modified beta-FeOOH rods on flame retardancy and smoke suppression of epoxy resin. *J. Appl. Polym. Sci.* **2021**, *138*, 16. [CrossRef]
14. Pan, Y.T.; Zhang, Z.D.; Yang, R.J. The rise of MOFs and their derivatives for flame retardant polymeric materials: A critical review. *Compos. Part B Eng.* **2020**, *199*, 15. [CrossRef]
15. Nabipour, H.; Wang, X.; Song, L.; Hu, Y. Metal-organic frameworks for flame retardant polymers application: A critical review. *Compos. Part A Appl. Sci. Manuf.* **2020**, *139*, 10. [CrossRef]
16. Huang, R.; Guo, X.Y.; Ma, S.Y.; Xie, J.X.; Xu, J.Z.; Ma, J. Novel Phosphorus-Nitrogen-Containing Ionic Liquid Modified Metal-Organic Framework as an Effective Flame Retardant for Epoxy Resin. *Polymers* **2020**, *12*, 108. [CrossRef]
17. Gu, L.Q.; Qiu, C.; Qiu, J.H.; Yao, Y.W.; Sakai, E.; Yang, L.T. Preparation and Characterization of DOPO-Functionalized MWCNT and Its High Flame-Retardant Performance in Epoxy Nanocomposites. *Polymers* **2020**, *12*, 613. [CrossRef]
18. Im, J.S.; Lee, S.K.; In, S.J.; Lee, Y.S. Improved flame retardant properties of epoxy resin by fluorinated MMT/MWCNT additives. *J. Anal. Appl. Pyrolysis* **2010**, *89*, 225–232. [CrossRef]
19. Yu, T.; Jiang, N.; Li, Y. Functionalized multi-walled carbon nanotube for improving the flame retardancy of ramie/poly(lactic acid) composite. *Compos. Sci. Technol.* **2014**, *104*, 26–33. [CrossRef]
20. *GB/T2408-2008*; Plastics—Determination of Burning Characteristics—Horizontal and Vertical Test. National Standardization Technical Committee Plastic: Beijing, China, 2008.
21. *GB/T2406.2-2009*; Plastics—Determination of Burning Behaviour by Oxygen Index—Part 2: Ambient—Temperature Test. National Standardization Technical Committee Plastic: Beijing, China, 2009.
22. *ASTM D7309*; Standard Test Method for Determining Flammability Characteristics of Plastics and Other Solid Materials Using Microscale Combustion Calorimetry. ASTM International: West Conshohocken, PA, USA, 2019.
23. Ma, J.; Guo, X.Y.; Ying, Y.P.; Liu, D.H.; Zhong, C.L. Composite ultrafiltration membrane tailored by MOF@GO with highly improved water purification performance. *Chem. Eng. J.* **2017**, *313*, 890–898. [CrossRef]
24. Howell, B.A. Thermal Degradation of Organophosphorus Flame Retardants. *Polymers* **2022**, *14*, 4929. [CrossRef]

Disclaimer/Publisher's Note: The statements, opinions and data contained in all publications are solely those of the individual author(s) and contributor(s) and not of MDPI and/or the editor(s). MDPI and/or the editor(s) disclaim responsibility for any injury to people or property resulting from any ideas, methods, instructions or products referred to in the content.

polymers

MDPI

Article

Cage Nanofillers' Influence on Fire Hazard and Toxic Gases Emitted during Thermal Decomposition of Polyurethane Foam

Arkadiusz Głowacki [1,*], Przemysław Rybiński [1,*], Monika Żelezik [2] and Ulugbek Zakirovich Mirkhodjaev [3]

1 Institute of Chemistry, The Jan Kochanowski University, 25-406 Kielce, Poland
2 Institute of Geography and Environmental Sciences, Jan Kochanowski University, 25-406 Kielce, Poland; monika.zelezik@ujk.edu.pl
3 Department of Biophysics, National University of Uzbekistan, Tashkent 100095, Uzbekistan; u.z.mirkhodjaev@gmail.com
* Correspondence: roarek@wp.pl (A.G.); przemyslaw.rybinski@ujk.edu.pl (P.R.)

Abstract: Polyurethane (PUR), as an engineering polymer, is widely used in many sectors of industries. However, the high fire risks associated with PUR, including the smoke density, a high heat release rate, and the toxicity of combustion products limit its applications in many fields. This paper presents the influence of silsesquioxane fillers, alone and in a synergistic system with halogen-free flame-retardant compounds, on reducing the fire hazard of polyurethane foams. The flammability of PUR composites was determined with the use of a pyrolysis combustion flow calorimeter (PCFC) and a cone calorimeter. The flammability results were supplemented with smoke emission values obtained with the use of a smoke density chamber (SDC) and toxicometric indexes. Toxicometric indexes were determined with the use of an innovative method consisting of a thermo-balance connected to a gas analyzer with the use of a heated transfer line. The obtained test results clearly indicate that the used silsesquioxane compounds, especially in combination with organic phosphorus compounds, reduced the fire risk, as expressed by parameters such as the maximum heat release rate (HRRmax), the total heat release rate (THR), and the maximum smoke density (SDmax). The flame-retardant non-halogen system also reduced the amounts of toxic gases emitted during the decomposition of PUR, especially NOx, HCN, NH_3, CO and CO_2. According to the literature review, complex studies on the fire hazard of a system of POSS–phosphorus compounds in the PUR matrix have not been published yet. This article presents the complex results of studies, indicating that the POSS–phosphorous compound system can be treated as an alternative to toxic halogen flame-retardant compounds in order to decrease the fire hazard of PUR foam.

Keywords: PUR composites; silsesquioxane; fire hazard; smoke emission; toxicometric index

check for updates

Citation: Głowacki, A.; Rybiński, P.; Żelezik, M.; Mirkhodjaev, U.Z. Cage Nanofillers' Influence on Fire Hazard and Toxic Gases Emitted during Thermal Decomposition of Polyurethane Foam. *Polymers* **2024**, *16*, 645. https://doi.org/10.3390/polym16050645

Academic Editors: Fei Xiao, Fubin Luo and Ke Sun

Received: 15 January 2024
Revised: 9 February 2024
Accepted: 21 February 2024
Published: 27 February 2024

Copyright: © 2024 by the authors. Licensee MDPI, Basel, Switzerland. This article is an open access article distributed under the terms and conditions of the Creative Commons Attribution (CC BY) license (https://creativecommons.org/licenses/by/4.0/).

1. Introduction

A fire hazard is a result of multiple aspects, including a high released heat value and oxygen deprivation, as well as smoke and toxic gas emissions [1,2].

During a fire hazard, people are exposed not only to fire or smoke but also to a mixture of toxic gases, such as CO, CO_2, NOx, HCN, SOx, and HCl [3–7].

In the event of a room fire, the most dangerous products of thermal decomposition and combustion are created during the combustion of polyurethane-based foams. Polyurethane foams are used, among others, in the production of upholstered furniture. The research conducted so far indicates that even in the initial phase of the thermal decomposition of PUR foam, the concentrations of CO and CO_2 are so high that they can pose a lethal threat to people. During the decomposition of PUR foam, high amounts of HCN and NOx, which have been proven to have asphyxiate properties, are also emitted [8–10].

Polyurethane, as an engineering composite, is widely used in many sectors of industry. However, the high fire risks associated with PUR, including the smoke density, a high heat release rate, and the toxicity of combustion products, limit its applications in many fields [11,12].

As a consequence, research on how to reduce the flammability and toxicity properties of PUR is attracting more and more attention.

Nowadays, to reduce the fire hazard of PUR composites, an intumescent flame-retardant system, first of all, in the form of expandable graphite (EG), is widely used [13]. EG is halogen free and acts mainly in the condensed phase. Exposed to heat, expandable graphite forms a low-density, thermal insulating layer on the surface of the polymer that prevents the transfer of both heat and oxygen. Unfortunately, the incorporation of EG into the PUR matrix decreases the mechanical properties of PUR, mainly due to its poor adhesion to the PUR matrix and large particle size. Furthermore, the expandable graphite clearly increases smoke emission during the thermal decomposition of PUR composites [14–16].

Currently, in order to develop a new intumescent flame-retardant system, organic phosphorous compounds, alone and in synergistic action with silsesquioxanes are being studied. Compounds based on phosphorus have several advantages, such as low toxicity, a lack of release of poisonous halogen gases, and the production of low amounts of smoke during combustion. Phosphorus compounds, similar to graphite, can form a char layer that protects polymers from combustion heat. Unfortunately, the phosphorus compound's carbon layer created during thermal decomposition very often lacks a homogenous structure. For this reason, the action efficiency of phosphorus compounds in PUR matrices is worse in comparison with expandable graphite. In order to improve the homogenous as well as insulative properties of the carbon layer created during the thermal decomposition of PUR, phosphorus compounds are joined with silsesquioxanes [17].

Silsesquioxanes are polyhedral structures of the general formula $(RSiO1.5)n$, where R is virtually any organic substituent or a hydrogen atom, and n is an integer, in most cases equal to 6, 8, 10, or 12. The silsesquioxane core is regarded as the smallest obtainable fragment of crystalline silica. The dimension of the silsesquioxane molecule is within 3 nm. This supposes that crystalline silica may stabilize and increase the amount of homogeneous carbon residue after the thermal decomposition of polymer composites [18–22].

Therefore, the aim of this work is to examine the effectiveness of a synergic system, a phosphorus compound with silsesquioxane, in decreasing fire hazards with special consideration for the emission of toxic gases and smoke from PUR composites.

2. Materials and Methods

2.1. Materials

The subject of this study was polyurethane foam, which was synthesized with the use of polyol (BASF, Elastoflex W5165/140) and isocyanide (BASF, Izo 135/158) (diphenylmethane diisocyanate–MDI) in a 2:1 proportion.

As a flame retardant, an organic phosphorus compound was used, triphenyl phosphate (TPP), which was produced by Everkem, Italy. The phosphorus content was equal to 9.5% (Figure 1).

The silsesquioxanes used were methacrylpropyl POSS MA0735 (POSS MA), aminopropyl POSS AM0265 (POSS AM), and OL POSS OL1170 (POSS OL) (Figure 1).

2.2. Preparation of PUR Composites

Flexible polyurethane foams were produced at the laboratory scale using the one-stage method involving a two-component system, with an equivalent ratio of OH to NCO groups equal to 2:1. Component A, i.e., Elastoflex W5165/140, was mixed with the appropriate amount of modifier (5 wt.% POSS and 10 wt.% TPP vs. polyol). Component B was diphenylmethane diisocyanate (MDI) (Table 1).

Components A and B were mixed and poured into a Teflon open mold, and then foam was frothed. Flexible polyurethane foams, after having been left to the end of the growth process for 1 h, were then conditioned to stable mass at a temperature of 23 ± 2 °C and under relative humidity less than 50 ± 5%, according to the PN-EN ISO 291:2010 standard [23].

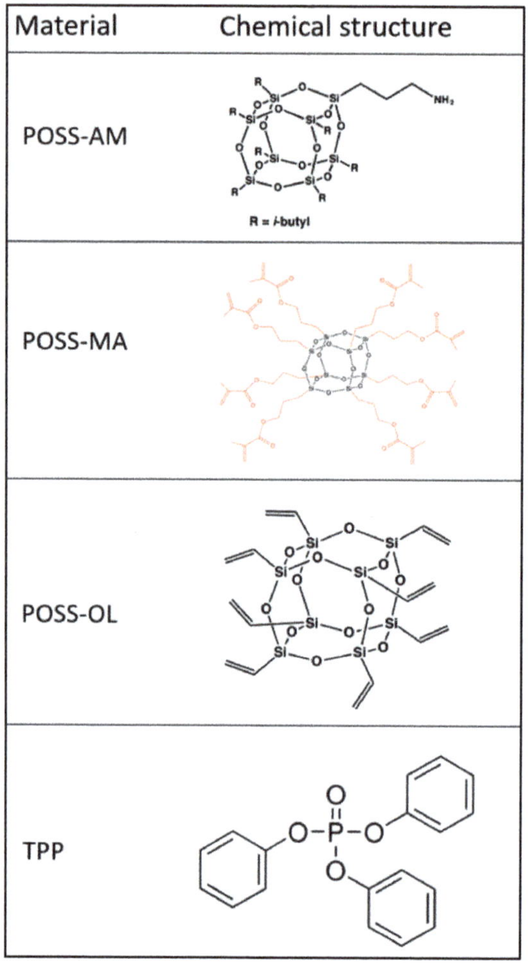

Figure 1. Chemical structure of used compounds.

Table 1. Composition of PUR composites in mass fraction (phr).

Composite	AM	MA	OL	TPP
PUR–AM	5			
PUR–MA		5		
PUR–OL			5	
PUR–TPP				10
PUR–AM–TPP	5			10
PUR–MA–TPP		5		10
PUR–OL–TPP			5	10

2.3. Methods

2.3.1. Scanning Electron Microscopy

Scanning electron microscopy (SEM) images were assessed by means of a Quanta 250 FEG electron microscope (FEI Company, Hillsboro, OR, USA) with electron gun and field emission (Schottky's emitter, Hillsboro, OR, USA).

2.3.2. Fourier-Transform Infrared Spectroscopy Analysis

Fourier-Transform Infrared Spectroscopy with attenuated total reflection (FTIR-ATR) was recorded on a PerkinElmer Spectrum (Waltham, MA, USA). Two FTIR spectrophotometer equipped with a single-reflection diamond ATR crystal on a ZeSe plate. Measurements were recorded using a spectrum computer program. The parameters were 4 scans from 400–4000 cm^{-1} in the transmittance mode with a resolution of 4 cm^{-1}.

2.3.3. Thermogravimetry Analysis

The thermograms of thermogravimetric analysis (TGA) were recorded using Netzsch STA 449 F3 Jupiter (Selb, Germany) in a temperature range from 25 to 650 °C, for samples 5 ± 1 mg placed in the open Al_2O_3 pan. The measurements were performed with a gas flow of 40/20 μL/min in nitrogen/oxygen and a heating rate of 10 °C/min. The results obtained were processed using the Proteus Thermal Analysis 8.0.3 computer program. The parameters considered during thermogravimetric analysis include the temperature at which the sample loses 5% of weight (T_5), the temperature at which the sample loses 50% of weight (T_{50}), the temperature at the maximum rate of sample decomposition (TR_{MAX}), the range of the combustion temperature of the residue after thermal decomposition of the sample (ΔT_s) and the residue at a temperature of 600 °C (P_{600})

2.3.4. Toxicity

The studies on the release of the toxic combustion products of PUR composites were carried out using a TG-gas analyzer coupled system (Netzsch TG 209 F1 Libra (Selb, Germany)) coupled with a real-time analyzer Bruker Omega 5 gas analyzer (Billerica, MA, USA). The measurement of toxic gas evolution analysis were performed at a temperature range from 30 °C to 650 °C with a gas flow of 40/20 μL/min in nitrogen/oxygen and a heating rate of 10 °C/min, for the 5 ± 1 mg samples. The gases were analyzed using a FTIR MIR detector (400–4000 cm^{-1}) (Billerica, MA, USA). Spectra were registered every 7–8 s (10 scans). Then, the spectra were converted using the Opus GA computer program, version 8.7.41 to the value of the gas emission concentration [ppm]. The gases recorded were CO_2, CO, HCl, NH_3, NO and NO_2.

The emission gas rate in ppm was converted to toxic gases release concentrations in g/m^3 using the ideal gas law.

$$PV = nRT$$

where

P—air/gas pressure in atm (1 atm = 1013.25 hPa),

V—volume of gas,

n—number of particles moles in a gas,

R—gas constant (0.08206 $\frac{L\times atm}{mol\times K}$), and

T—Kelvin temperature scale (T °C + 273).

By transforming the above formula, the number of moles per liter of air in 1 atm. in 25 °C is converted:

$$\frac{n}{V} = \frac{P}{R \times T} = \frac{1\ \text{atm}}{298\ \text{K} \times 0.08206\ \frac{L\times atm}{mol\times K}} = 0.0409\frac{\text{mol}}{\text{L}}$$

Then, the volume concentration (g/m^3) of toxic combustion products is converted into mass per volume of air:

$$Concentration \frac{g}{m^3} = \frac{0.0409 \times C[ppm] \times M}{1000}$$

where

C—emission gas rate in [ppm], and
M—converted gas molar mass.

2.3.5. Microcalorimetry PCFC

The flammability of the PUR, PUR–AM, PUR–MA, PUR–OL, PUR–TPP, PUR–AM–TPP, PUR–MA–TPP and PUR–OL–TPP was tested using PCFC (pyrolysis combustion flow calorimeter) produced by Fire Testing Technology Ltd. (East Grinstead, UK). The procedure was performed in according to the ASTM D 7309 standard [24]. The pyrolizer's temperature was 650 °C, and the combustor's heat was 900 °C. The composites were heated at a 1 °C/s rate. The test was performed in conditions of nitrogen/oxygen (80/20 cc/min). During the test, the following parameters were recorded: the heat release rate, the maximum (peak) value of PCFC HRR (W/g), the time to HRR peak, total HR (kJ/g), and the heat release capacity (J/gK).

2.3.6. Flammability

The PUR composites were tested using a cone calorimeter, produced by Fire Testing Technology Ltd., according to the PN-EN ISO 5660 standard [25]. The samples, all having the dimensions 100 × 100 × 50 mm, were tested in a horizontal position with a heat radiant flux density of 35 kW/m^2. During the test, the following parameters were recorded: initial sample weight, sample weight during testing, final sample weight, time to ignition (TTI), total heat released (THR), effective combustion heat (EHC), the average weight loss rate (MLR), and the heat release rate (HRR).

2.3.7. The Smoke Density

The smoke density was recorded using a Fire Testing Technology Smoke Density Chamber (SDC) according to the PN-EN ISO 5659-2 standard [26]. The samples, all having dimensions 75 × 75 × 15 mm, were tested with a heat radiant flux density of 25 kW/m^2. During the test, the following parameters were recorded: initial sample weight, weight during the test, final sample weight, maximum specific optical density ($D_s$max), specific optical density after 4 min of testing Ds(4), area under the specific optical density curve (VOF4) and light attenuation coefficient after testing.

3. Result and Discussion

3.1. Surface Morphology of PUR Composites

The test results obtained by SEM clearly indicate that the PUR composites are characterized by a porous structure. Free spaces filled with air during the thermal decomposition of PUR not only increase the efficiency of exothermic combustion reactions through the diffusion of oxygen into the reaction environment, but also catalyze degradation reactions and thermal destruction of PUR chains (Figure 2A,B).

Introducing an organophosphate compound in the form of TPP to the polyurethane foam matrix significantly reduces the size of free spaces in the PUR structure (Figure 2C,D). The results obtained by the SEM analysis clearly indicate that TPP is homogeneously distributed in the polyurethane matrix, and thus limits the degree of porosity of the tested PUR foam.

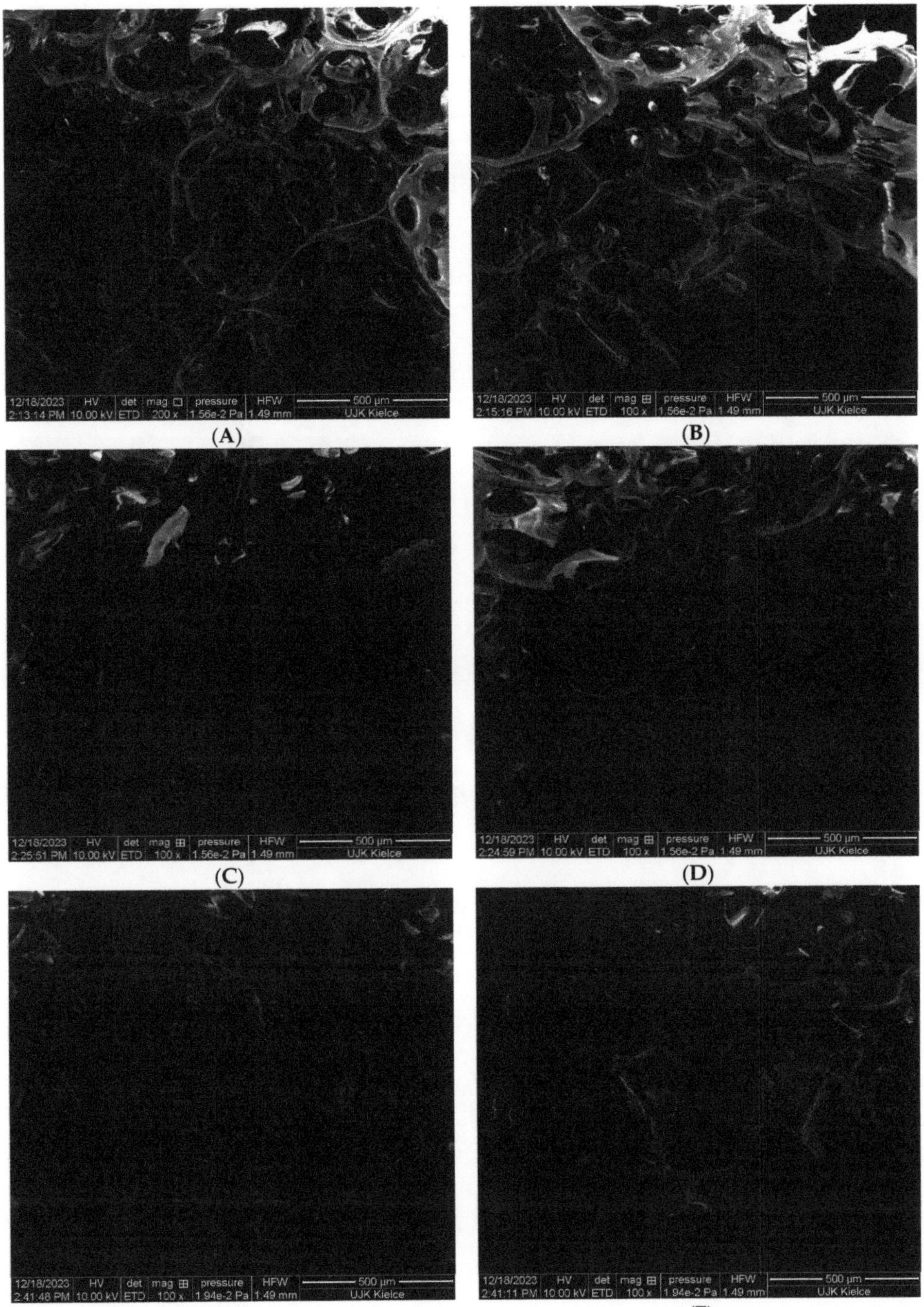

Figure 2. *Cont.*

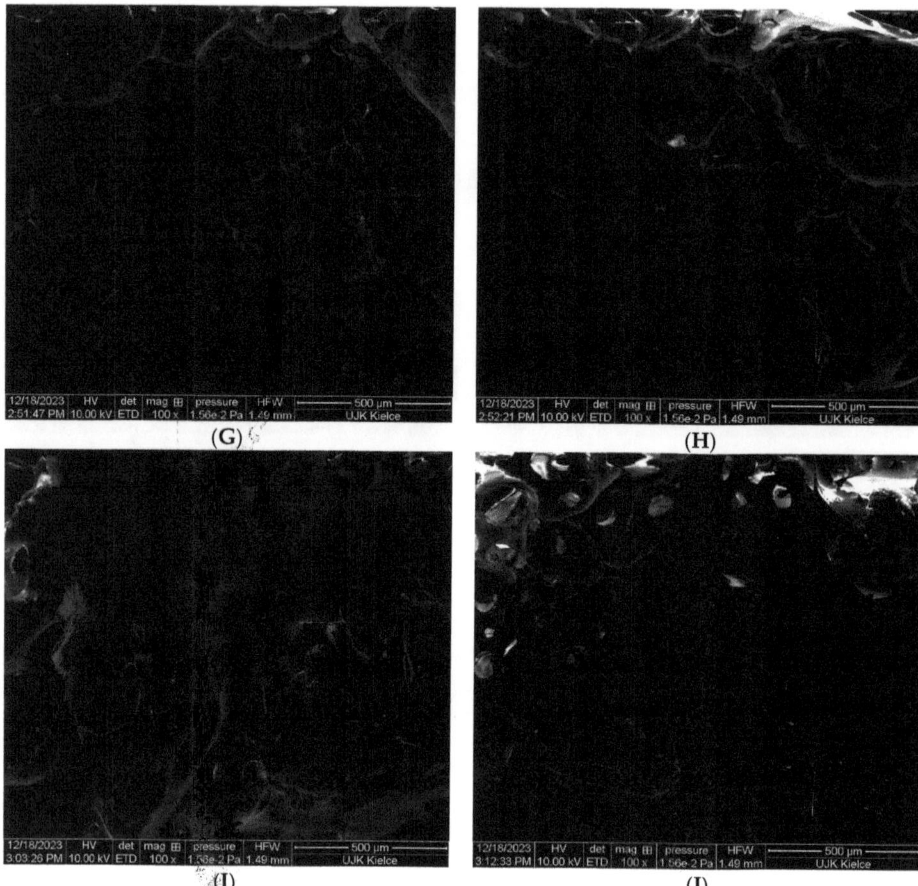

Figure 2. SEM images of PUR composites: (**A**,**B**) reference PUR composite; (**C**,**D**) PUR composite containing TPP; (**E**,**F**) PUR composite containing an AM–TPP system; (**G**,**H**) PUR composite containing an TPP–MA system; (**I**,**J**) PUR composite containing an AM–TPP OL–TPP system.

A significant reduction in the porosity of the tested PUR foam was achieved for the TPP–POSS system. Regardless of the type of POSS used, both the number and size of free spaces in the PUR structure were significantly reduced in relation to the pure PUR as well as in the PUR–TPP composite [27,28].

The analysis of the obtained SEM images clearly indicates that the best effect in terms of reducing the porosity of the PUR foam was achieved for the TPP–OL system. In the case of the PUR–TPP–OL (Figure 2I,J), the porous structure of PUR practically disappeared.

The structural studies of the PUR–POSS composites by the IR spectroscopy revealed the characteristic absorbance bands. The band at 3337 cm^{-1} corresponded to the stretching vibrations of N-H (Figure 3) [29]. The peak at 2868 and 2868 cm^{-1} corresponded to the symmetric and asymmetric stretching vibration of C-H in hydrocarbon chains, respectively. The peak with a maximum of 1710 cm^{-1} derived from stretching vibrations was associated with the occurrence of the carbonyl group. In the IR spectra, the following signals were also visible: signals at 1538 and 1509 cm^{-1} corresponded to the bending vibrations of the N-H group, whereas the signal at 1094 cm^{-1} was associated with the stretching vibrations of C-O-C groups [30,31].

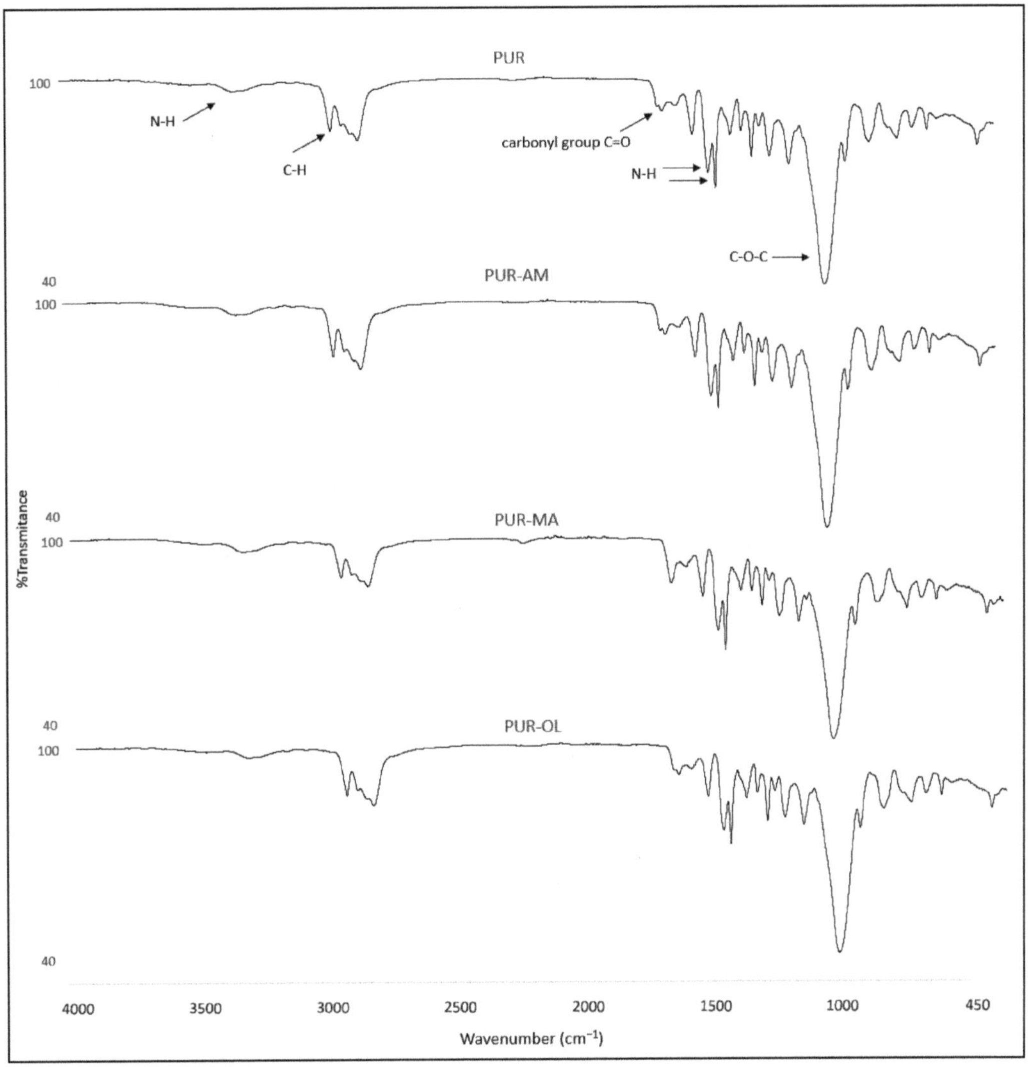

Figure 3. FTIR spectra of PUR, PUR–AM, PUR–MA and PUR–OL composites.

New signals at 1489, 1186 and 959 cm^{-1} were registered for the PUR composites containing TPP. The peak at 1489 cm^{-1} and 959 cm^{-1} corresponded to a vibration aromatic ring, whereas the peak at 1186 cm^{-1} corresponded to the valence vibration of the P-O-C group (Figure 4) [32,33].

Additional signals were not recorded for the FTIR spectrum of the PUR composites containing TPP with POSS (AM, MA, OL). Also, no loss of signals was observed for the spectrum of the PUR−TPP composites. Therefore, it was found that TPP does not react with polyol and isocyanate during the PUR synthesis process, or even POSS compounds added at this stage.

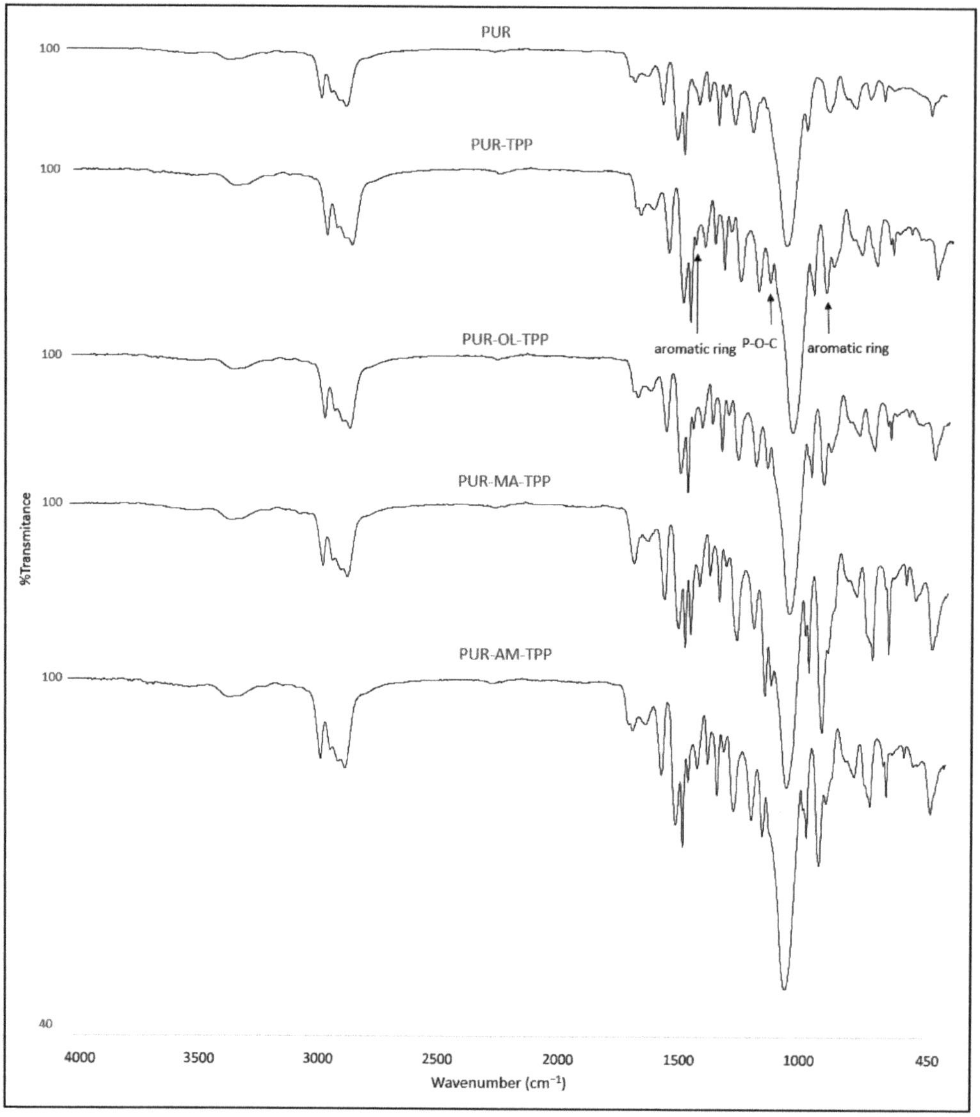

Figure 4. FTIR spectra of the PUR, PUR–TPP and PUR–OL–TPP composites.

3.2. Thermal Analysis and Flammability of PUR Composites

The results of the thermal analysis indicate that POSS compounds have an ambiguous effect on the thermal parameters of PUR composites (Table 2). The POSS–AM as well as POSS–MA increase the value of the T_5 and T_{50} parameters, and significantly reduce the value of the dm/dt parameter, especially in the case of POSS–MA (Figure 5A). Limiting the rate of thermal decomposition dm/dt is of key importance from the point of view of reducing the fire hazard of the tested materials. The lower the value of the dm/dt parameter, the lower the intensity decomposition of a given composite, and thus the less flammable composite enters the flame zone.

Table 2. Thermal analysis results of PUR composites.

Composite	T_5 (°C)	T_{50} (°C)	T_{RMAX} (°C)	dm/dt (%/min)	P_{TD} (%)	ΔT_s (°C)	P_{600} (%)
PUR	295	365	365	10.1	20.1	460–600	1.45
PUR–AM	271	345	330	9.3	23.7	475–600	3.60
PUR–MA	305	385	370	8.3	31.6	475–600	10.81
PUR–OL	295	350	340	10.6	27.2	450–600	7.20
PUR–TPP	260	350	345	9.6	25.5	235–600	0.15
PUR–AM–TPP	270	375	360	6.7	26.6	450–600	4.54
PUR–MA–TPP	270	385	395	6.1	25.2	465–600	4.27
PUR–OL–TPP	195	345	345	7.4	23.6	455–600	4.71

It is worth noting that the value of the dm/dt parameter was reduced by nearly 18% compared to the reference sample in the case of the PUR–MA composite, which suggests that the intensity of the combustion process (flame feeding) and the emission of toxic gaseous destructs are significantly lower in the case of the POSS–MA sample when compared to the PUR reference sample (Table 2).

All tested POSS compounds (AM, MA, OL) cause an increase in both residues after thermal decomposition, the PTD parameter, as well as residues at a temperature of 600 °C, the P_{600} parameter. A clear increase in the value of the P_{TD} and P_{600} parameters indicates that POSS compounds can catalyze the formation of a ceramic boundary layer during the decomposition of the PUR composite. This layer, due to its insulating properties, primarily limits the transfer of heat from the flame to the interior of the sample, as well as the transfer of liquid and gaseous destructs to the flame [34,35].

The application of an organophosphate flame-retardant compound into the PUR matrix in the form of TPP also had an ambiguous effect on the thermal stability of the PUR–TPP composite (Figure 5B). In the presence of TPP, thermal stability parameters such as T_5, T_{50} and T_{RMAX} were not improved, while the dm/dt parameter had a value close to the reference sample (PUR composite).

However, it should be noted that the residue parameter after thermal decomposition (P_{TD} parameter) was increased, which indicates that TPP generates a carbon residue of a potential insulating nature during thermal decomposition. The aromatic rings enrich the TPP structure with carbon (Figure 1). Nevertheless, the obtained test results clearly indicate that the carbon residue after thermal decomposition was almost completely burned at ΔT = 435–600 °C. The P600 parameter was only 0.15% in the case of the PUR–TPP composite.

In the case of PUR composites containing the synergistic POSS–TPP system, the amount of residue after thermal decomposition (P_{TD} parameter) was practically no different in relation to the residue after thermal decomposition of the PUR–TPP composite (Figure 5C).

Nevertheless, the P_{600} parameter value for the composites containing the POSS–TPP system was significantly higher when compared to the PUR–TPP composite. These were 4.54, 4.27, and 4.71 for the PUR–AM–TPP, PUR–MA–TPP, and PUR–OL–TPP composites, respectively (Table 2).

The increase in the P_{600} parameter for the composites containing the POSS–TPP system indicates that POSS compounds stabilize the carbon residue through physical interactions, i.e., the formation of a ceramic coating on the carbon surface and as a result of a chemical reaction based on the formation of thermally stable Si-C silicon carbide [36].

It is also worth noting that those composites which contained the POSS–TPP system, especially the PUR–MA–TPP ones, were characterized by higher values of the T_{50} and T_{RMAX} parameters, along with a simultaneously reduced value of the dm/dt parameter, compared to the PUR–TPP reference sample.

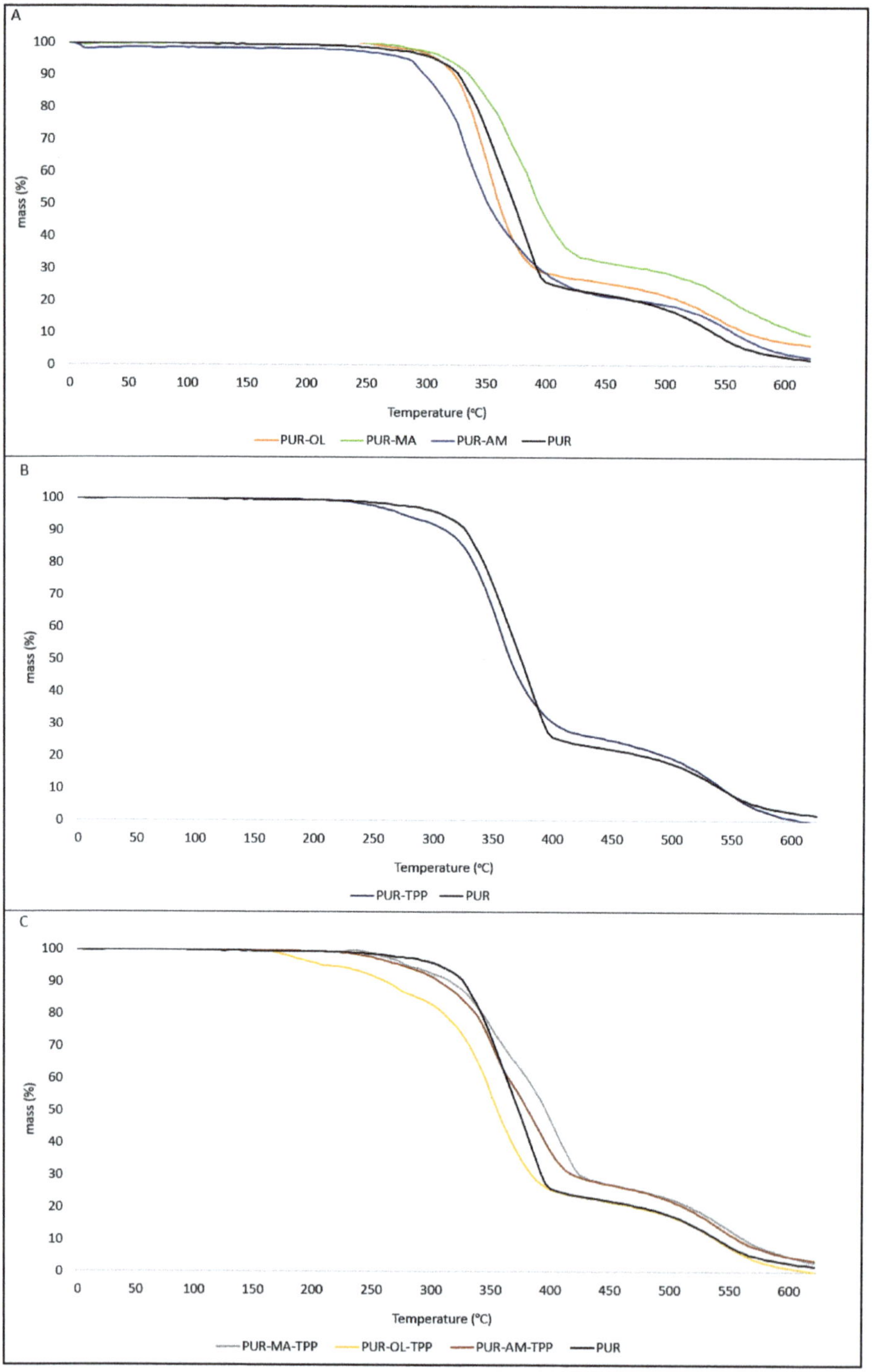

Figure 5. TG thermograms of PUR composites: (**A**) PUR, PUR–AM, PUR–MA, and PUR–OL; (**B**) PUR and PUR–TPP; (**C**) PUR, PUR–AM–TPP, PUR–MA–TPP, and PUR–OL–TPP.

The test results obtained using the PCFC method were corelating positively with those obtained when using the thermal analysis method [37,38]. Among the tested POSS compounds, the PUR–MA composite had the highest reduction value for both HRR_{MAX} and THR parameters. It cannot be excluded that despite the lack of changes in the IR spectrum, the POSS–MA was applied to the structure of the PUR foam through polar groups located at the end of the alkyl chains or bound to the PUR through reversible intermolecular interactions [39,40].

POSS–AM and POSS–OL also have a beneficial effect on reducing the flammability of PUR composites; however, the reduction in the values of HRR_{MAX} and THR parameters was observed for a much smaller extent than in the case of the composite containing POSS–MA (Figure 6).

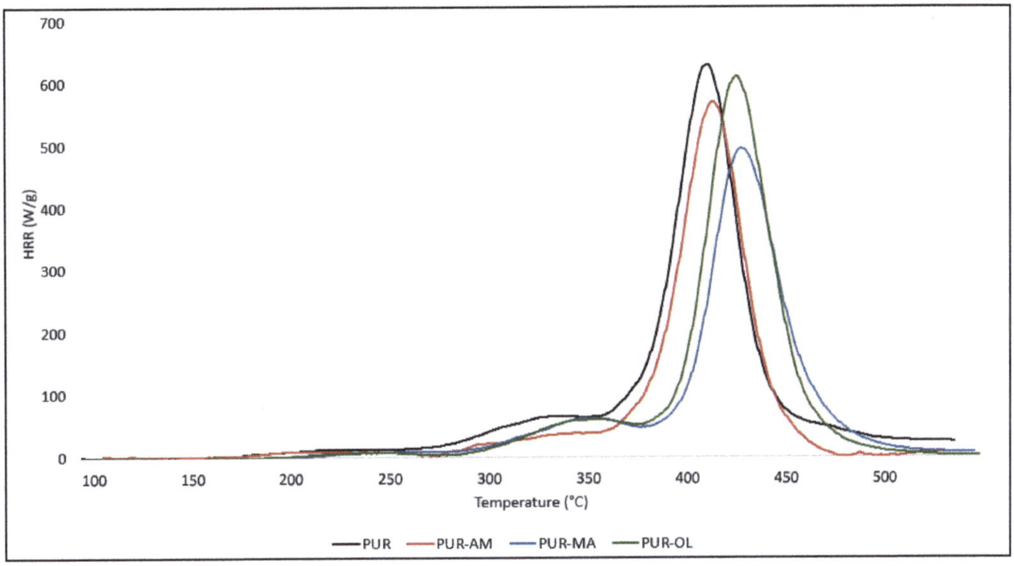

Figure 6. HRR curves of PUR composites: PUR, PUR–AM, PUR–MA, and PUR–OL.

The test results presented in Table 3 indicate that triphenyl polyphosphorus (TPP) is much more effective in reducing the flammability of PUR compared to POSS compounds. In the presence of TPP, the HRR_{MAX} parameter was reduced by 38%. It is also worth noting that the TPP significantly reduced the value of the heat capacity parameter, i.e., HRC. Reducing the HRC parameter value by 37.8% significantly limits the susceptibility of the modified PUR foam to both maintaining combustion processes and fire development (Table 3) (Figure 7).

The highest degree of fire hazard reduction in the tested PUR foams, expressed by means of the HRR_{MAX}, THR, and HRC parameter values, was obtained for the synergistic effect of POSS compounds and TPP. The results presented in Table 3 indicate that all tested composites, i.e., PUR–AM–TPP, PUR–MA–TPP and PUR–OL–TPP, are characterized by similar values of HRR_{MAX}, $THRR_{MAX}$, THR, and HRC (Figure 8).

Considering that the PUR composites containing the synergistic POSS–TPP system are characterized by the greatest reduction in flammability in comparison to the PUR reference sample, they were next subjected to flammability tests in real conditions (Table 4).

Table 3. Flammability test results of PUR composites obtained with the use of the PCFC method.

Composite	HRR_{MAX} (W/g)	$THRR_{MAX}$ (°C)	THR (kJ/g)	HRC (J/gK)
PUR	629.0	415	26.3	578
PUR–AM	570.5	410	22.8	532
PUR–MA	495.1	420	23.1	461
PUR–OL	610.1	415	26.7	578
PUR–TPP	390.1	415	24.8	354
PUR–AM–TPP	341.1	420	24.3	315
PUR–MA–TPP	375.5	420	23.5	349
PUR–OL–TPP	354.4	420	24.0	331

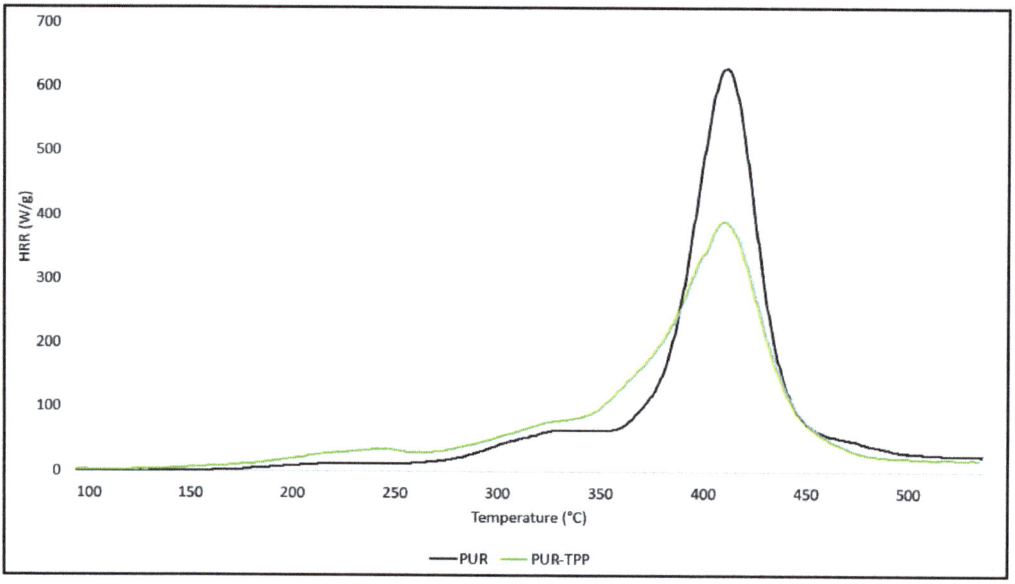

Figure 7. HRR curves of PUR composites: PUR, PUR–TPP.

The test results obtained using the cone calorimetry method confirmed the effectiveness of the POSS–TPP system in the flammability reduction processes of the tested PUR foams. In real conditions, it is worth noting that the POSS–TPP system, especially in the form of AM–TPP and OL–TPP, significantly reduced the total heat released (THR parameter) of the tested PUR composites. In the case of AM–TPP, the THR parameter value was reduced by 70.1%, while it was reduced by 68.7% in the case of OL–TPP compared to the reference sample (PUR) (Table 4).

It should also be emphasized that the MA–TPP system effectively reduced the MARHE parameter informing about the intensity of fire. The PUR–MA–TPP composite, similarly to the PUR–OL–TPP one, is characterized by a significant reduction time to flameout compared to the PUR–TPP composite. Reducing the extinguishing time of the sample, without increasing the HRR_{MAX} value, reduces the fire hazard of the tested PUR composites.

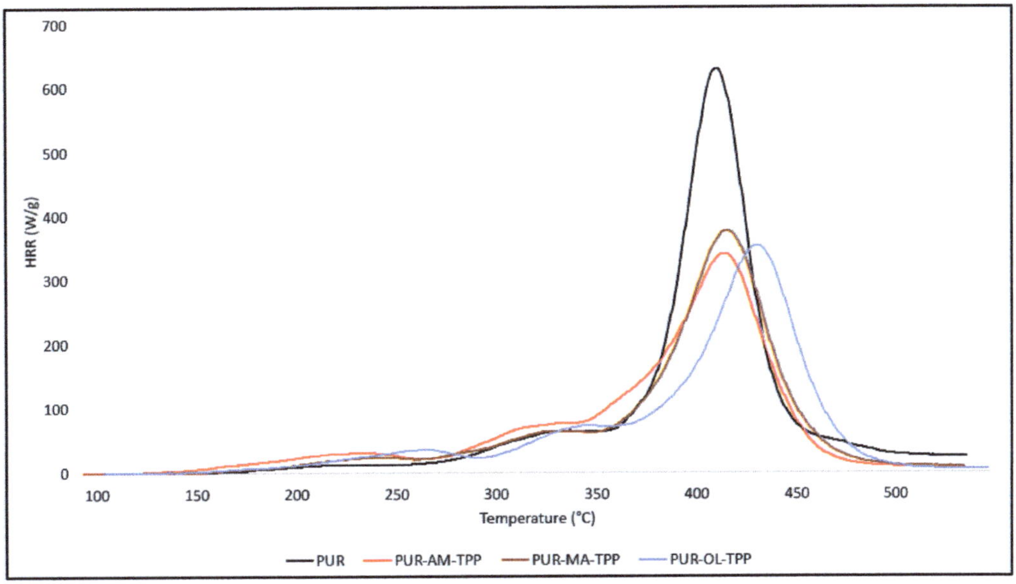

Figure 8. HRR curves of PUR composites: PUR, PUR–AM–TPP, PUR–MA–TPP, and PUR–OL–TPP.

Table 4. Flammability test results of PUR composites obtained using the cone calorimeter method.

Composite	PUR	PUR–TPP	PUR–AM–TPP	PUR–MA–TPP	PUR–OL–TPP
t_i (s)	30	40	41	79	32
$t_{f\text{-}o}$ (s)	944	1098	1094	205	315
HRR (kW/m^2)	62.70	43.46	109.9	47.77	67.4
HRR_{max} (kW/m^2)	145.92	97.42	162.1	101.3	131.7
$tHRR_{max}$ (s)	115	85	120	145	100
THR (MJ/m^2)	60.5	46.1	17.9	49.0	18.9
EHC (MJ/kg)	17.25	13.61	14.10	13.81	13.8
EHC_{max} (MJ/kg)	67.35	72.60	74.31	74.55	65.28
MLR (g/s)	0.032	0.028	0.069	0.031	0.043
MLR_{max} (g/s)	0.151	0.122	0.204	0.102	0.132
AMLR ($g/m^2{\cdot}s$)	3.74	3.51	12.27	4.34	9.15
FIGRA ($kW/m^2{\cdot}s$)	1.27	1.14	1.35	0.69	1.317
MARHE (kW/m^2)	81.55	56.89	93.13	51.46	80.15

3.3. Emission of Smoke and Toxic Gases

Smoke is defined as an aerosol of solid or liquid particles generated during the pyrolysis process or thermo-oxidative decomposition of an organic fuel. The amount of smoke generated depends on many factors, including the chemical composition of the fuel, the availability of oxygen, the intensity of the heat flux, as well as the conditions of the combustion process (flameless or flame combustion) [41].

Reducing the amount of smoke emitted directly translates into increased fire safety. It is common knowledge that smoke hinders the orientation of people caught in a fire; and therefore, it is the main factor causing disorientation and panic. It is also a carrier of toxic organic destructs, especially those coming from the group of dioxins and polycyclic aromatic hydrocarbons [42].

The obtained test results clearly indicate that the flame-retardant system used for the tested PUR–POSS–TPP composites has a beneficial effect on reducing their smoke production (Ds_{MAX} parameter, Table 5).

Table 5. Smoke emission of PUR composites.

Composite	D_{sMAX}	TD_{sMAX}	Ds(4)	VOF_4
PUR	300.1	600	109.4	211.3
PUR–TPP	263.5	600	107.2	217.3
PUR–AM–TPP	203.8	600	74.22	153.3
PUR–MA–TPP	203.4	600	96.29	188.7
PUR–OL–TPP	212.9	600	89.12	202.9

In the case of composites containing AM–TPP and MA–TPP, the value of the maximum optical density of smoke was reduced by over 32% in comparison to the PUR reference (Table 5).

Also, the amount of smoke emitted after 4 min of the test, the Ds(4) parameter, is much lower for the composites containing POSS–TPP than for the PUR reference sample.

It should also be emphasized that the PUR–TPP composite is also characterized by lower values of the Ds_{MAX} and Ds(4) parameters compared to the PUR reference composite. It is probable that the char formed during the thermal decomposition of TPP of good insulating properties limits the decomposition of the composite and reduces smoke emissions.

Silsesquioxanes have a beneficial effect on the reduction in toxic gas emissions generated during PUR decomposition, especially when they occur in a synergistic system with TPP (Table 5) (Figure 9).

Applying the silsesquioxane compounds to the PUR foam matrix results primarily in the reduction in carbon monoxide emissions. It is worth noting that the amount of hydrogen cyanide emitted was also reduced. However, POSS compounds had an ambiguous effect on the CO_2 and NO_2 emission values. In the case of the PUR–OL composite, a significant increase in the amount of NO_2 emitted was recorded.

Applying the organophosphate compound TPP to the PUR matrix results in reduced CO, CO_2, and HCN emissions compared to the PUR reference sample (Table 6). Nevertheless, TPP, just like POSS–OL, clearly affects the amount of nitrogen dioxide (NO_2) emissions. Considering that both TPP and POSS–OL do not contain nitrogen, it is possible that TPP and POSS–OL accelerate the oxidative decomposition of the urethane bond.

The synergistic POSS–TPP system clearly reduces the amount and toxicity of gases emitted during the decomposition of polyurethane composites. The POSS–TPP system not only reduces the amount of CO and CO_2, but also the amount of HCN and NO_2 emissions during the decomposition of PUR composites (Figure 10).

The reduction in toxic gas emissions in the POSS–TPP system results primarily from the formation of a uniform carbon layer during the thermal decomposition of the PUR composite. The main sources of carbon in the PUR–POSS–TPP composite are the aromatic structures of TPP, which, during thermal decomposition of the composite, create a three-dimensional carbon residue with a large specific surface area, also chemically stabilized by the POSS decomposition products. The reduction in the emission values of the determined toxic gases results primarily from their impeded diffusion through the boundary layer, as well as the effect of adsorption processes taking place in the boundary layer.

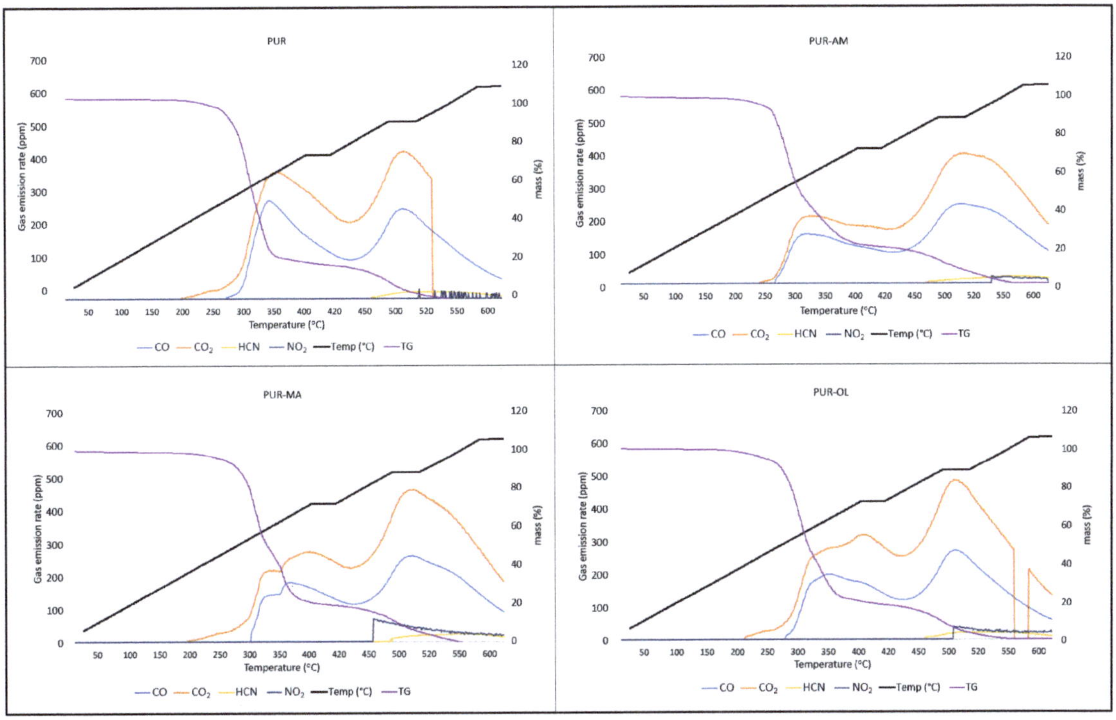

Figure 9. TG-gas analyzer—real-time gas emission results for PUR, PUR–AM, PUR–MA, and PUR–OL.

Table 6. Concentration of toxic gases emitted during thermal oxidation decomposition of PUR composites.

Composite	Gas Concentration (g/m^3)				
	CO	CO_2	HCN	NO_2	Total
PUR	71.52	182.28	2.22	0.972	256.9
PUR–AM	68.49	175.67	1.89	1.054	247.0
PUR–MA	69.20	199.50	2.04	1.412	272.1
PUR–OL	69.21	192.64	2.15	6.03	269.9
PUR–TPP	66.65	154.86	2.04	14.03	237.5
PUR–AM–TPP	64.90	145.40	1.98	11.24	238.4
PUR–MA–TPP	69.51	154.67	1.94	10.37	236.5
PUR–OL–TPP	64.19	161.18	1.74	14.47	241.5

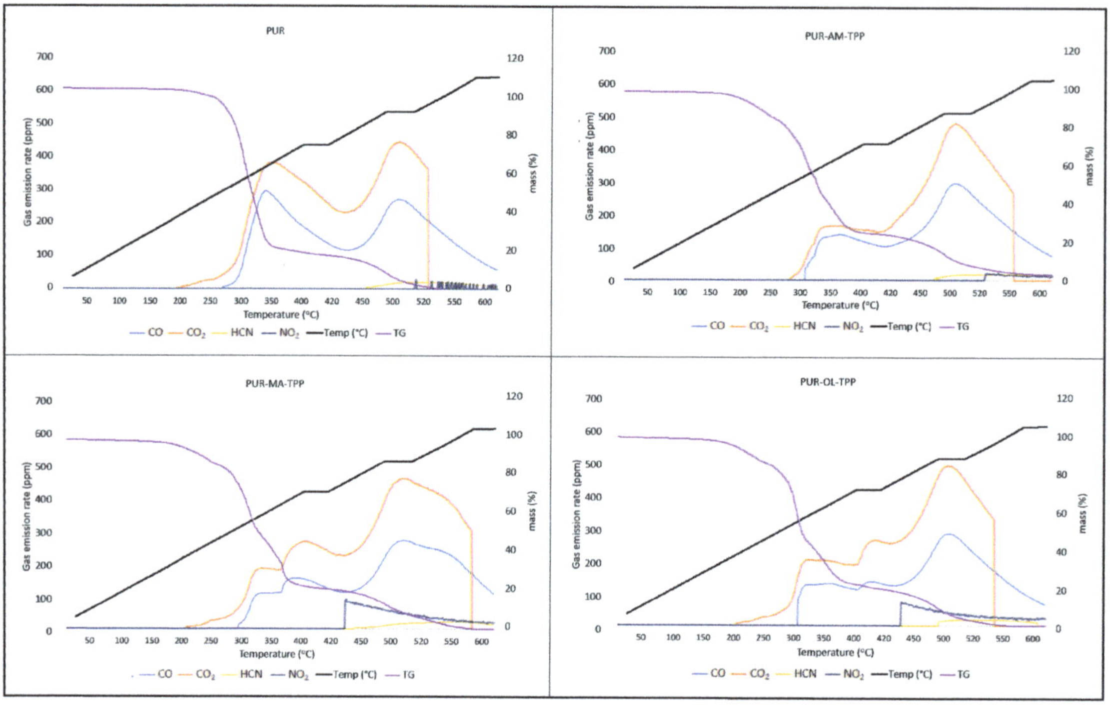

Figure 10. TG-gas analyzer—real-time gas emission results of PUR, PUR–AM–TPP, PUR–MA–TPP, and PUR–OL–TPP.

4. Summary

This paper presented the influence of silsesquioxane fillers, along with a synergistic system and halogen-free flame-retardant compounds, on reducing the fire hazard of polyurethane foams.

The test results obtained by the SEM clearly indicate that the PUR composite is characterized by a porous structure. The best effect in terms of reducing the porosity of the PUR foam was achieved for the TPP–OL system. In the case of the PUR–TPP–OL system, the porous structure of PUR practically disappeared.

The thermal analysis results indicated that POSS compounds have an ambiguous effect on the thermal parameters of PUR composites.

All the tested POSS compounds (AM, MA, and OL) cause an increase in both residues after the thermal decomposition of the PTD parameter, as well as residues at a temperature of 600 °C when it comes to the P_{600} parameter. A clear increase in the value of the P_{TD} and P_{600} parameters indicates that POSS compounds can catalyze the formation of a ceramic boundary layer during the decomposition of the PUR composite.

The P_{600} parameter value for the composites containing the POSS–TPP system is significantly higher when compared to the PUR–TPP composite. The increase in the P_{600} parameter for the composites containing the POSS–TPP system indicates that POSS compounds stabilize the carbon residue through physical interactions, i.e., the formation of a ceramic coating on the carbon surface and because of a chemical reaction based on the formation of thermally stable Si-C silicon carbide.

The PUR–MA composite has the highest reduction value for both HRR_{MAX} and THR. It cannot be excluded that despite the lack of changes in the IR spectrum, the POSS–MA was applied to the structure of the PUR foam through polar groups located at the end of the alkyl chains or bound to the PUR through reversible intermolecular interactions.

The obtained test results clearly indicate that the flame-retardant system used for the tested PUR–POSS–TPP composites has a beneficial effect on reducing their smoke production.

Silsesquioxanes have a beneficial effect on the reduction in toxic gas emissions generated during PUR decomposition, especially when they occur in a synergistic system with TPP.

The synergistic POSS–TPP system clearly reduces the amount and toxicity of gases emitted during the decomposition of polyurethane composites. The POSS–TPP system reduces not only the amount of CO and CO_2, but also the amount of HCN and NO_2 emissions during the decomposition of PUR composites.

Author Contributions: Conceptualization, P.R. and A.G.; methodology, P.R., A.G. and U.Z.M.; investigation, P.R., U.Z.M., M.Ż. and A.G.; formal analysis, P.R., U.Z.M., M.Ż. and A.G.; data curation, P.R., U.Z.M. and A.G.; writing—original draft preparation, P.R. and A.G.; writing—review and editing, P.R., A.G. and U.Z.M. All authors have read and agreed to the published version of the manuscript.

Funding: The work was founded by the National Centre for Research and Development POIR 01.01.01-00-0238/21.

Institutional Review Board Statement: Not applicable.

Data Availability Statement: Data are contained within the article.

Conflicts of Interest: The authors declare no conflict of interest.

Abbreviations

PUR	polyurethane
POSS	polyhedral oligomeric silsesquioxanes
MA	methacrylpropyl POSS
AM	aminopropyl POSS
OL	vinyl POSS
TPP	triphenyl phosphate
SEM	scanning electron microscope
FTIR	Fourier-Transform Infrared Spectroscopy
ATR	attenuated reflection
TGA	thermogravimetry analysis
TG	thermogravimeter
T_5	temperature 5% sample weight loss, °C
T_{50}	temperature 50% sample weight loss, °C
T_{RMAX}	temperature at the maximum rate of sample decomposition, °C
dm/dt	rate of sample decomposition, wt.%/min
P_{TD}	residue after thermal decomposition, %
ΔT_s	range of combustion temperature of the residue after thermal decomposition of sample, °C
P_{600}	residue at T = 600 °C
MIR	mid-infrared
PCFC	pyrolysis combustion flow calorimeter
HRR	heat release rate, W/g
HR Capacity	heat release capacity, J/g·K
Cone	cone calorimeter method
t_i	time to ignition, s
$t_{f\text{-}0}$	time to flameout, s
THRR	temperature at the maximum heat release rate, °C
EHC	effective combustion heat, MJ/kg
AMLR	average weight loss rate, $g/m^2 \cdot s$
THR	total heat release, MJ/m^2
HRR_{max}	maximum heat release rate, kW/m^2
$tHRR_{max}$	time to maximum heat release rate, s
MARHE	maximum average heat release rate, kW/m^2
FIGRA	fire growth rate index, $kW/m^2 \cdot s$
$D_s$max	maximum specific optical density

Ds(4)	specific optical density after 4 min of testing
VOF4	are under the specific optical density curve
Dc	light attenuation coefficient after testing

References

1. Chen, X.; Jiang, Y.; Jiao, C. Smoke suppression properties of ferrite yellow on flame retardant thermoplastic polyurethane based on ammonium polyphosphate. *J. Hazard. Mater.* **2014**, *266*, 114–121. [CrossRef]
2. Hirschler, M.M. Fire Hazard and Toxic Potency of the Smoke from Burning Materials. *J. Fire Sci.* **1987**, *5*, 289–307. [CrossRef]
3. Woolley, W.D. Smoke and Toxic Gas Production from Burning Polymers. *J. Macromol. Sci. Part A—Chem.* **1982**, *17*, 1–33. [CrossRef]
4. Krawiec, P.; Warguła, Ł.; Małozięć, D.; Kaczmarzyk, P.; Dziechciarz, A.; Czarnecka-Komorowska, D. The Toxicological Testing and Thermal Decomposition of Drive and Transport Belts Madeof Thermoplastic Multilayer Polymer Materials. *Polymers* **2020**, *12*, 2232. [CrossRef]
5. Rybiński, P.; Syrek, B.; Żukowski, W.; Bradło, D. Impact of Basalt Filler and Ceramizable Additives on the Toxicity of Gaseous Products Emitted from Thermal Decomposition of Silicone Rubber Composites. *Materials* **2019**, *12*, 3478. [CrossRef] [PubMed]
6. Wang, H.; Tian, Y.; Li, J.; Chen, X. Experimental study on thermal effect and gas release laws of coal-polyurethane cooperative spontaneous combustion. *Sci. Rep.* **2021**, *11*, 1994. [CrossRef] [PubMed]
7. Cheung, W.K.; Zeng, Y.; Lin, S.; Huang, X. Modelling carbon monoxide transport and hazard from smouldering for building fire safety design analysis. *Fire Saf. J.* **2023**, *140*, 103895. [CrossRef]
8. McKenna, S.T.; Hull, T.R. The fire toxicity of polyurethane foams. *Fire Sci. Rev.* **2016**, *5*, 3. [CrossRef]
9. Jiao, L.; Xiao, H.; Wang, Q.; Sun, J. Thermal degradation characteristics of rigid polyurethane foam and the volatile products analysis with TG-FTIR-MS. *Polym. Degrad. Stab.* **2013**, *98*, 2687–2696. [CrossRef]
10. Salasinska, K.; Borucka, M.; Leszczyńska, M.; Zatorski, W.; Celiński, M.; Gajek, A.; Ryszkowska, J. Analysis of flammability and smoke emission of rigid polyurethane foams modified with nanoparticles and halogen-free fire retardants. *J. Therm. Anal. Calorim.* **2017**, *130*, 131–141. [CrossRef]
11. Asefnejad, A.; Khorasani, M.T.; Behnamghader, A.; Farsadzadeh, B.; Bonakdar, S. Manufacturing of biodegradable polyurethane scaffolds based on polycaprolactone using a phase separation method: Physical properties and in vitro assay. *Int. J. Nanomed.* **2011**, *2375*, 2375–2384. [CrossRef]
12. Somarathna, H.M.C.C.; Raman, S.N.; Mohotti, D.; Mutalib, A.A.; Badri, K.H. The use of polyurethane for structural and infrastructural engineering applications: A state-of-the-art review. *Constr. Build. Mater.* **2018**, *190*, 995–1014. [CrossRef]
13. Duquesne, S.; Le Bras, M.; Bourbigot, S.; Delobel, R.; Camino, G.; Eling, B.; Lindsay, C.; Roels, T. Thermal degradation of polyurethane and polyurethane/expandable graphite coatings. *Polym. Degrad. Stab.* **2001**, *74*, 493–499. [CrossRef]
14. Modesti, M.; Lorenzetti, A.; Simioni, F.; Camino, G. Expandable graphite as an intumescent flame retardant in polyisocyanurate–polyurethane foams. *Polym. Degrad. Stab.* **2002**, *77*, 195–202. [CrossRef]
15. Chen, Y.; Luo, Y.; Guo, X.; Chen, L.; Xu, T.; Jia, D. Structure and Flame-Retardant Actions of Rigid Polyurethane Foams with Expandable Graphite. *Polymers* **2019**, *11*, 686. [CrossRef]
16. Acuña, P.; Li, Z.; Santiago-Calvo, M.; Villafañe, F.; Rodríguez-Perez, M.; Wang, D.-Y. Influence of the Characteristics of Expandable Graphite on the Morphology, Thermal Properties, Fire Behaviour and Compression Performance of a Rigid Polyurethane Foam. *Polymers* **2019**, *11*, 168. [CrossRef]
17. Hu, X.; Wang, D.; Wang, S. Synergistic effects of expandable graphite and dimethyl methyl phosphonate on the mechanical properties, fire behavior, and thermal stability of a polyisocyanurate–polyurethane foam. *Int. J. Min. Sci. Technol.* **2013**, *23*, 13–20. [CrossRef]
18. Ayandele, E.; Sarkar, B.; Alexandridis, P. Polyhedral Oligomeric Silsesquioxane (POSS)-Containing Polymer Nanocomposites. *Nanomaterials* **2012**, *2*, 445–475. [CrossRef] [PubMed]
19. Zhou, H.; Ye, Q.; Xu, J. Polyhedral oligomeric silsesquioxane-based hybrid materials and their applications. *Mater. Chem. Front.* **2017**, *1*, 212–230. [CrossRef]
20. Rybiński, P.; Syrek, B.; Bradło, D.; Żukowski, W. Effect of POSS Particles and Synergism Action of POSS and Poly-(Melamine Phosphate) on the Thermal Properties and Flame Retardance of Silicone Rubber Composites. *Materials* **2018**, *11*, 1298. [CrossRef]
21. Rybiński, P.; Bradło, D.; Żukowski, W.; Syrek, B. Determination of toxic products emissions of polymers thermal decomposition using fluidised bed reactor and FTIR analysis. *Polym. Test.* **2019**, *79*, 106040. [CrossRef]
22. Lewicki, J.P.; Pielichowski, K.; Jancia, M.; Hebda, E.; Albo, R.L.F.; Maxwell, R.S. Degradative and morphological characterization of POSS modified nanohybrid polyurethane elastomers. *Polym. Degrad. Stab.* **2014**, *104*, 50–56. [CrossRef]
23. *PN-EN ISO 291:2010*; Technical Terminology: Translation of Units of Measurement and Other Basic Physical Quantities. Polish Committee for Standardization: Warsaw, Poland, 2010.
24. *ASTM D7309-14*; Standard Test Method for Determining Flammability Characteristics of Plastics and Other Solid Materials Using Microscale Combustion Calorimetry. ASTM International: West Conshohocken, PA, USA, 2014.
25. *PN-EN ISO 5660-1:2017*; Reaction-to-Fire Tests—Heat Release, Smoke Production and Mass Loss Rate—Part 1: Heat Release Rate (Cone Calorimeter Method with a Nominal Heat Flux of 50 kW/m^2). Polish Committee for Standardization: Warsaw, Poland, 2017.

26. *PN-EN ISO 5659-2:2012*; Plastics—Smoke Generation—Part 2: Determination of Optical Density by a Single-Chamber Test. Polish Committee for Standardization: Warsaw, Poland, 2012.
27. Verdolotti, L.; Di Caprio, M.R.; Lavorgna, M.; Buonocore, G.G. Polyurethane Nanocomposite Foams. *Polyurethane Polym.* **2017**, *9*, 277–310.
28. Hebda, E.; Bukowczan, A.; Michałowski, S.; Wroński, S.; Urbaniak, P.; Kaczmarek, M.; Hutnik, E.; Romaniuk, A.; Wolun-Cholewa, M.; Pielichowski, K. Examining the influence of functionalized POSS on the structure and bioactivity of flexible polyurethane foams. *Mater. Sci. Eng. C* **2019**, *108*, 110370. [CrossRef]
29. Kwiatkowski, K.; Nachman, M. The Abrasive Wear Resistance of the Segmented Linear Polyurethane Elastomers Based on a Variety of Polyols as Soft Segments. *Polymers* **2017**, *9*, 705. [CrossRef]
30. Członka, S.; Strąkowska, A.; Kairytė, A. Application of Walnut Shells-Derived Biopolyol in the Synthesis of Rigid Polyurethane Foams. *Materials* **2020**, *13*, 2687. [CrossRef]
31. Hatchett, D.W.; Kodippili, G.; Kinyanjui, J.M.; Benincasa, F.; Sapochak, L. FTIR analysis of thermally processed PU foam. *Polym. Degrad. Stab.* **2005**, *87*, 555–561. [CrossRef]
32. Yin, X.; Dong, C.; Luo, Y. Effects of hydrophilic groups of curing agents on the properties of flame-retardant two-component waterborne coatings. *Colloid Polym. Sci.* **2017**, *295*, 2423–2431. [CrossRef]
33. Tang, Q.; Gao, K. Structure analysis of polyether-based thermoplastic polyurethane elastomers by FTIR, 1H NMR and 13C NMR. *Int. J. Polym. Anal. Charact.* **2017**, *22*, 569–574. [CrossRef]
34. Mariappan, T.; Zhou, Y.; Hao, J.; Wilkie, C.A. Influence of oxidation state of phosphorus on the thermal and flammability of polyurea and epoxy resin. *Eur. Polym. J.* **2013**, *49*, 3171–3180. [CrossRef]
35. Bourbigot, S.; Turf, T.; Bellayer, S.; Duquesne, S. Polyhedral oligomeric silsesquioxane as flame retardant for thermoplastic polyurethane. *Polym. Degrad. Stab.* **2009**, *94*, 1230–1237. [CrossRef]
36. Bao, X.; Nangrejo, M.R.; Edirisinghe, M.J. Preparation of silicon carbide foams using polymeric precursor solutions. *J. Mater. Sci.* **2000**, *35*, 4365–4372. [CrossRef]
37. Ramgobin, A.; Fontaine, G.; Penverne, C.; Bourbigot, S. Thermal Stability and Fire Properties of Salen and Metallosalens as Fire Retardants in Thermoplastic Polyurethane (TPU). *Materials* **2017**, *10*, 665. [CrossRef]
38. Rybiński, P.; Syrek, B.; Marzec, A.; Szadkowski, B.; Kuśmierek, M.; Śliwka-Kaszyńska, M.; Mirkhodjaev, U.Z. Effects of Basalt and Carbon Fillers on Fire Hazard, Thermal, and Mechanical Properties of EPDM Rubber Composites. *Materials* **2021**, *14*, 5245. [CrossRef]
39. Tan, H.; Li, J.; Guo, M.; Du, R.; Xie, X.; Zhong, Y.; Fu, Q. Phase behavior and hydrogen bonding in biomembrane mimicing polyurethanes with long side chain fluorinated alkyl phosphatidylcholine polar head groups attached to hard block. *Polymer* **2005**, *46*, 7230–7239. [CrossRef]
40. Peyrton, J.; Avérous, L. Structure-properties relationships of cellular materials from biobased polyurethane foams. *Mater. Sci. Eng. R Rep.* **2021**, *145*, 100608. [CrossRef]
41. Butler, K.M.; Mulholland, G.W. Generation and Transport of Smoke Components. *Fire Technol.* **2004**, *40*, 149–176. [CrossRef]
42. Rybiński, P.; Mirkhodjaev, U.Z.; Żukowski, W.; Bradło, D.; Gawlik, A.; Zamachowski, J.; Żelezik, M.; Masłowski, M.; Miedzianowska, J. Effect of Hybrid Filler, Carbon Black–Lignocellulose, on Fire Hazard Reduction, including PAHs and PCDDs/Fs of Natural Rubber Composites. *Polymers* **2023**, *15*, 1975. [CrossRef] [PubMed]

Disclaimer/Publisher's Note: The statements, opinions and data contained in all publications are solely those of the individual author(s) and contributor(s) and not of MDPI and/or the editor(s). MDPI and/or the editor(s) disclaim responsibility for any injury to people or property resulting from any ideas, methods, instructions or products referred to in the content.

polymers

Article

Synergistic Modification of Polyformaldehyde by Biobased Calcium Magnesium Bi-Ionic Melamine Phytate with Intumescent Flame Retardant

Shike Lu [1,2], Xueting Chen [2], Bin Zhang [2], Zhehong Lu [3], Wei Jiang [3], Xiaomin Fang [1,2,*], Jiantong Li [1,*], Baoying Liu [1], Tao Ding [1,2] and Yuanqing Xu [1,2]

1 Henan Engineering Research Center of Functional Materials and Catalytic Reaction, Henan University, Kaifeng 475001, China; lushike1025@163.com (S.L.); liubaoying666@163.com (B.L.); dingtao@henu.edu.cn (T.D.); xuyuanqing@henu.edu.cn (Y.X.)

2 College of Chemistry and Molecular Sciences, Henan University, Kaifeng 475001, China; cxt17861100039@163.com (X.C.); 13938523364@163.com (B.Z.)

3 National Special Superfine Powder Engineering Research Center of China, Nanjing University of Science and Technology, Nanjing 210014, China; 15736870133@163.com (Z.L.); superfine_jw@126.com (W.J.)

* Correspondence: xmfang@henu.edu.cn (X.F.); jiantongli@163.com (J.L.)

Abstract: Intumescent flame retardants (IFRs) are mainly composed of ammonium polyphosphate (APP), melamine (ME), and some macromolecular char-forming agents. The traditional IFR still has some defects in practical application, such as poor compatibility with the matrix and low flame-retardant efficiency. In order to explore the best balance between flame retardancy and mechanical properties of flame-retardant polyformaldehyde (POM) composite, a biobased calcium magnesium bi-ionic melamine phytate (DPM) synergist was prepared based on renewable biomass polyphosphate phytic acid (PA), and its synergistic system with IFRs was applied to an intumescent flame-retardant POM system. POM/IFR systems can only pass the V-1 grade of the vertical combustion test (UL-94) if they have a limited oxygen index (LOI) of only 48.5%. When part of an IFR was replaced by DPM, the flame retardancy of the composite was significantly improved, and the POM/IFR/4 wt%DPM system reached the V-0 grade of UL-94, and the LOI reached 59.1%. Compared with pure POM, the PkHRR and THR of the POM/IFR/4 wt%DPM system decreased by 61.5% and 51.2%, respectively. Compared with the POM/IFR system, the PkHRR and THR of the POM/IFR/4 wt%DPM system were decreased by 20.8% and 27.5%, respectively, and carbon residue was increased by 37.2%. The mechanical properties of the composite also showed a continuous upward trend with the increase in DPM introduction. It is shown that the introduction of DPM not only greatly reduces the heat release rate and heat release amount of the intumescent flame-retardant POM system, reducing the fire hazard, but it also effectively improves the compatibility between the filler and the matrix and improves the mechanical properties of the composite. It provides a new approach for developing a new single-component multifunctional flame retardant or synergist for intumescent flame-retardant POM systems.

Keywords: polyformaldehyde; phytate; synergistic effect; flame retardancy

check for updates

Citation: Lu, S.; Chen, X.; Zhang, B.; Lu, Z.; Jiang, W.; Fang, X.; Li, J.; Liu, B.; Ding, T.; Xu, Y. Synergistic Modification of Polyformaldehyde by Biobased Calcium Magnesium Bi-Ionic Melamine Phytate with Intumescent Flame Retardant. *Polymers* **2024**, *16*, 614. https://doi.org/10.3390/polym16050614

Academic Editors: Fei Xiao, Fubin Luo and Ke Sun

Received: 27 December 2023
Revised: 1 February 2024
Accepted: 8 February 2024
Published: 23 February 2024

Copyright: © 2024 by the authors. Licensee MDPI, Basel, Switzerland. This article is an open access article distributed under the terms and conditions of the Creative Commons Attribution (CC BY) license (https://creativecommons.org/licenses/by/4.0/).

1. Introduction

POM is a milky, opaque, crystalline, linear thermoplastic resin without a side chain whose structural formula is $(CH_2O)_n$. It has the advantages of good rigidity and hardness, excellent fatigue resistance, creep resistance, and chemical resistance, self-lubrication, a high thermal deformation temperature, stable mechanical properties, and good surface gloss, so it is widely used in electronic and electrical, light industry, machinery, building materials, and other fields [1]. However, because POM is flammable, there is a great fire hazard. Owing to its special chain molecular structure and ultra-high oxygen content, once ignited, it will continue to release a large amount of heat and toxic smoke, causing great harm to humans and the environment. Therefore, flame-retardant modification of POM is one of the difficulties and points of concern for many scientific researchers and related enterprises [2]. At present, adding flame retardants to polymers by blending is the simplest and most efficient way to alleviate this problem. Commonly used flame retardants include halogenated flame retardants, inorganic flame retardants, phosphorus and nitrogen flame retardants, silicon flame retardants, and intumescent flame retardants [3,4]. Intumescent flame retardants (IFR) have the advantages of being halogen-free, have low toxicity, are environmentally friendly, and are considered a class of flame retardants with the greatest application potential after halogen-based flame retardants [5–7]. However, IFR additives have poor compatibility with polymer matrices, low flame-retardant efficiency, and the additive amount is often above 30 wt% to achieve a certain flame-retardant effect, which makes the mechanical properties of the composite material greatly reduced [8].

According to the literature research, the commonly used strategies to solve the above problems mainly include the following: (1) search for novel efficient acid/char sources of IFR [9]; (2) surface modification of common IFR ingredient, such as APP [10]; (3) preparation of single-component multifunctional flame retardants [11]; (4) add synergistic agents [12–14], etc. Phytic acid (PA) is a natural organic phosphorus compound acid source extracted from plant seeds. It has a series of characteristics such as high storage capacity in nature, ultra-high phosphorus content, and excellent metal ion complexation ability, and it can react with organic amine compounds. These are great qualities for it to become a high-efficiency acid source in IFRs and to prepare single-component multifunctional flame retardants and organometallic salt synergists [15–17]. Cheng et al. used phytic acid as an acid source and chitosan and biochar as carbon sources to prepare flame-retardant coatings and applied them to the flame-retardant modification of cotton fabrics. The modified cotton fabric exhibited good thermal degradation and thermal oxidation stability, which reduces the fire hazard [9]. Yang et al. synthesized a new one-component intumescent flame retardant in the form of microporous nanosheets called hexa-(4-aminophenoxy) cyclotriphosphonitrile–phytic acid (HACP-PA), which assembled the carbon source, acid source, and gas source into a molecular structure. After adding 5 wt% HACP-PA to polylactic acid (PLA), the V-0 grade of UL-94 was achieved, and the LOI value was increased to 24.2%. Compared with PLA, the total heat release and peak heat release rate of the PLA composite with 5 wt% HACP-PA were reduced by 5% and 15.3%, respectively. In addition, the total smoke volume was also significantly reduced by 31.0%, and combustible volatiles were significantly reduced by the incorporation of HACP-PA [18]. Gong et al. added nickel phytate (PA-Ni) as a synergist with an intumescent flame retardant to modify polylactic acid (PLA). By adding 4 wt% PA-Ni and 11 wt% IFR, the resulting PLA composite can achieve the V-0 test grade of UL94. Compared with pure PLA, the peak heat release rate of PLA/11IFR/4PA-Ni is reduced by 62.3%, and the carbon residue is significantly increased [19]. Zhan et al. used layered melamine phytates (MEL-PA) with intumescent flame retardants in flame-retardant polypropylene (PP). Studies have shown that the addition of MEL-PA can effectively improve the limiting oxygen index of PP composites, greatly reduce the heat and smoke release of PP/IFR, and increase the degree of graphitization of the carbon layer, which proves that MEL-PA and IFRs have a good synergistic effect on flame-retardant PP [20].

In this paper, a biobased calcium magnesium bi-ionic melamine phytate (DPM) was designed and prepared. It was employed to modify POM by compounding with the IFR, containing ammonium polyphosphate (APP), benzoxazine (BOZ), and melamine (ME) as the acid source, carbon source, and gas source, respectively; it was fabricated and optimized in our previous works [19,21]. The flame retardancy and mechanical properties of the intumescent flame-retardant POM composites with DPM were studied. It was found that DPM not only has an excellent synergistic flame-retardant effect with IFRs but also enhances the compatibility between fillers and substrates, improves the mechanical properties of composite materials, and has good application prospects. It provides a new approach for the development of biobased environmentally friendly intumescent flame retardants.

2. Experimental Section

2.1. Materials

Phytic acid (PA, 70% aqueous solution), magnesium hydroxide (99.5%), and calcium hydroxide (99.5%) were purchased from Xiya Chemical Co., Ltd, Chengdu, China. 4, 4-diaminodiphenyl sulfone (DDS, 99.5%) was obtained from Shanghai Meryer Technologies Co., Ltd., Shanghai, China. POM (POM, MC90), melamine (ME, 99%), and antioxidant (1010, industrial grade) were provided by Henan Kaifeng Longyu Chemical Co., Ltd, Kaifeng, China. APP (TF-201, crystal form II, $n \geq 1000$) was purchased from Shifang Taifeng New Flame Retardant Co., Ltd, Shifang, China. Benzoxazine (BOZ) was made in our laboratory [22].

2.2. Preparation of DPM

Firstly, 1.5 equivalent of calcium hydroxide and 1.5 equivalent of magnesium hydroxide were dispersed in deionized water (0.2 mol/L) at room temperature to prepare a calcium magnesium bimetallic ion suspension solution, and 1 equivalent of phytic acid and 4 equivalent of melamine were dissolved in deionized water separately to prepare a phytic acid solution (0.2 mol/L) and melamine suspension (0.2 mol/L). Then, 1 equivalent of DDS was dissolved in ethanol (0.5 mol/L) for use. The preparation diagram is shown in Figure 1.

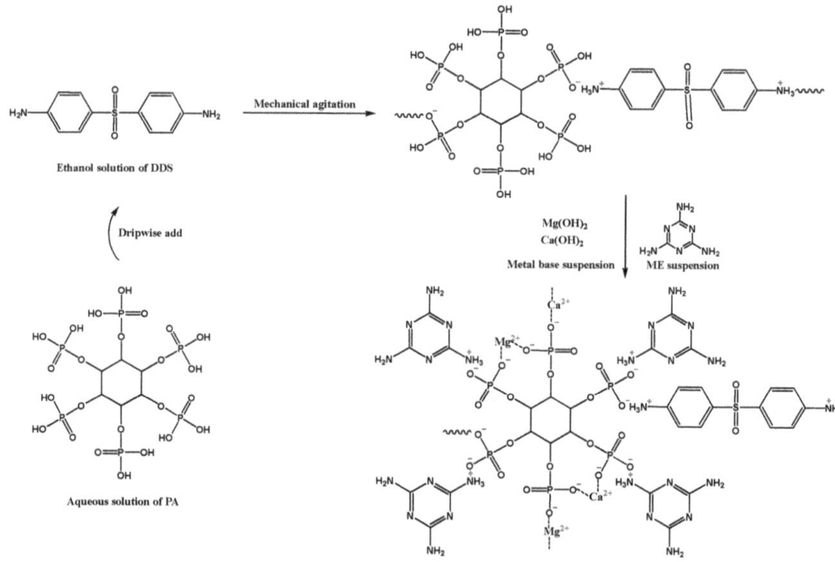

Figure 1. Preparation route and one of the possible structural schematic diagrams of DPM.

Secondly, the DDS ethanol solution was slowly and uniformly added to the phytic acid aqueous solution at room temperature. After stirring for 30 min, the calcium and magnesium bimetallic ion aqueous solution was added slowly and evenly to the above reaction solution. Then, it was stirred for another 30 min, the temperature was raised to 85 °C, and the melamine suspension was added dropwise. A large amount of precipitation was quickly generated, and the temperature was kept with stirring for 4 h until the reaction was completed.

Finally, DPM was obtained by filtering, washing, and drying, with a yield of 92.0%.

2.3. Preparation of POM Composites

POM pellets, intumescent flame retardant (IFR) (composed of APP/BOZ/ME), and DPM were dried at 80 °C in an electric blast drying oven for more than 4 h. After premixing all the ingredients with weighted amounts according to the design formula, the mixture was melt-blended, extruded, and granulated by a twin-screw extruder at 165 °C to 175 °C. The obtained pellets were further dried at 80 °C for more than 6 h. After thorough drying, they were injected by an injection molding machine to make standard bars for further testing. The plasticizing temperature of the injection molding machine was 170 °C to 180 °C.

2.4. Characterization

Fourier transform infrared spectroscopy (FTIR) analysis was performed by the Nicolet 170sx Fourier Infrared spectrometer (Bruker Spectrometer, Saarbrücken, Germany), using potassium bromide tablets with an optical test range of 400 cm^{-1} to 4000 cm^{-1} and a resolution of 4 cm^{-1}.

The vertical combustion performance (UL-94) was tested on the horizontal vertical combustion tester (Suzhou Testech Testing Instrument Technology Co., Ltd., Suzhou, China) according to the GB/T 2408-2008 [23] test standard, with a spline size of 125 × 12.5 × 3.2 mm.

The limiting oxygen index (LOI) was tested in accordance with GB/T 2406.2-2009 [24] standard on the smart oxygen index fume hood integrated machine (Suzhou Testech Testing Instrument Technology Co., Ltd., Suzhou, China), and the sample size was 80 × 10 × 4 mm.

The cone calorimetry (Fire Testing Technology Ltd., East Grinstead, UK) was measured using the ISO 5660-1 2015 [25] standard to study the combustion behavior at a thermal radiant flux of 50 kw/m^2 and a temperature of 700 °C.

The thermal degradation gas infrared analysis test was carried out in a combination of a thermogravimetric analyzer (TGA/SDTA 851, Mettler-Toledo, Zurich, Switzerland) and Fourier infrared tester (INVENIOS, Bruker, Saarbrücken, Germany), heated from room temperature to 800 °C in a nitrogen atmosphere at a heating rate of 10 °C/min and a gas flow rate of 50 mL/min.

The surface topography of the coke slag after cone calorimetry was analyzed by scanning electron microscopy (SEM, JEOL JSM-7610F, UK). The test voltage was 5 kV. The scanning electronic microscope was equipped with an energy-dispersive X-ray spectrometer (EDS) for elemental analysis using elemental mapping with a voltage of 8 kV and a detection limit of 0.01%.

Elemental analysis was performed by X-ray fluorescence spectroscopy (XRF, S2 RANGER, Bruker) with a test range of 11Na-92U and a content range of ppm-100%.

The degree of graphitization of the carbon layer was measured by a laser microscope Raman spectrometer (Renishaw in Via, Renishaw, London, UK). The excitation wavelength was 532 nm and the spectral range was 1000–2000 cm^{-1}.

The mechanical properties of the samples were tested by an electronic universal testing machine (TCS-2000, GOTECH Testing Machines Inc., Taiwan) at room temperature. The tensile test was conducted according to the GB/T 1040.1-2006 [26] standard, the tensile rate was 50 mm/min, and the sample size was 150 × 10 × 4 mm. The bending test was conducted according to the GB/T 9341-2008 [27] standard, the drop rate was 2 mm/min, and the sample size was 80 × 10 × 4 mm.

According to the standard GB/T 1043-2008 [28], a sample with a size of 80 × 10 × 4 mm was used to open a V-shaped notch with a depth of 2 mm in the center, and the notch impact test was carried out with the impact testing machine (ZBC-8400-C, GOTECH Testing Machines Inc., Taiwan).

3. Results and Discussion

3.1. Characterization of DPM

The FTIR spectra of DPM are shown in Figure 2. According to the spectrum, DPM retained a -CH vibration absorption peak near 2900 cm^{-1}, and O=P and O-P characteristic absorption peaks appeared near 1640 cm^{-1} and 1100 cm^{-1}, respectively. It can be proved that DPM retains the basic structure of phytic acid [29]. The characteristic peaks of -NH_2 and triazine rings in ME appear near 1510 cm^{-1} and 800 cm^{-1}, and the characteristic absorption peaks of a benzene ring and -SO_2- groups of DDS appear at 3200–3500 cm^{-1} and 1100 cm^{-1} wave numbers [30]. In addition, the characteristic peak of -OH near 3700 cm^{-1} was not clearly reflected in DPM, indicating that the successful grafting of ME and DDS and the full reaction of $Mg(OH)_2$ and $Ca(OH)_2$ as raw materials had been achieved.

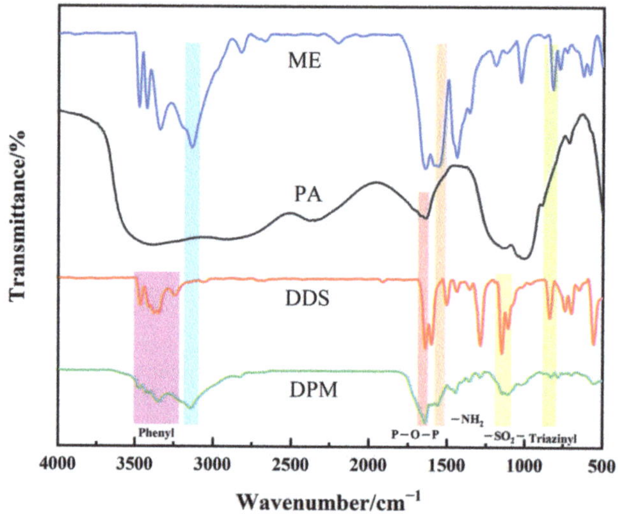

Figure 2. FTIR spectra of DPM.

Further combined with EDS map test analysis (Figure 3), the atomic percentages of P elements and S, Ca, and Mg elements in DPM are 3.18%, 0.37%, 0.75%, and 0.89%, respectively, and the ratio between these four major elements is 1:0.12:0.24:0.28, which is roughly similar to the theoretical value (1:0.17:0.25:0.25), and it can be seen from the figure that the distribution of elements in DPM is very uniform. Finally, according to the results of XRF analysis (Table 1), it can be seen that the ratio of major elements P, S, Ca, and Mg in DPM is 1:0.29:0.41:0.29, which is roughly consistent with the theoretical value (1:0.17:0.25:0.25). Combined with FITR, EDS maps, and XRF analysis results, it can be inferred that DPM was successfully prepared.

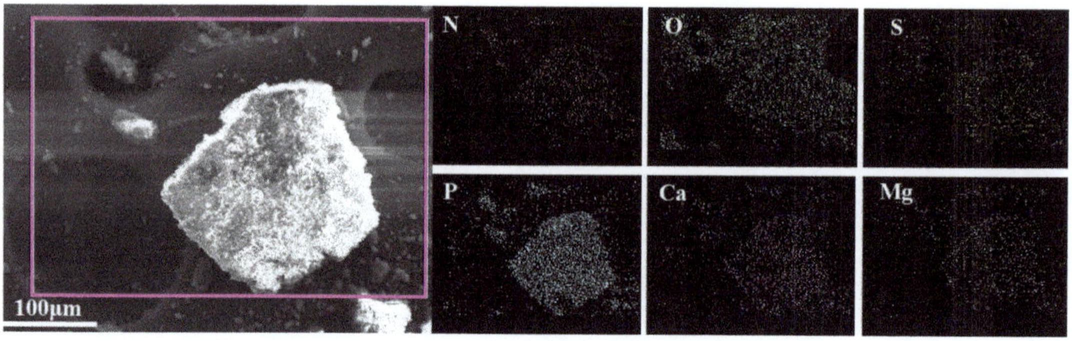

Figure 3. SEM images and EDS map results of DPM.

Table 1. Results of XRF analysis of DPM.

Component	Weight/%	Element	Atomic Ratio (Relative to the P Element) *
P_2O_5	55.09	P	1.00
SO_3	17.88	S	0.29
CaO	17.64	Ca	0.41
MgO	9.10	Mg	0.29
Cl	0.17	--	--
Rest	0.16	--	--

Note: "*" represents the ratio of the atoms of each element to the P element in DPM.

Figure 4 shows the TG and DTG diagrams of DPM. As can be seen from the figure, the initial decomposition temperature of DPM ($T_{-5\%}$) is 229.7 °C, the main maximum weight loss rate temperature (T_{max}) has two intervals, T_{max1} and T_{max2}, which are 305.2 °C and 537.2 °C, respectively, and the residual amount at 800 °C is 43.6%.

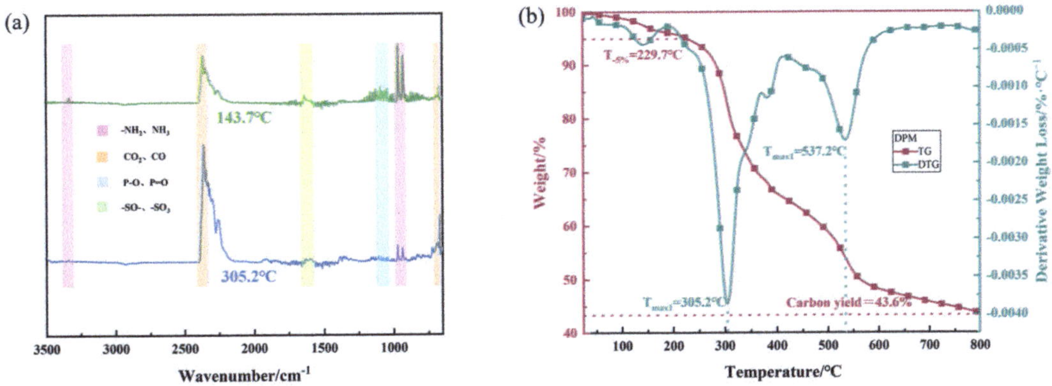

Figure 4. FTIR spectra of DPM thermal decomposition products at different temperatures (**a**) and thermogravimetric curves (**b**).

Figure 5 shows the real-time infrared spectrum of gas released by DPM under nitrogen thermogravimetric conditions. It can be seen from the three-dimensional figure that under the maximum decomposition rate temperature (305.2 °C) (Figure 5), the asymmetric stretching vibration peak and bending vibration peak of carbon dioxide appear near 2349 cm^{-1} and 667 cm^{-1}. N-H bending vibration absorption peaks emerge near 1630 cm^{-1} and 950 cm^{-1}. There are P-O stretching vibrations and bending vibrations in the vicinity of 1600–1740 cm^{-1}. The characteristic absorption peak of the sulfonic acid group was found near 1000–1300 cm^{-1}. In addition, at 143.7 °C, the peak value near 950 cm^{-1} is higher, and the corresponding NH_3 or other amino volatiles may be generated from the early release of partially exposed amino groups in the DPM structure. The results showed that CO_2, a small amount of amines, a very small quantity of phosphate–oxygen radicals, and sulfonic acid groups were released by thermal decomposition of DPM at this temperature.

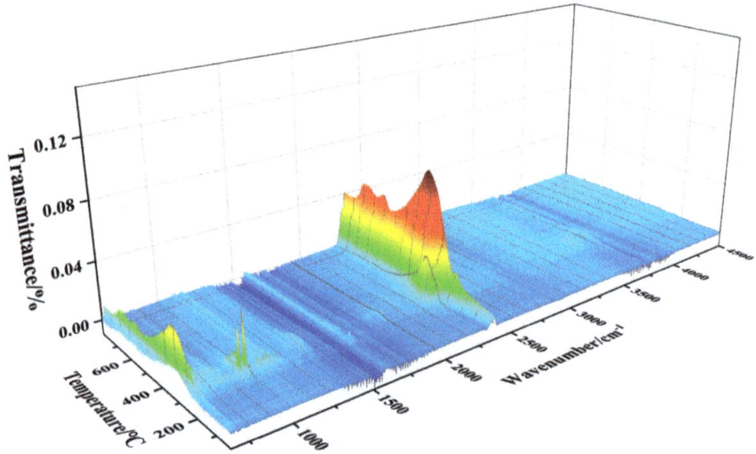

Figure 5. FTIR 3D diagram of thermal decomposition products of DPM at real-time temperature.

3.2. Flame-Retardancy Analysis

Vertical combustion (UL-94) and limited oxygen index (LOI) tests are the basic and intuitive ways to observe the flame-retardant properties of polymeric materials. The UL-94 grade is mainly judged according to the total burning time (Tall) after two ignites. The shorter the T_{all}, the more difficult the material is to ignite. The LOI indicates the minimum oxygen concentration that can support the combustion of the material, and the higher the LOI, the better the flame-retardant performance.

Table 2 shows the UL-94 grade and LOI values of the DPM and IFR synergistically flame-retardant POM composites. Pure POM is extremely flammable, and the combustion process is accompanied by a large number of droplets; the UL-94 test cannot pass any grade, and the LOI is only 15%. After the addition of the IFR, the UL-94 test time of the POM/IFR system has been significantly reduced; its UL-94 reached V-1 grade, and the LOI has been significantly increased to 48.5%. When the amount of DPM is increased from 1% to 8%, the combustion performance of each system is further improved. The test results show that with the introduction of DPM synergism, the UL-94 and LOI of POM composites show a trend of first increase and then decrease. The UL-94 of the POM/IFR/4 wt%DPM and POM/IFR/5 wt%DPM systems can reach the V-0 grade with an LOI of 59.1% and 57.6%, respectively. In contrast, the POM/IFR/4 wt%DPM system shows the best flame retardancy.

Table 2. Flame-retardancy data of POM composites.

Sample	UL-94 (3.2 mm)					LOI/%
	t_1/s	t_2/s	T_{All}/s	Dripping	Grade	
POM	--	--	--	Yes	NR	15
POM/IFR	12.4	221.3	233.7	No	V-1	48.5
POM/IFR/1 wt%DPM	5	135.2	140.2	No	V-1	53.2
POM/IFR/2 wt%DPM	0	75.2	75.2	No	V-1	54.3
POM/IFR/3 wt%DPM	0	70.5	70.5	No	V-1	55.5
POM/IFR/4 wt%DPM	0	36.2	36.2	No	V-0	59.1
POM/IFR/5 wt%DPM	0	41.6	41.6	No	V-0	57.6
POM/IFR/6 wt%DPM	0	110.2	110.2	No	V-1	57.3
POM/IFR/7 wt%DPM	0	120.3	120.3	No	V-1	56.9
POM/IFR/8 wt%DPM	0	123.2	123.2	No	V-1	56.3

Note: The total additive amount of flame retardants and synergists in all systems was kept at 30 wt%, where the additive amount of DPM replaced the corresponding amount of IFR. t_1 and t_2 represent the total burning time of the first and second ignition, respectively; T_{all} is the total time of ten fires after two ignites.

According to the results in Table 2, we subsequently selected four representative systems, namely pure POM, POM/IFR, POM/IFR/1 wt%DPM, and POM/IFR/4 wt%DPM, to further investigate their combustion behavior and mechanism of action.

3.3. Investigation of Combustion Behavior

Cone calorimetry (CONE) simulates real fire conditions, giving us a lot of information about the combustion process. Due to the comprehensiveness and reliability of the data, it has been widely used in the flame-retardancy evaluation of various composite materials, such as various engineering plastics, foam materials, and wood materials. The characteristic data of ignition time (TTI), heat release rate (HRR), total heat release (THR), effective heat of combustion (EHC), total smoke release (TSP), and specific extinction area (SEA) can be obtained [31,32].

Figure 6 shows the curves of HHR, THR, smoke release rate (SPR), and TSP of the POM composites, respectively, and the specific data are shown in Table 3. From the test data, pure POM is extremely flammable, the ignition time (TTI) is 43 s, and after ignition, it burns violently to completion without any residue. The peak heat release rate (PkHRR) is 335.55 kW/m^2, and the average heat release amount (AvHRR) is 233.15 kW/m^2. The THR was 133.08 MJ/m^2, the mean effective heat of combustion (MeanEHC) was 14.54 MJ/kg, the mean mass loss rate (AvMLR) was 19.71 g/(m^2·s), and the SEA, TSP, and carbon residue were all zero. All composites with added flame retardants have shorter TTI than pure POM, which is caused by the premature decomposition of the IFR before the POM matrix at high temperatures. The PkHRR, AvHRR, and THR of POM/IFR composites with 30%IFR were reduced to 163.17 kW/m^2, 42.49 kW/m^2, 89.47 MJ/m^2, with a MeanEHC of 10.51 MJ/kg and an AvMLR of 4.13 g/(m^2·s). Compared with pure POM, SEA, TSP, and carbon residue increased to 86.30 m^2/kg, 7.57 m^2 and 11.3%, respectively.

Table 3. The main data of CONE test of composite materials.

Sample	POM	POM/IFR	POM/IFR/1 wt%DPM	POM/IFR/4 wt%DPM
TTI (S)	43	23	22	23
PkHRR (kW/m^2)	335.55	163.17	132.79	129.17
PFI	7.8	7.1	6.0	5.6
AvHRR (kW/m^2)	233.15	42.49	41.22	32.71
THR (MJ/m^2)	133.08	89.47	95.45	64.90
MeanEHC (MJ/kg)	14.54	10.51	11.45	8.33
SEA (m^2/kg)	0.00	86.30	159.40	109.31
AvMLR (g/(m^2·s))	19.71	4.13	3.87	3.7
TSP (m^2)	0.00	7.57	9.2	6.8
Residue (%)	0.0	11.3	15.0	15.5

When 1% DPM was introduced, the PkHRR, AvHRR, THR, MeanEHC, and AvMLR of the composite were further reduced, and the SEA, TSP, and carbon residues were increased to a certain extent. With the increase in DPM introduction, the test results of the composite materials were improved. The PkHRR, AvHRR, THR, MeanEHC, and AvMLR of the POM/IFR/4 wt%DPM system were further reduced to 129.17 kW/m^2, 32.71 kW/m^2, 64.90 MJ/m^2, 8.33 MJ/kg, and 3.7 g/(m^2·s). Compared with pure POM and POM/IFR systems, the THR of the POM/IFR/4 wt%DPM system was reduced by 51.2% and 27.5%, respectively, and the SEA and TSP were also decreased to 109.31 m^2/kg and 6.8 m^2, respectively; the carbon residue was increased to 15.5%. The above data all show that the introduction of DPM greatly improves the flame-retardant efficiency of the IFR, significantly reduces the amount of heat released during combustion, improves the amount of carbon residue of composite materials, and strengthens the flame-retardant effect of the condensed phase. At the same time, it also inhibits the combustion of the material in the gas phase, resulting in the release of a large amount of refractory smoke during the combustion process, which plays the role of reducing the amount of oxygen, combustible gas, and heat released during the combustion process. These results indicate that DPM does not only have a flame-retardant effect in the gas phase but also promotes the condensed phase flame-retardant effect of the crosslinked carbon of flame-retardant POM composites and is a good flame-retardant synergist for the intumescent flame-retardant POM system.

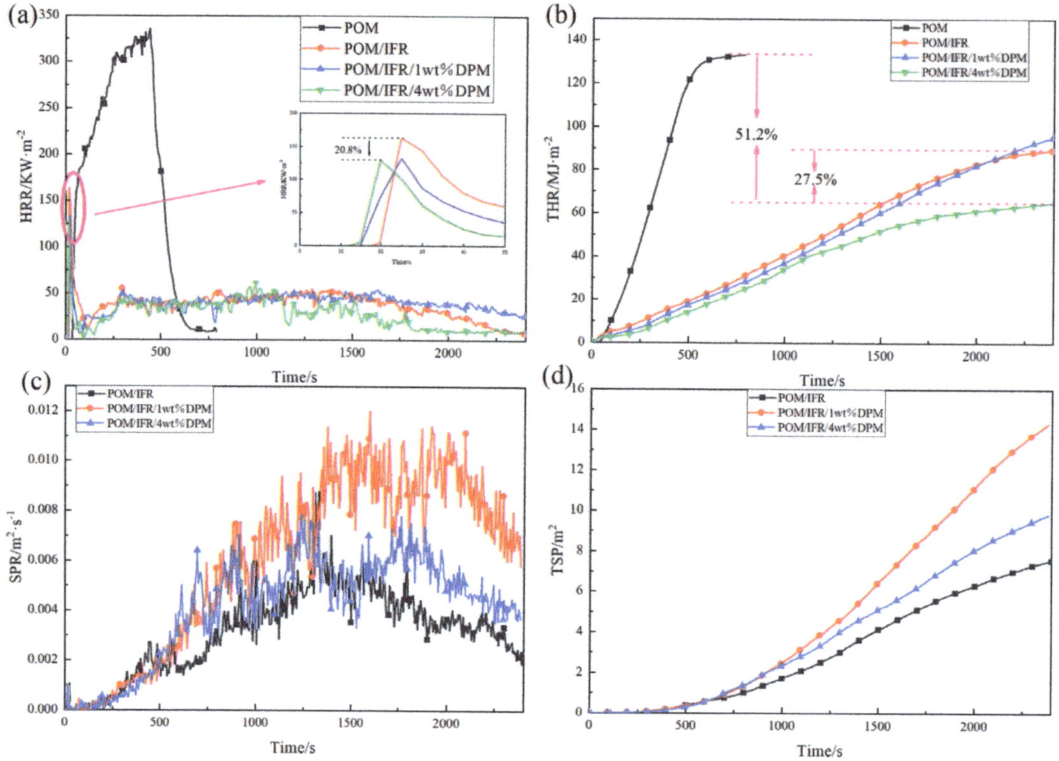

Figure 6. HRR curve (**a**), THR curve (**b**), SPR curve (**c**), and TSP curve (**d**) of composite material obtained by CONE test.

3.4. Carbon Residue Analysis

The digital photos, SEM, and LRS of residual carbon after the combustion of flame-retardant POM composites are shown in Figure 7. After pure POM is ignited, it burns violently until complete without any carbon residue generation. From the digital photos, the carbon residue of the POM/IFR system has increased significantly, reaching 11.3%, but the surface carbon layer is relatively loose, fragmentary, soft, and without any luster. After the introduction of DPM, the amount of residual carbon in the system is further increased; the POM/IFR/1 wt%DPM system reaches 15.0%, the POM/IFR/4 wt%DPM system reaches 15.5%, and the quality of the carbon layer is obviously improved, is more compact, and the surface shows a certain metallic luster. In particular, the carbon layer generated by the POM/IFR/4 wt%DPM system after combustion appears as a whole, and the external carbon layer basically has no damaged carbon residue, almost completely covering and protecting the internal carbon layer, and the amount of carbon residue has further increased. Thus, the carbon layer shows excellent quality and can play an efficient barrier role.

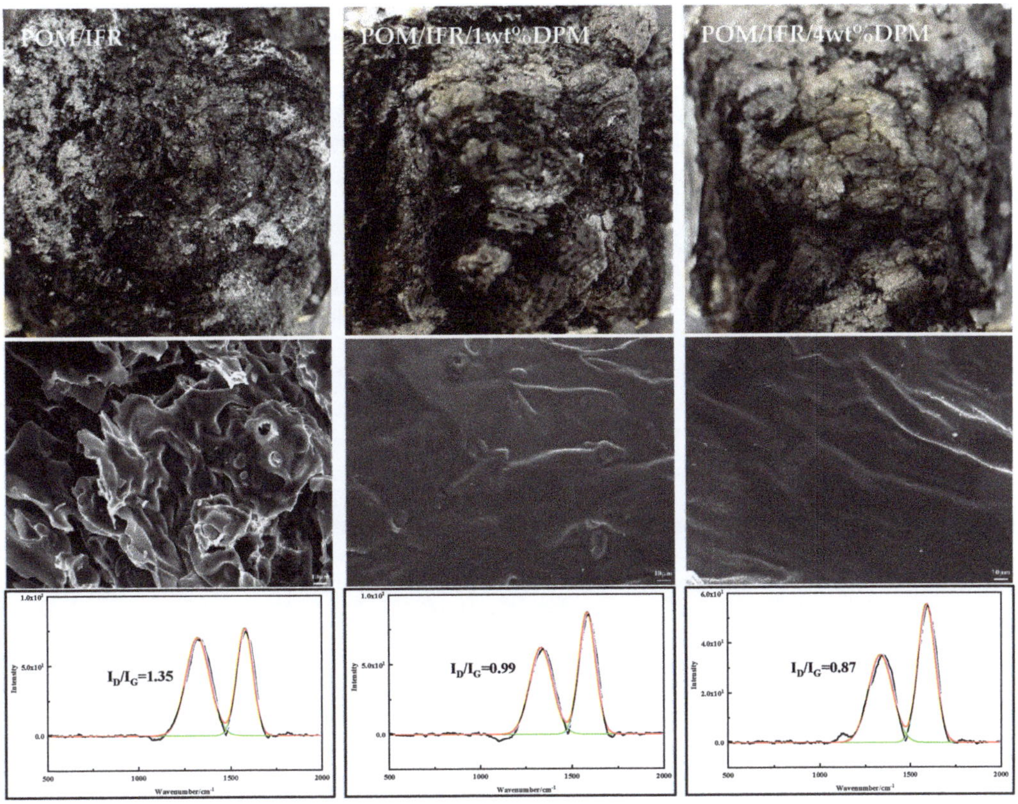

Figure 7. Digital photos, SEM images, and LRS maps of carbon residue after combustion of composite materials.

It can be seen more obviously from the SEM diagram of carbon residue (Figure 7) that the surface of the carbon layer of the POM/IFR system is broken and there are many carbon slag and holes on the surface that cannot effectively block the transfer of oxygen, heat, and combustible gas mixture, so the POM/IFR system does not achieve the ideal flame-retardant effect. After the introduction of DPM, the carbon layer of the POM/IFR/1 wt%DPM system is compact and continuous, showing a certain metallic luster, but there are still a few holes and broken carbon residue. The POM/IFR/4 wt%DPM system has a good improvement, and the carbon layer is completely and continuously arranged in a compact and dense manner, basically without broken carbon residue and holes, which can exert effective heat insulation and oxygen insulation, block the diffusion and transfer of combustible gas mixture during combustion, and play a good shielding and protecting role in the interior of the system. Therefore, the POM/IFR/4 wt%DPM system shows outstanding flame retardancy.

In order to further analyze the carbon layer quality, the carbon layer after the CONE test was analyzed by Raman spectroscopy (LRS) test, as shown in Figure 7. The LRS of the carbon layer mainly has two peaks, one near 1350 cm^{-1} (peak D) and the other near 1580 cm^{-1} (peak G), which represents sp^2 hybridization of the ordered carbon structure with in-plane stretching vibrations [33,34]. The graphitization degree of the carbon layer is usually expressed by the peak area ratio (I_D/I_G) of peak D to peak G; the smaller the I_D/I_G value, the higher the degree of graphitization and the better the quality of the carbon layer. The I_D/I_G values of carbon residue in the POM/IFR, POM/IFR/1 wt%DPM, and POM/IFR/4 wt%DPM systems were 1.35, 0.99, and 0.87, respectively. The graphitization degree of the carbon layer in the POM/IFR/4 wt%DPM system was the highest. The results show that the introduction of DPM in the flame-retardant POM can effectively improve the degree of graphitization of the carbon layer, obtain a higher quality carbon layer, effectively provide heat insulation, and protect the interior of the composite material, resulting in excellent flame retardancy of the composite material. The above analysis shows that the comprehensive flame-retardant effect of the POM/IFR/4 wt%DPM system is the best, which is basically consistent with the combustion test data (UL-94 reaches V-0 grade, LOI up to 59.1%).

3.5. Thermogravimetric Infrared Analysis

Figure 8 shows the thermogravimetric analysis (TG and DTG) curves of POM and flame-retardant POM composites under a nitrogen atmosphere, and the specific data are shown in Table 4. The initial decomposition temperature $T_{-5\%}$ of pure POM is 300.7 °C, the maximum decomposition rate temperature T_{max} is 338.8 °C, and no carbon residue was generated at 600 °C. Compared with POM, the $T_{-5\%}$ of the POM/IFR system decreased significantly to 261.7 °C, which was caused by the advanced decomposition due to the introduction of the IFR; T_{max} was 269.1 °C, which was also significantly advanced, and the carbon residue increased to 17% at 600 °C. The results show that although the introduction of the IFR can reduce the thermal stability of the composite, it can effectively improve the carbon formation property of the composite. After the introduction of DPM, the $T_{-5\%}$ and T_{max} of the POM/IFR/1 wt%DPM and POM/IFR/4 wt%DPM systems are basically the same as those of POM/IFR systems, except that their residual carbon content increases to 21.8% and 23.5%, respectively, at 600°C, both of which are higher than the theoretical carbon residue content of their respective systems. It can be shown that the introduction of DPM does not play a single role but has an excellent synergistic effect with the IFR, which can more efficiently promote the carbon formation of the flame-retardant POM composite.

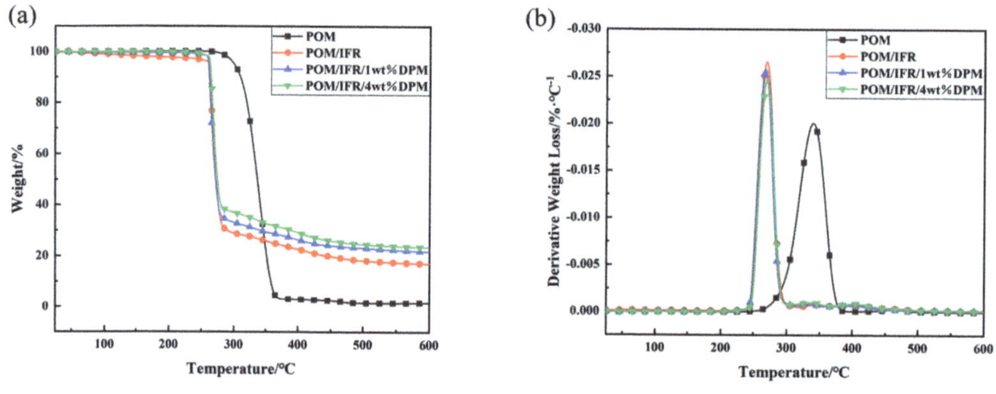

Figure 8. TG (**a**) and DTG (**b**) curves of composite materials.

Table 4. Thermogravimetric data of composite materials.

Sample	$T_{-5\%}$/°C	$T_{-10\%}$/°C	$T_{-50\%}$/°C	T_{max}/°C	Actual Carbon Residue (600 °C)/%
POM	300.7	310.1	336.8	338.8	0.0
POM/IFR	261.7	263.1	271.5	269.1	17.0
POM/IFR/1 wt%DPM	260.1	261.2	271.1	266.8	21.8 (17.3 *)
POM/IFR/4 wt%DPM	262.6	263.9	274.1	268.8	23.5 (18.6 *)

Note: The data with "*" in brackets are the theoretical carbon residues of the sample, and its calculation formula is $Y = Y_{POM/IFR} \times W_{POM/IFR} + Y_{DPM} \times W_{DPM}$ (where "Y" represents the carbon residue amount and "W" represents the corresponding mass fraction).

Figure 9 shows the thermogravimetric infrared (TG-IR) analysis results of pure POM and flame-retardant POM composites. As can be seen from the three-dimensional figure, the gas products during the combustion of POM and the flame-retardant POM system are basically similar. The temperature range of pure POM gas release is about 250~375 °C, while the temperature range of the POM/IFR and POM/IFR/4 wt%DPM systems is greatly shortened to around 240~260 °C and 245~265 °C, respectively. According to the FTIR spectra (Figure 10), the gas products of pure POM at the maximum decomposition rate are mainly carbonyl, carbonyl compounds (1715 cm^{-1}–1770 cm^{-1}), and hydrocarbon compounds (2640 cm^{-1}–2940 cm^{-1}). However, with the introduction of the IFR and DPM, the gas products of the POM/IFR and POM/IFR/4 wt%DPM systems show weak characteristic absorption peaks near 3500 cm^{-1}; this may be the water released by the intumescent flame-retardant POM system during the carbonization crosslinking reaction under high-temperature conditions, which can dilute the combustible volatile matter and oxygen concentration to a certain extent and improve the flame retardancy of the composite material. Compared with the POM/IFR system, there is no obvious difference in the gas generated by the POM/IFR/4 wt%DPM system. Combined with the cone calorimetric analysis of each system and the carbon residue analysis, it can be seen that mutual reactions occurred among DPM, IFR, and POM in the combustion process. DPM mainly plays a catalytic role in the condensed phase, leading to rapid carbonization and more high-quality char residue [29]. The ME grafted on phytate still retains some -NH_2 groups, and some of the amine refractory gas mixture generated during combustion is further oxidized into nitrogen oxide or nitrogen and water at high temperatures, thus diluting the concentration of combustible volatiles in the combustion zone and playing a certain flame-retardant role in the gas phase [30].

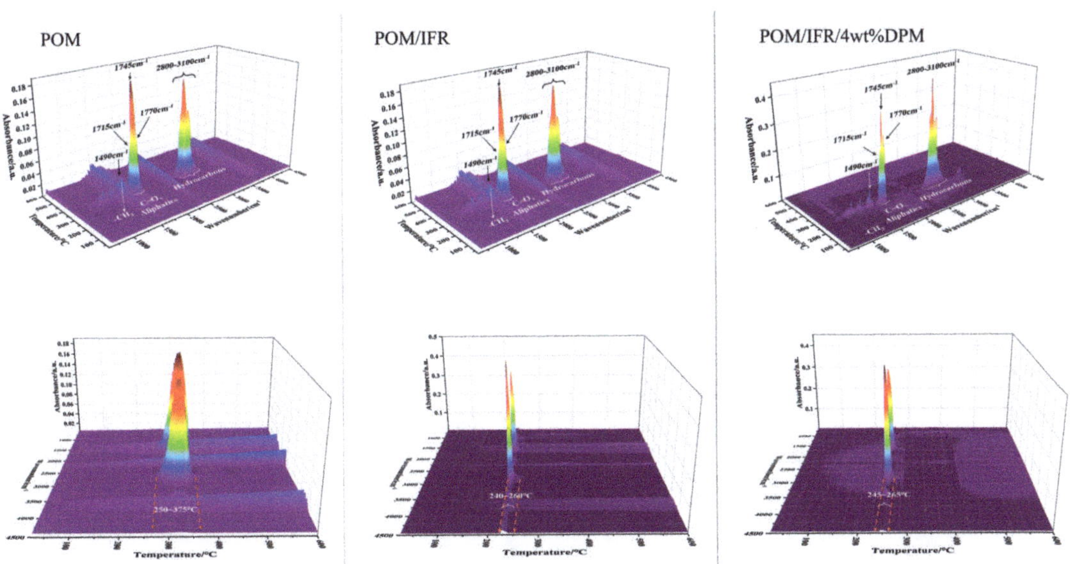

Figure 9. Real-time TG-IR 3D image of composite materials.

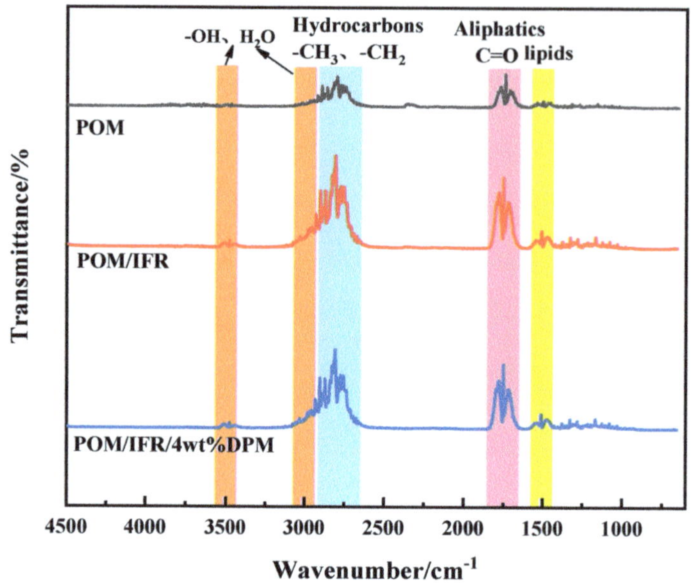

Figure 10. Infrared spectrum of gas released by composite at real maximum thermogravimetric rate.

3.6. Flame-Retardant Mechanism Analysis

The mechanism of the DPM-synergistic IFR to flame-retardant POM is shown in Figure 11. First, at the initial stage of combustion, the acidic substance released by the thermal decomposition of APP reacts with the charring agent BOZ for esterification and promotes the ring-opening and crosslinking of BOZ to interweave with the polymer matrix to form a three-dimensional network carbon layer [33]. Ca^{2+} and Mg^{2+} in DPM can not only catalyze the esterification reaction between the acid and carbon source but also strengthen the carbon layer, which is consistent with the flame-retardant mechanism of phytic acid metal salts reported in the literature [29]. Secondly, DPM can also release a certain amount of amino compounds, ·SO, $\cdot PO_2$, and ·PO free radicals when heated to capture highly active free radicals maintaining combustion such as ·H and ·OH in the gas phase. Finally, when ME is decomposed by heat, a large amount of water vapor, CO_2, and other nonflammable gases are generated [30]. While diluting the concentration of flammable gas and oxygen, the molten carbon layer is rapidly intumescent and foamed, forming a dense intumescent carbon foam layer, which has a shielding effect and thus achieves a good flame-retardant effect. DPM has an excellent synergistic effect with the IFR. The introduction of it can strengthen the quality of the carbon layer and improve its shielding effect. In addition, when heated, DPM will release refractory gas and free radical catchers to strengthen the dilution effect and quenching effect. Therefore, DPM can be used as a good flame-retardant synergist for the intumescent flame-retardant POM system.

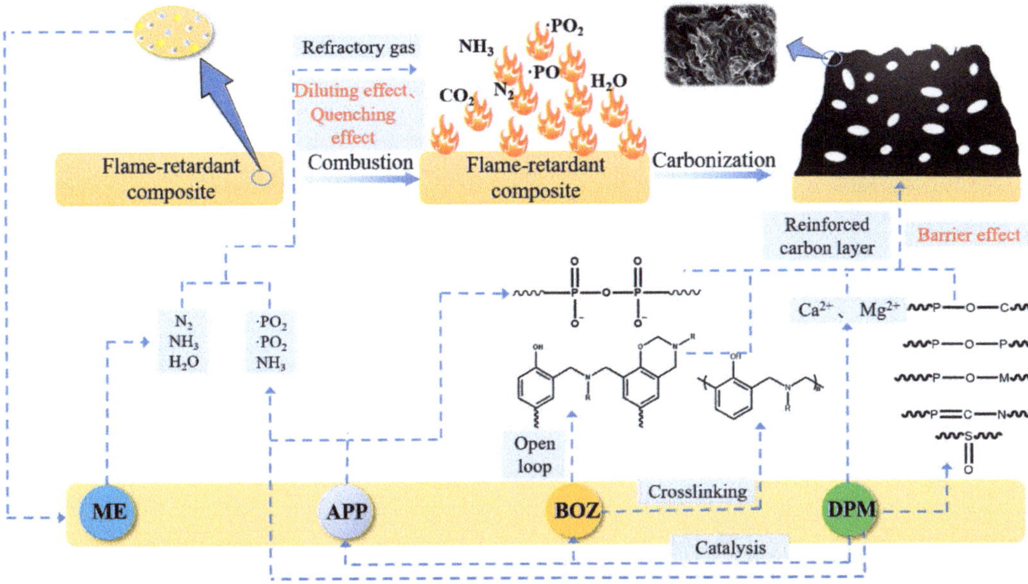

Figure 11. Flame-retarding mechanism diagram of DPM-synergistic IFR flame-retarding POM.

3.7. Mechanical Property Analysis

Table 5 shows the test results of mechanical properties of pure POM and flame-retardant POM composites. From the test data, pure POM has excellent mechanical properties. However, after the introduction of the IFR, the mechanical properties of the POM/IFR system deteriorated sharply. The notch impact strength, bending strength, and tensile strength decreased to 2.78 kJ/m^2, 48.04 MPa, and 36.20 MPa, respectively, which decreased by 50.5%, 31.7%, and 42.9% compared with pure POM. The compatibility between the IFR and the matrix is not good. When DPM is introduced into the intumescent flame-retardant POM system, the mechanical properties of the composite are comprehensively improved, showing a continuous upward trend with the increase in DPM addition. The notch impact strength, bending strength, and tensile strength of POM/IFR/4 wt%DPM reached 3.28 kJ/m^2, 51.24 MPa, and 38.68 MPa, respectively; of note, the notch impact strength increased by 18.0% compared with POM/IFR, with excellent flame retardancy and good mechanical properties.

Table 5. Mechanical data of composite materials.

Sample	Notched Impact Strength (kJ/m^2)	Bending Modulus (MPa)	Bending Strength (MPa)	Tensile Strength (MPa)
POM	5.62 ± 0.06	2172.29 ± 22.57	70.37 ± 0.13	63.3.15
POM/IFR	2.78 ± 0.56	2707.04 ± 16.36	48.04 ± 0.34	36.20 ± 0.59
POM/IFR/1 wt%DPM	3.13 ± 0.21	2882.26 ± 17.23	48.25 ± 0.52	37.25 ± 0.32
POM/IFR/2 wt% DPM	3.25 ± 0.17	3058.51 ± 32.30	50.04 ± 0.63	37.87 ± 0.41
POM/IFR/3 wt% DPM	3.22 ± 0.05	2998.92 ± 42.26	50.32 ± 0.77	38.13 ± 0.33
POM/IFR/4 wt% DPM	3.28 ± 0.10	3067.59 ± 17.22	51.24 ± 0.11	38.68 ± 0.22
POM/IFR/5 wt% DPM	3.31 ± 0.08	3079.59 ± 25.21	52.39 ± 0.21	39.98 ± 0.15
POM/IFR/6 wt% DPM	3.32 ± 0.06	2968.26 ± 32.18	53.39 ± 0.54	40.03 ± 0.12
POM/IFR/7 wt% DPM	3.27 ± 0.12	3033.28 ± 32.09	53.44 ± 0.82	40.55 ± 0.52
POM/IFR/8 wt% DPM	3.28 ± 0.03	3200.23 ± 24.86	53.84 ± 1.85	41.00 ± 0.28

In order to further explore the reasons why DPM improves the mechanical properties of composite materials, the impact cross-sections of pure POM and flame-retardant POM composites were scanned by SEM, as shown in Figure 12. After the sections were enlarged 1000 times, it can be seen that the pure POM impact section is very rough, showing obvious ductile fracture traces without any longitudinal cracks and holes. After the addition of the IFR, the impact cross-section of the POM/IFR system presents a large height difference, which is full of particles with uneven particle size, and there are cracks and holes at the joints with the matrix, and the interface separation of the two phases is serious, which indicates the poor compatibility of the IFR and POM matrix, leading to serious deterioration of the mechanical properties of the POM/IFR composites. Compared with the POM/IFR system, the height difference of the impact section of the POM/IFR/4 wt%DPM system has been significantly reduced, and the surface is rougher, but there are still some additive particles with uneven particle size on the surface, and there are still small cracks and holes at the two joints. With the increase in DPM addition and the decrease in IFR addition, in the POM/IFR/8 wt%DPM system, the impact fracture and hole phenomenon improved, and the two joints are tighter without obvious cracks, which also makes the composite perform better in terms of mechanical properties. The improvement in the mechanical properties of flame-retardant POM composites by DPM may be due to the special structure of DPM as a chain macromolecular organometallic amine salt synergist, which improves the compatibility between POM and filler, contributes to the dispersion of the intumescent flame-retardant system during processing, and improves the interface interaction between the filler and the matrix.

Figure 12. SEM image of composite section.

4. Conclusions

In this paper, DPM was prepared and characterized. The effects of its synergy on the flame retardancy and mechanical properties of POM were discussed under the condition that the total additive amount of the IFR and DPM remained at 30 wt%. The results were as follows:

(1) Through the super-strong metal complexation ability of PA and the reaction with organic amine compounds, the cationic unsaturated state of calcium and magnesium bi-ionic melamine phytates were connected with diamine DDS to prepare DPM. The reaction conditions were simple and mild, and the yield was very high (92%).

(2) The use of DPM as a synergist in the intumescent flame-retardant POM system can effectively improve the flame retardancy of POM composite materials, especially the POM/IFR/4 wt%DPM system. Its UL-94 test passed the V-0 grade and the LOI reached the highest value, 59.1%, while the THR decreased by 51.2% compared with pure POM. Compared with the POM/IFR system, its THR decreased by 27.5%, and the carbon residue increased significantly to 15.5%. Moreover, the synergistic system of DPM and IFR cannot only improve the carbonization ability of the system, leading to the formation of a dense carbon layer, but it can also inhibit the combustion of the gas phase. The gas phase and the condensed phase cooperate to perform an efficient flame-retardant effect so that the flame retardancy of the DPM synergistic flame-retardant system is significantly improved.

(3) DPM as a synergist in the intumescent flame-retardant POM system can effectively improve the mechanical properties of the composite materials. Compared with the POM/IFR system, the mechanical properties of the composite materials basically show a continuously rising trend with the increase in DPM addition, and the SEM analysis of the section also proves that DPM can improve the interface interaction of POM and filler and enhance their compatibility.

Author Contributions: Conceptualization, X.F.; Methodology, T.D. and W.J.; Software, B.Z.; Validation, X.C.; Formal analysis, J.L.; Investigation, Z.L.; Resources, X.F.; Data curation, B.L.; Writing—original draft, S.L.; Writing—review & editing, S.L.; Project administration, W.J. and X.F.; Funding acquisition, Y.X. All authors have read and agreed to the published version of the manuscript.

Funding: This project was financially supported by the National Natural Science Foundation of China (No. 51703051), the International Science and Technology Cooperation Project of Henan Province, China (No. 232102521020), and the Key Project of the Education Department of Henan Province (21A430005).

Data Availability Statement: Data are contained within the article.

Conflicts of Interest: The authors declare that they have no conflicts of interest.

References

1. Lu, S.Y.; Hamerton, I. Recent developments in the chemistry of halogen-free flame retardant polymers. *Prog. Polym. Sci.* **2002**, *27*, 1661–1712. [CrossRef]
2. Yuan, L.; Wang, Z.; Qi, W. Effects of magnesium hydroxide and its synergistic systems on the flame retardance of polyformaldehyde. *J. Appl. Polym. Sci.* **2012**, *125*, 968–974.
3. Andrzejewski, J.; Skórczewska, K.; Kloziński, A. Improving the toughness and thermal resistance of polyoxymethylene/poly(lactic acid) blends: Evaluation of structure-properties correlation for reactive processing. *Polymers* **2020**, *12*, 307. [CrossRef] [PubMed]
4. Kang, X.; Liu, Y.; Chen, N.; Feng, W.; Liu, B.; Xu, Y.; Li, J.; Ding, F.; Fang, X. Influence of modified ammonium polyphosphate on the fire behavior and mechanical properties of polyformaldehyde. *J. Appl. Polym. Sci.* **2021**, *138*, 50156. [CrossRef]
5. Zhao, W.; Li, B.; Xu, M.; Yang, K.; Lin, L. Novel intumescent flame retardants: Synthesis and application in polycarbonate. *Fire Mater.* **2013**, *37*, 530–546. [CrossRef]
6. Hu, X.M.; Wang, D.M. Enhanced fire behavior of rigid polyurethane foam by intumescent flame retardants. *J. Appl. Polym. Sci.* **2013**, *129*, 238–246. [CrossRef]
7. Su, X.; Yi, Y.; Tao, J.; Qi, H. Synergistic effect of zinc hydroxystannate with intumescent flame retardants on fire retardancy and thermal behavior of polypropylene. *Polym. Degrad. Stab.* **2012**, *97*, 2128–2135. [CrossRef]
8. Wang, J.; Guo, Y.; Zhao, S.; Huang, R.Y.; Kong, X.J. A novel intumescent flame retardant imparts high flame retardancy to epoxy resin. *Polym. Adv. Technol.* **2020**, *31*, 932–940. [CrossRef]
9. Cheng, X.; Shi, L.; Fan, Z.; Yu, Y.; Liu, R. Bio-based coating of phytic acid, chitosan, and biochar for flame retardant cotton fabrics. *Polym. Degrad. Stab.* **2022**, *199*, 109898. [CrossRef]
10. Sun, Y.; Yuan, B.; Shang, S.; Zhang, H.; Shi, Y.; Yu, B.; Qi, C.; Dong, H.; Chen, X.; Yang, X. Surface modification of ammonium polyphosphate by supramolecular assembly for enhancing fire safety properties of polypropylene. *Compos. Part B Eng.* **2020**, *181*, 107588. [CrossRef]
11. Xiao, D.; Li, Z.; Gohs, U.; Wagenknecht, U.; Voit, B.; Wang, D.Y. Functionalized allylamine polyphosphate as a novel multifunctional highly efficient fire retardant for polypropylene. *Polym. Chem.* **2017**, *8*, 6309–6318. [CrossRef]
12. Fan, M.; Feng, N.; Zhang, Y.; Wang, Z.; Qu, M.; Zhang, G. Synergistic effects of aluminium hypophosphite on the flame retardancy and thermal degradation behaviours of a novel intumescent flame retardant thermoplastic vulcanisate composite. *Plast. Rubber Compos.* **2019**, *48*, 270–280. [CrossRef]

13. Liu, S.; Fang, Z.; Yan, H.; Chevali, V.S.; Wang, H. Synergistic flame retardancy effect of graphene nanosheets and traditional retardants on epoxy resin. *Compos. Part A Appl. Sci. Manuf.* **2016**, *89*, 26–32. [CrossRef]
14. Zhou, Q.; Gong, K.; Zhou, K.; Zhao, S.; Shi, C. Synergistic effect between phosphorus tailings and aluminum hypophosphite in flame-retardant thermoplastic polyurethane composites. *Polym. Adv. Technol.* **2019**, *30*, 2480–2487. [CrossRef]
15. Chen, K.; Cheng, J.; Wu, B.; Liu, C.; Guo, J. Synergistic effects of strontium carbonate on a novel intumescent flame-retardant polypropylene system. *Polym. Adv. Technol.* **2021**, *32*, 3018–3027. [CrossRef]
16. Sheng, Y.; Chen, Y.; Bai, Y. Catalytically synergistic effects of novel $LaMnO_3$ composite metal oxide in intumescent flame-retardant polypropylene system. *Polym. Compos.* **2014**, *35*, 2390–2400. [CrossRef]
17. Feng, J.; Zhang, X.; Ma, S.; Xiong, Z.; Zhang, C.; Jiang, Y.; Zhu, J. Syntheses of metallic cyclodextrins and their use as synergists in a poly (vinyl alcohol)/intumescent flame retardant system. *Ind. Eng. Chem. Res.* **2013**, *52*, 2784–2792. [CrossRef]
18. Yang, W.; Zhang, H.; Hu, X.; Liu, Y.; Zhang, S.; Xie, C. Self-assembled bio-derived microporous nanosheet from phytic acid as efficient intumescent flame retardant for polylactide. *Polym. Degrad. Stab.* **2021**, *191*, 109664. [CrossRef]
19. Gong, W.; Fan, M.; Luo, J.; Liang, J.; Meng, X. Effect of nickel phytate on flame retardancy of intumescent flame retardant polylactic acid. *Polym. Adv. Technol.* **2021**, *32*, 1548–1559. [CrossRef]
20. Zhan, Y.; Yuan, B.; Shang, S. Synergistic effect of layered melamine-phytate and intumescent flame retardant on enhancing fire safety of polypropylene. *J. Therm. Anal. Calorim.* **2022**, *147*, 285–295. [CrossRef]
21. Yang, Z.; Chen, X.; Lu, S.; Wang, Z.; Li, J.; Liu, B.; Fang, X.; Ding, T.; Xu, Y. Synergistic Flame Retardant Properties of Polyoxymethylene with Surface Modified Intumescent Flame Retardant and Calcium Carbonate. *Polymers* **2023**, *15*, 537. [CrossRef]
22. Lu, Z.; Feng, W.; Kang, X.; Wang, J.; Xu, H.; Li, J.; Wang, W.; Liu, B.; Fang, X. Flame retardant effect and mechanism of benzoxazine as synergist in intumescent flame-retardant polyoxymethylene. *Polym. Adv. Technol.* **2020**, *31*, 2512–2525. [CrossRef]
23. Standardization Administration of the People's Republic of China. *GB/T 2408-2008*; Plastics—Determination of Burning Characteristics—Horizontal and Vertical Test. Standards Press of China: Beijing, China, 2008; pp. 8–14.
24. Standardization Administration of the People's Republic of China. *GB/T 2406.2-2009*; Plastics—Determination of Burning Behaviour by Oxygen Index—Part 2: Ambient-Temperature Test. Standards Press of China: Beijing, China, 2009; pp. 6–15.
25. International Organization for Standardization. *ISO5660-1:2015*; Reaction-to-Fire Tests—Heat Release, Smoke Production and Mass Loss Rate. International Organization for Standardization: Geneva, Switzerland, 2015; pp. 3–31.
26. Standardization Administration of the People's Republic of China. *GB/T 1040.1-2006*; Plastics—Determination of Tensile Properties—Part 1: General Principles. Standards Press of China: Beijing, China, 2006; pp. 8–24.
27. Standardization Administration of the People's Republic of China. *GB/T 9341-2008*; Plastics—Determination of Flexural Properties. Standards Press of China: Beijing, China, 2008; pp. 8–14.
28. Standardization Administration of the People's Republic of China. *GB/T 1043-2008*; Plastics—Determination of Charpy Impact Properties—Part 1: Non-Instrumented Impact Test. Standards Press of China: Beijing, China, 2008; pp. 8–14.
29. Yang, W.; Tawiah, B.; Yu, C.; Qian, Y.F.; Wang, L.L.; Yuen, A.C.Y.; Zhu, S.E.; Hu, E.Z.; Chen, T.B.Y.; Yu, B.; et al. Manufacturing, mechanical and flame retardant properties of poly(lactic acid) biocomposites based on calcium magnesium phytate and carbon nanotubes. *Compos. Part A Appl. Sci. Manuf.* **2018**, *110*, 227–236. [CrossRef]
30. Li, W.X.; Zhang, H.J.; Hu, X.P.; Yang, W.X.; Cheng, Z.; Xie, C.Q. Highly efficient replacement of traditional intumescent flame retardants in polypropylene by manganese ions doped melamine phytate nanosheets. *J. Hazard. Mater.* **2020**, *398*, 123001. [CrossRef] [PubMed]
31. Kang, X.; Lu, Z.; Feng, W.; Wang, J.; Fang, X.; Xu, Y.; Wang, Y.; Liu, B.; Ding, T.; Ma, Y.; et al. A novel phosphorous and silicon-containing benzoxazine: Highly efficient multifunctional flame-retardant synergist for polyoxymethylene. *Adv. Compos. Hybrid Mater.* **2021**, *4*, 127–137. [CrossRef]
32. Mun, S.Y.; Lee, S.Y.; Lim, H.M. Flame Retardant Properties of Basalt Fiber Reinforced Epoxy Composite with Inorganic Fillers. *Compos. Res.* **2019**, *32*, 368–374.
33. Leng, Y.M.; Zhao, X.; Fu, T.; Wang, X.L.; Wang, Y.Z. Bio-based flame-retardant and smoke-suppressing wood plastic composites enabled by phytic acid tyramine salt. *ACS Sustain. Chem. Eng.* **2022**, *10*, 5055–5066. [CrossRef]
34. Liu, D.; Zhao, W.; Cui, Y.; Zhang, T.; Ji, P. Influence of the Chemical Structure on the Flame Retardant Mechanism and Mechanical Properties of Flame-Retardant Epoxy Resin Thermosets. *Macromol. Mater. Eng.* **2022**, *307*, 2200169. [CrossRef]

Disclaimer/Publisher's Note: The statements, opinions and data contained in all publications are solely those of the individual author(s) and contributor(s) and not of MDPI and/or the editor(s). MDPI and/or the editor(s) disclaim responsibility for any injury to people or property resulting from any ideas, methods, instructions or products referred to in the content.

polymers MDPI

Article

Thermal Stability of Highly Filled Cellulosic Biocomposites Based on Ethylene–Vinyl Acetate Copolymer

Pavel Gennadievich Shelenkov [1,2], Petr Vasilievich Pantyukhov [1,3,*], Svetlana Vladimirovna Aleshinskaya [2], Alexander Andreevich Maltsev [1], Zubarzhat Rafisovna Abushakhmanova [1,3], Anatoly Anatolievich Popov [1,3], Jose Javier Saavedra-Arias [4] and Matheus Poletto [5]

1 Department of Biological and Chemical Physics of Polymers, Emanuel Institute of Biochemical Physics, Russian Academy of Sciences, 119334 Moscow, Russia; shell1183@mail.ru (P.G.S.); aam.0205@yandex.ru (A.A.M.); zubarzhat.akh@gmail.com (Z.R.A.); anatoly.popov@mail.ru (A.A.P.)
2 Joint Research Center, LLC "Metaclay Research and Development", 143026 Moscow, Russia; aleshinskaya@gmail.com
3 Higher Engineering School "New Materials and Technologies", Plekhanov Russian University of Economics, 115054 Moscow, Russia
4 Department of Physics, Universidad Nacional, Heredia 40101, Costa Rica; jose.saavedra.arias@una.ac.cr
5 Postgraduate Program in Engineering of Processes and Technologies, University of Caxias do Sul, Caxias do Sul 95070-560, Brazil; mpolett1@ucs.br
* Correspondence: p.pantyukhov@gmail.com

Abstract: The effect of plant-based fillers on thermal resistance in highly filled biocomposites based on ethylene–vinyl acetate copolymer (EVA) was studied. Wood flour and microcrystalline cellulose were used as fillers. It was shown that the introduction of microcrystalline cellulose into EVA did not affect the thermal stability of the polymer matrix. In contrast, the introduction of wood flour into EVA led to a significant increase in the thermal stability of the entire biocomposite. Oxidation induction time increased from 0 (pure EVA) to 73 min (EVA + wood flour biocomposites). The low-molecular weight phenolic compounds contained in wood flour are likely able to diffuse into the polymer matrix, exerting a stabilizing effect. The discovered stabilizing effect is a positive development for expanding the possibilities of technological processing of biocomposites, including multiple processing.

Keywords: highly filled biocomposite; ethylene–vinyl acetate copolymer (EVA); wood flour; microcrystalline cellulose; thermal-oxidative stabilization; natural antioxidants

check for updates

Citation: Shelenkov, P.G.; Pantyukhov, P.V.; Aleshinskaya, S.V.; Maltsev, A.A.; Abushakhmanova, Z.R.; Popov, A.A.; Saavedra-Arias, J.J.; Poletto, M. Thermal Stability of Highly Filled Cellulosic Biocomposites Based on Ethylene–Vinyl Acetate Copolymer. *Polymers* **2024**, *16*, 2103. https://doi.org/10.3390/polym16152103

Academic Editors: Fei Xiao, Fubin Luo and Ke Sun

Received: 8 May 2024
Revised: 9 June 2024
Accepted: 13 June 2024
Published: 24 July 2024

Copyright: © 2024 by the authors. Licensee MDPI, Basel, Switzerland. This article is an open access article distributed under the terms and conditions of the Creative Commons Attribution (CC BY) license (https://creativecommons.org/licenses/by/4.0/).

1. Introduction

EVA-based composites, filled with natural dispersed fillers, such as cellulose, starch, and wood flour were characterized in earlier works [1–3]. These composites are widely used, including for the creation of biodegradable products. The usage of a natural filler can reduce the cost of a composite based on synthetic plastics [4]. It is known that recycling composites based on synthetic polymers leads to thermal oxidation of the matrix polymer. To prevent thermal-oxidative destruction, antioxidants and heat stabilizers are usually added to the composition. The role of antioxidants (inhibitors) is to break the active chain due to interaction with the peroxide radical [5]. The most widely used synthetic antioxidant is Irganox 1010 (BASF Corporation, Charlotte, NC, USA). However, recent studies showed that synthetic antioxidants were poorly compatible with polymers; they can dissolve in water, diffuse onto the surface of the polymer, and evaporate over an elongated period [6,7]. One of the alternatives is the usage of natural additives as antioxidants and heat stabilizers. Poletto (2020) [8] examined the influence of natural oil additives on the physicochemical parameters and thermal stability of mixtures based on recycled polypropylene with wood flour. It was stated that there is an influence of natural oil additives on the physicochemical parameters and thermal stability of biocomposites based on recycled polypropylene with wood flour. The obtained results confirmed that when mixing wood flour with 2 wt.% of octane oil, the thermal destruction temperature of the polypropylene/wood flour biocomposite increased from 300 °C to 312 °C. At the same time, the strength and flexural modulus of elasticity increased noticeably. That work concluded that natural oils improve interfacial adhesion between wood flour and polypropylene matrix. Another work by Vorobyova and Prykhod (2019) [9] displayed the antioxidant effect of various organic fillers, such as dried and shredded buckwheat husk sowing (*Fagopyrum esculentum*), carposome of crab-of-the-woods (*Laetiporus sulphureus*), carposome of chaga mushroom (*Inonotus obliquus*), and thallus of lichen of oakmoss (*Evernia prunastri*), in composites with low-density polyethylene. The work studied the antioxidant effect of both the fillers themselves and their extracts on the polyethylene matrix. The results showed that the most effective antioxidant was an extract of oakmoss (*E. prunastri*), which increased the induction period of polyethylene oxidation by more than 10 times. Cerruti et al. (2009) [10] described the effect of the extract from tomato peel and seeds on the stabilization of polypropylene. The authors argued that lycopene, a carotenoid pigment found in large quantities in tomatoes, is a promising antioxidant for polymers. The effectiveness of other substances contained in plants, including quercetin, α-tocopherol, and cyclodextrin, enhanced polyethylene stabilization; the addition of these substances significantly increased the induction period of the oxidation [11]. The stabilizing effect of flavonoids (chrysin, quercetin, hesperidin, naringin, silibinin) under the influence of UV irradiation and temperature on polypropylene was also studied [12]. In general, a review of the literature data has shown that the use of natural additives as antioxidants in mixtures with synthetic polymers is an effective method of stabilization. However, in a majority of these works, it was not the raw plant particles used, but it was their extracts, where polyphenolic compounds exist in a concentrated form. When a small amount of plant particles is introduced, the effect of thermal stabilization of the polymer matrix is negligible. In highly filled biocomposites (over 50 wt.% of vegetable filler), the effect of thermal stabilization may be more obvious, but this research has not been carried out yet. In previous works devoted to the preparation and investigation of highly filled biocomposites, it was found that the content of VA in the EVA macromolecule had an influence on elongation at the break of the entire biocomposite; the higher the VA content, the higher the elongation [13–15]. Also, the effect of thermal stabilization of the polymer matrix due to the introduction of wood flour was accidentally discovered, and our current research investigates this effect more deeply. The use of natural, biodegradable vegetable fillers opens additional advantages, such as thermal stability and waste reduction, for the prospective use of biocomposites. Thus, the main objective of this work was to study the thermal stability of highly filled biocomposites based on wood flour compared to the ones with pure cellulose. It was also important to discover the correlations between the VA content in the EVA macromolecule or molecular mass of the EVA macromolecule and the effect of thermal stabilization by wood flour. For that reason, five different grades of EVA were used.

2. Materials and Methods

2.1. Materials

Five different grades of EVA produced by LG Chem (Seoul, Republic of Korea), differing in vinyl acetate content and melt flow index, were used as polymer matrices (Table 1).

Wood flour (WF) of deciduous wood provenance, provided by "Novotop" (Moscow, Russia), and microcrystalline cellulose (MCC) grade 101, produced by "Progress" (Kemerovo, Russia), were used as fillers. The chemical composition of the fillers, known from the literature, is presented in Table 2. The fillers were sifted through a sieve with a mesh diameter of 100 microns and dried at 105 °C for 24 h in an oven before mixing with the polymer.

Table 1. Basic characteristics of EVA grades (provided by LG Chem in technical datasheets).

EVA Grade	Content of Vinyl Acetate Groups [wt.%]	[mol.%]	Melt Flow Index (MFI) at 190 °C/2.16 kg [g/10 min]
28005	28	11	5
28025	28	11	25
28150	28	11	150
15006	15	5	6
19150	19	7	150

Table 2. Chemical composition of the fillers.

Filler	Cellulose [wt.%]	Lignin [wt.%]	Pentosanes [wt.%]	Polyuronic Acid [wt.%]	Reference
Wood flour (WF)	46	20	29	5	[16]
Microcrystalline cellulose (MCC)	100	-	-	-	-

Preparation of Biocomposites

Mixing of EVA with natural fillers was performed via heated mixing rolls UBL6175BL (Dongguan, China) with the temperature of the rolls set at 130 °C and 150 °C, and the rotation speed was 8 rpm. As a result, biocomposites based on five grades of EVA with two fillers (MCC and WF) were obtained. The content of the fillers was 50 wt%. The obtained biocomposites were molded by thermohydraulic press GOTECH GT-7014-H30C (Taichung, Taiwan) at 140 °C and 40 kgf/cm^2 over 1 min. The thickness of the resulting flat sheets varied from 0.4 mm to 0.5 mm.

2.2. Methods

(1) Determination of oxidation induction temperature (dynamic OIT)

Tests were carried out following ISO 11357-6:2018 [17]. The sample was heated at a constant rate in an oxygen atmosphere until the oxidation reaction was detected on the thermal curve. The onset of oxidation was indicated by a sharp increase in generated heat, observed via differential scanning calorimeter (DSC). The test was carried out using DSC 214 Polyma NETZSCH (Selb, Germany), at a heating rate of 10 °C/min. The oxygen flow rate was 50 mL/min, standard 40 μL aluminum crucibles without a lid were used, the masses of the samples varied in the range of 10 ± 3 mg, each measurement was carried out at least 2 times, and the average value was used for calculations.

(2) Determination of oxidation induction time (isothermal OIT)

Tests were carried out following the same ISO standard, using the same calorimeter, and crucibles as described above. For isothermal analysis, the sample was heated to 200 °C at 20 °C/min in an inert gas (nitrogen—100 mL/min) until 200 °C, kept for 5 min at this temperature, and then the gas was switched to oxygen (100 mL/min). The test was continued at a constant temperature of 200 °C until the oxidation reaction occurred.

(3) Thermogravimetric analysis (TGA)

The thermal degradation was studied using a thermogravimetric analyzer TGA/DSC3+ Mettler Toledo (Greifensee, Switzerland) following ISO 11358-1:2022 [18]. About 25–30 mg of the crushed sample was placed into a 150 μL crucible made of aluminum oxide. The measurements were carried out in closed crucibles, and the crucibles' lids had a hole (made by the manufacturer). The measurement was carried out in atmospheric air (100.0 mL/min): 15 min at 30 °C, then heated from 30 °C to 850 °C at a rate of 20.00 K/min.

(4) Fourier-transform infrared spectroscopy (FTIR)

To assess the effect of vinyl acetate content on the interaction of natural filler with EVA, studies were carried out using an FT-803 Simex IR-Fourier spectrometer (Novosibirsk,

Russia) with diamond crystal, employing the method of attenuated total reflectance (ATR). The temperature was 23 ± 2 °C, and the wavelength range was $4000 \leq V \leq 600$ cm^{-1}. The change in the peaks for EVA + WF biocomposites was recorded in the range of 1600–1650 cm^{-1} (peaks 1590 cm^{-1} and 1650 cm^{-1}). It was seen that 1590 cm^{-1} is sensitive to aromatic compounds and phenols from lignin in wood flour. The value 1650 cm^{-1} is the peak of oxidized phenols (benzophenones) and also is the peak of the double bond in vinyl acetate [19].

3. Results and Discussion

The results of TGA indicated that biocomposites with WF were more stable under heating than biocomposites with MCC. The TG curves (solid lines) and their first derivatives (DTG curves, dashed lines) for biocomposites based on EVA 19150 are illustrated in Figure 1. A similar nature of the curves was discovered for biocomposites based on all studied EVA grades (presented in Supplementary Materials). There are two peaks in the DTG curve of pure EVA. The first one (371 °C) corresponds to the degradation of the side vinyl acetate chain and the release of acetic acid. The second one (484 °C) characterizes the decomposition of the polymer backbone. Although biocomposites with WF began to lose weight earlier than the biocomposites with MCC, this decrease was insignificant. However, the peak of maximum weight loss (DTG curve) for the biocomposite with WF relative to the biocomposite with MCC was shifted by 16 degrees (from 358 to 374 °C) to a higher temperature. In addition, the DTG peak intensity of the biocomposite with WF was significantly lower than that of the biocomposite with MCC. Further analysis of the 19150-MCC biocomposite demonstrates that at the first stage of degradation, indicated by a peak at 358 °C, the filler was mainly destroyed (the peak for pure MCC was also detected at 358 °C). The weight loss of the biocomposite at 400 °C was 43% with an MCC content of 50%. Thus, at 400 °C, almost the entire MCC was destroyed. The biocomposite 19150-WF at 400 °C lost 35% of its weight, which indicated that only 15% of the filler remained. Although biocomposites with WF began to lose weight earlier than biocomposites with MCC, they better resisted thermal destruction at higher temperatures.

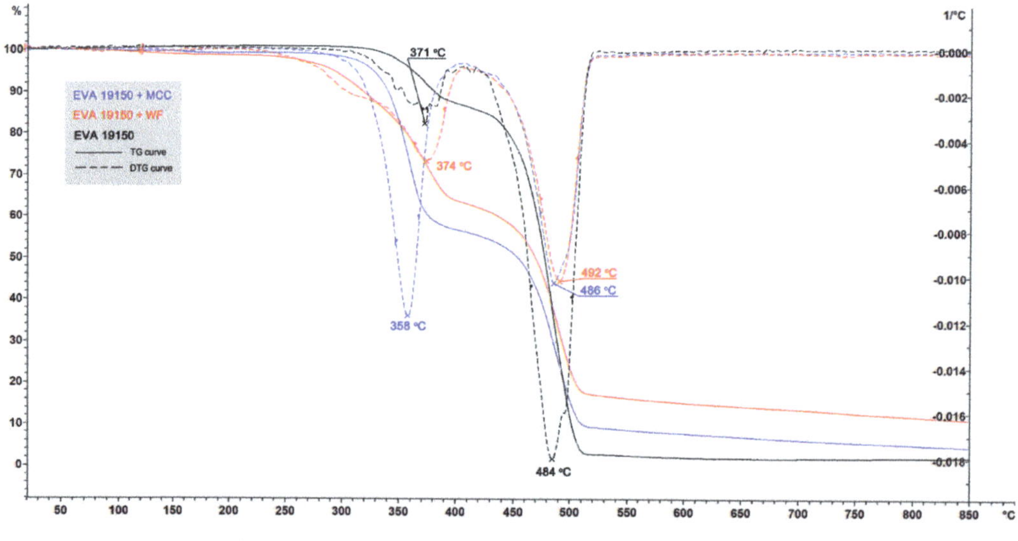

Figure 1. Thermogravimetric analysis (TGA) of biocomposites based on EVA 19150 matrix and pure EVA 19150 matrix.

Table 3 demonstrates the temperatures of thermal destruction of biocomposites on polymer matrices with different vinyl acetate contents. For comparison, the values of pure fillers and pure polymer matrices are also given. Biocomposites with WF showed higher thermal stability than biocomposites with MCC with all polymer matrices. All biocomposites with MCC had a peak in the temperature region where MCC destruction occurs (~358 °C), and the peaks of biocomposites with WF were determined by the destruction of matrices, sometimes even shifting them to the right (toward higher temperatures) along the abscissa axis. Thus, it can be hypothesized that WF has a thermal stabilization effect on polymer matrices.

Table 3. Onset, peak, and end temperatures at DTG curves (the first derivatives of TGA).

Composition	T_{onset}, °C	T_{peak}, °C	T_{end}, °C
WF	253	354	756
MCC	311	358	374
EVA 15006	330	374/481	517
EVA 15006 + WF	266	373/492	515
EVA 15006 + MCC	317	359/489	513
EVA 19150	313	372/484	513
EVA 19150 + WF	261	374/492	515
EVA 19150 + MCC	306	358/486	514
EVA 28005	329	367/487	513
EVA 28005 + WF	259	371/488	835
EVA 28005 + MCC	308	356/490	514

For the confirmation of the effect of thermal stabilization of the polymer matrix uncovered by TGA, additional investigations of biocomposites on thermal oxidation were carried out. The samples were studied using a differential scanning calorimeter under pure oxygen, so the onset temperatures of the thermodegradation processes were lower than in thermogravimetric analysis (which occurred in an air atmosphere). Figure 2 shows the kinetic dependences of the oxidative induction of EVA-MCC biocomposites depending on temperature: 100–0, 50–50, and 0–100 wt.%, using EVA with different vinyl acetate content.

MCC had a significantly longer period of thermal stability than EVA. The onset of thermal oxidation for EVA and the EVA-MCC biocomposite are very close. This result indicates the absence of chemical interaction between EVA and MCC. Therefore, MCC, which is more stable to oxidation, does not impact EVA in their biocomposites. This initial conclusion was confirmed by the study of the molecular structure of EVA-MCC using the FTIR method. The spectra of EVA-MCC biocomposites (Figure 3) are almost identical to the pure polymers. Therefore, it can be concluded that MCC in the EVA matrix is an inert filler. It should be noted that the onset of EVA oxidation shifts to a higher temperature with increasing concentrations of VA; with a content of 15% VA, the temperature of the onset of oxidation was about 200 °C, and with an increase in VA above 19%, oxidation began above 220 °C. These data look contradictory since it was shown [20] that the introduction of acetate functional groups into the polyolefin chain led to a decrease in thermal stability. However, in that work, thermal stability was assessed by the TGA method in an inert atmosphere based on weight loss, and the dynamic OIT method operates with a change in enthalpy. At the onset of thermal oxidation, the mass of the sample may not decrease; it decreases already with the formation of volatile products, so the results obtained by different methods may not converge.

A different effect of the filler on the thermal stability was found in composites based on EVA with WF. WF begins to oxidize at sufficiently low temperatures, in the region of the thermal stability of EVA. This can be explained by the chemical composition of WF. Unlike MCC, in addition to cellulose, WF contains lignin, hemicelluloses, and extractives. Some of them are more thermostable than cellulose (lignin), and others are less thermostable (hemicelluloses). Therefore, oxidation does not start for all wood components at the same time. TGA curves of pure fillers (Figure 4) indicated an earlier onset and later completion (a wider range) of thermal destruction of WF compared to MCC.

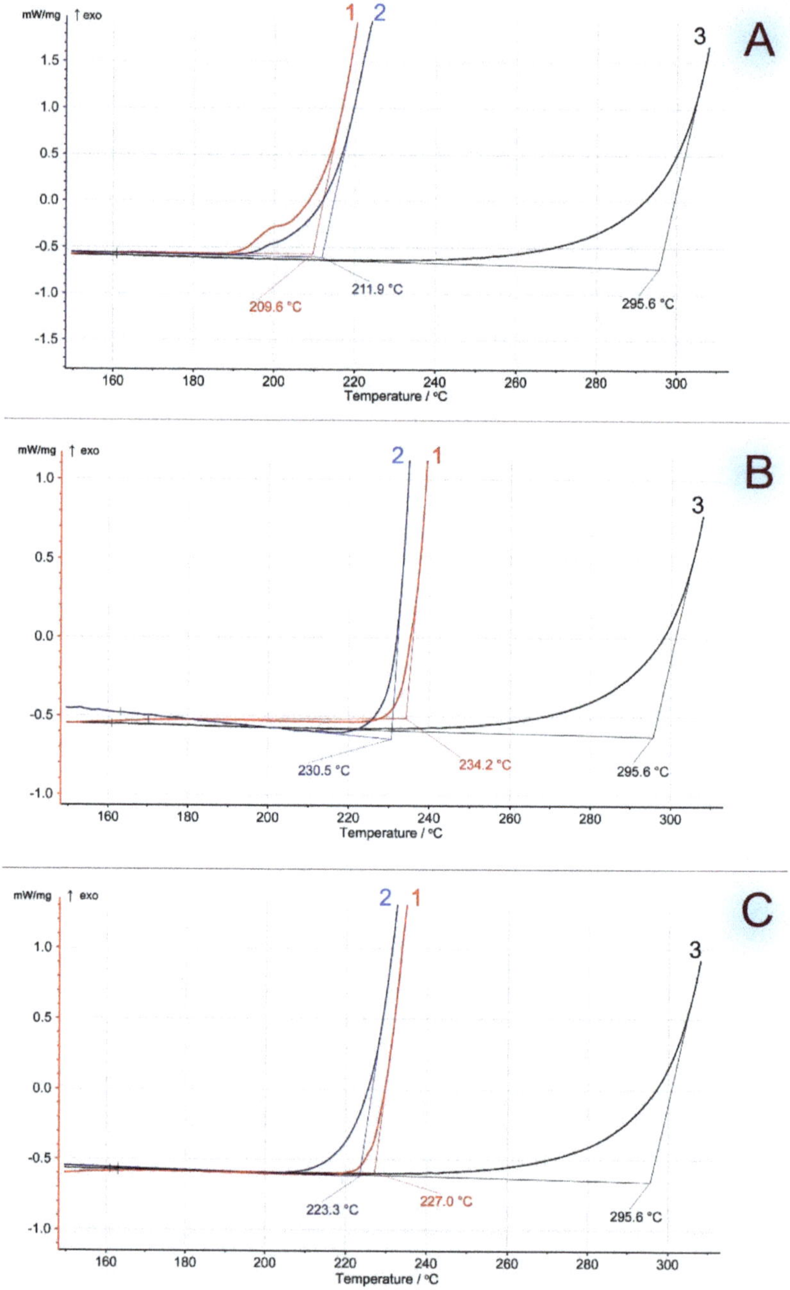

Figure 2. Oxidation induction temperature (dynamic OIT) of MCC-based biocomposites. Curves: 1—EVA, 2—EVA + MCC, 3—MCC. (**A**)—EVA with 15% of VA; (**B**)—EVA with 19% of VA; (**C**)—EVA with 28% of VA.

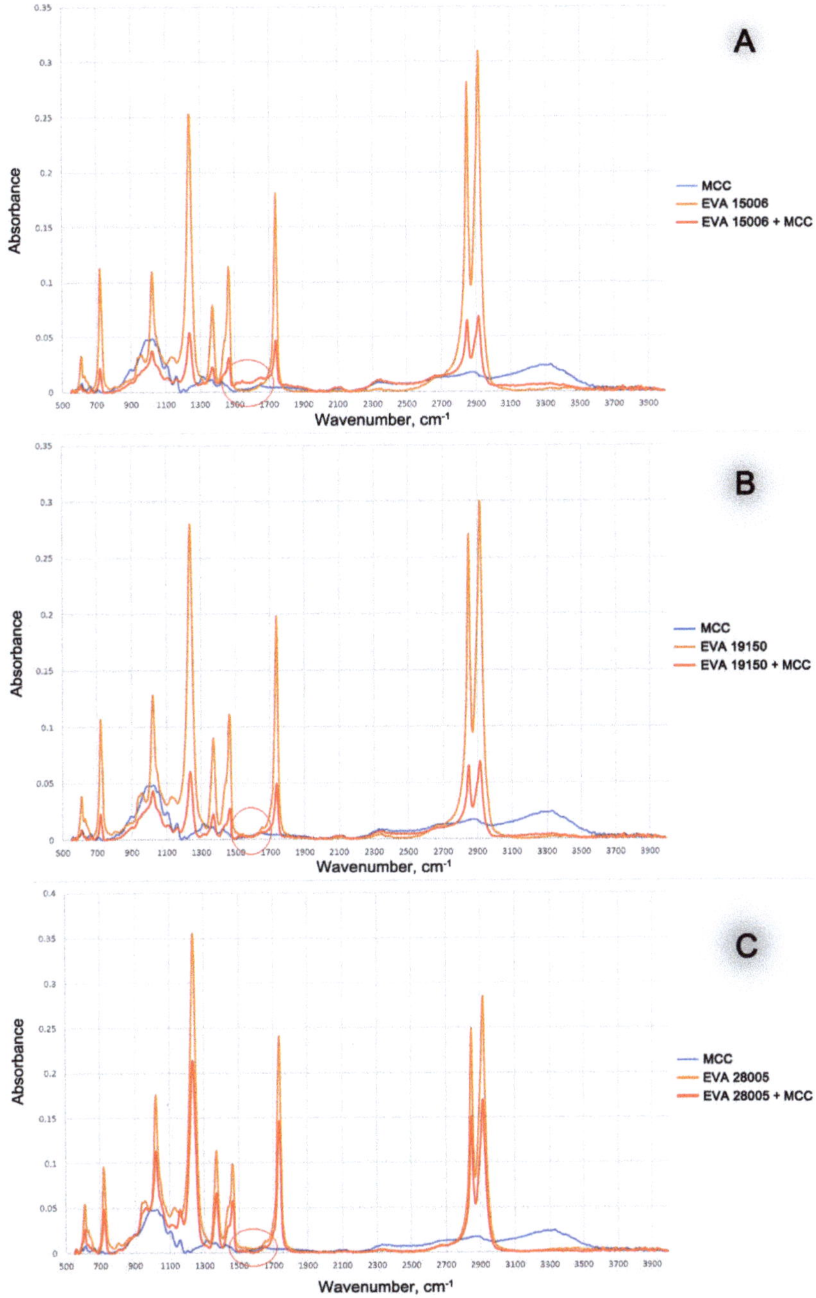

Figure 3. FTIR spectra of MCC-based biocomposites. FTIR spectra of pure EVA and MCC are also presented as a reference. (**A**)—EVA with 15% of VA; (**B**)—EVA with 19% of VA; (**C**)—EVA with 28% of VA.

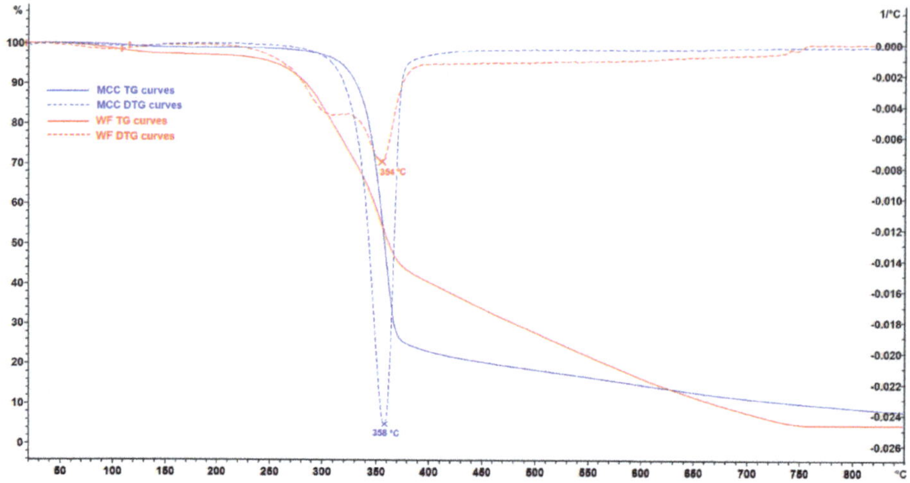

Figure 4. TG (solid) and DTG (dotted) curves of the fillers. Wood flour (WF)—red curves, microcrystalline cellulose (MCC)—blue curves.

Figure 5 presents the results of oxidation induction temperature determination (dynamic OIT) for EVA-WF biocomposites: 100–0, 50–50, and 0–100 wt.% using EVA with a different content of VA. The EVA-WF biocomposite exhibits a thermal stabilization effect. It begins to oxidize much later at higher temperatures than the individual components of the mixture. In terms of the thermal oxidation onset temperature, it significantly exceeded the performance of both EVA and WF. Low-molecular weight substances can diffuse into the EVA matrix from WF at the processing temperature under shear deformation, and this is the underlying reason why they can increase the stability of EVA against oxidation. Previous work showed the possibility of such a diffusion [21]. The model medium was liquid hexadecane (models the Kuhn segment in polyethylene); at 160 °C, polyphenolic substances (270 nm in the UV spectrum) and chlorophyll (410 nm and 670 nm) diffused into hexadecane from plant-based fillers. In another work, it was shown that the addition of plant components containing polyphenolic substances (birch leaves, mixed herbs hay) into a polyethylene matrix inhibits the oxidation process at a melt temperature of 130 °C [22]. In the paper [23], the effect of thermal stabilization of LDPE with the following agricultural waste as fillers was shown: grape pomace waste, wood shavings, turmeric waste, coffee grounds, and orange peel waste. Adding 4 wt.% of grape pomace waste led to an increase in the thermal oxidation temperature to 60 °C. This corresponds to the addition of 1 wt.% of synthetic thermal stabilizer Irganox 1010. At the same time, reprocessing of the composites led to an increased effect of thermal stabilization due to a more complete diffusion of polyphenols from fillers into the polymer.

Probably, during the process of compounding, phenols (including tannins, dihydroquercetin), which are natural antioxidants, diffuse into the polymer melt from wood flour. Diffusing into the polymer matrix, they stabilize it from oxidative degradation. At the same time, the polymer matrix protects plant particles of fillers from contact with atmospheric oxygen. This synergistic effect causes increased thermal stability of biocomposites with wood flour and other plant fillers containing polyphenolic antioxidants.

It is also possible that physicochemical interactions between polar groups of wood and EVA occurs, probably resulting in hydrogen bonds, which may explain the higher thermal stability of the biocomposite.

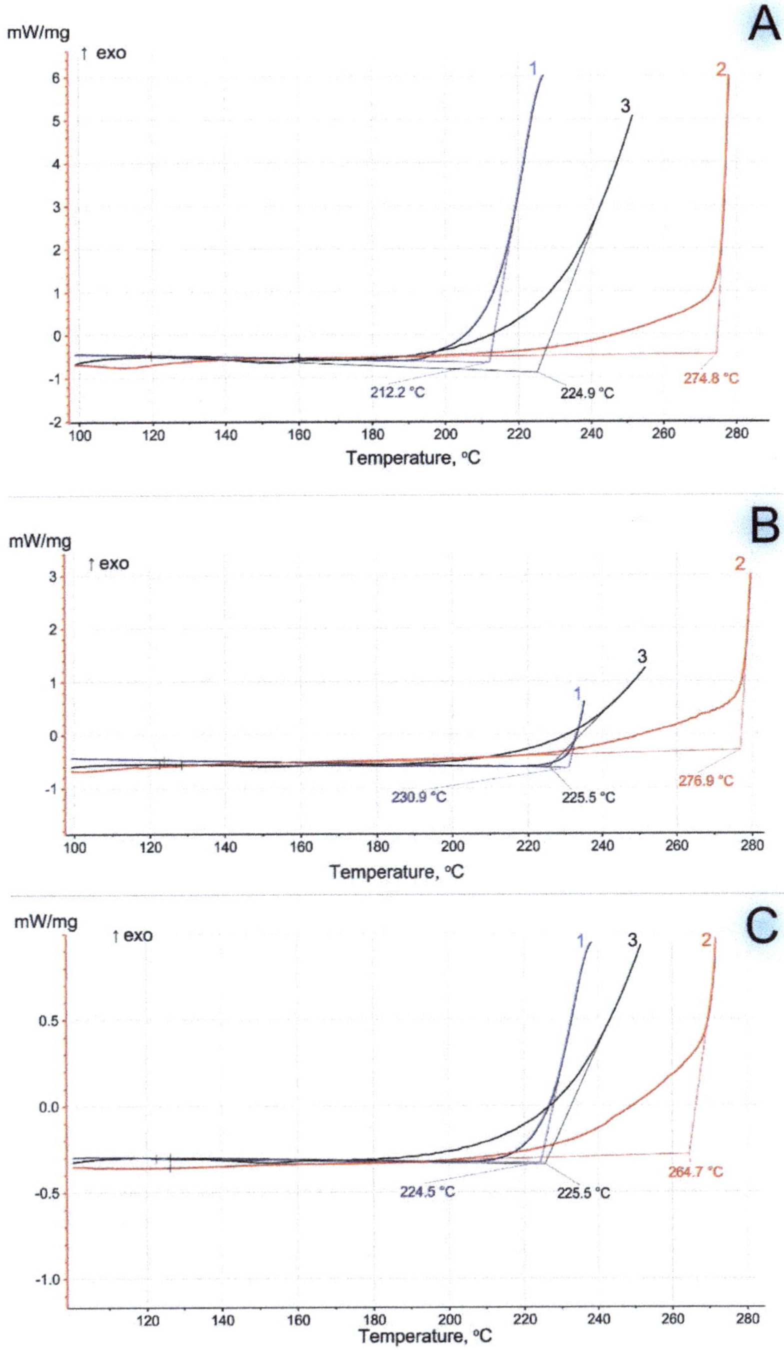

Figure 5. Oxidation induction temperature (dynamic OIT) of WF-based biocomposites. Curves: 1—EVA, 2—EVA + WF, 3—WF. (**A**)—EVA with 15% of VA; (**B**)—EVA with 19% of VA; (**C**)—EVA with 28% of VA.

The slope of the curves for biocomposites (EVA + WF) is less steep than that for pure WF (Figure 5). The tangent of the thermal oxidation curve's slope characterizes the oxidation rate of pure WF and biocomposites made of EVA with WF. For biocomposites, the tangent slope of the curves for EVA + WF is three-times less than that for pure WF (Table 4). It can be concluded that the diffusion of antioxidants from WF into the polymer matrix of EVA has a significant effect on the inhibition of thermal-oxidative degradation.

Table 4. The slope angle of thermal oxidation curves (dynamic OIT).

Composition	Tangent of the Inclination Angle in the Range of 220–240 °C
Wood flour (WF)	0.029 ± 0.001
EVA (15 wt.% of VA) + WF	0.010 ± 0.001
EVA (19 wt.% of VA) + WF	0.009 ± 0.001
EVA (28 wt.% of VA) + WF	0.008 ± 0.001

With an increase in the amount of VA in EVA, the oxidation rate decreases. This can be explained by the fact that with increasing VA content, the chemical affinity of EVA to the antioxidants contained in WF increases. This leads to better solubility of the antioxidants in the polymer matrix, and finally, it leads to a slowdown in oxidation processes. The yield of antioxidants was confirmed by studying the structure of biocomposites using FTIR. Figure 6 shows the IR spectra of EVA + WF biocomposites. In contrast to biocomposites with MCC, in the IR spectra of biocomposites with WF, a peak appears in the region of 1600–1650 cm^{-1} and it increases with increasing VA content in EVA. A more detailed examination of the spectra shows that this peak has two maxima, one of which is numbered (1) 1590 cm^{-1}, which duplicates the peak of wood flour, and the second (2) 1650 cm^{-1} is a small peak in the EVA spectrum. The peak at 1590 cm^{-1} indicates aromatic compounds such as phenols from wood flour lignin, while the peak at 1650 cm^{-1} indicates oxidized phenols (benzophenones) and at the same time indicates the peak of the double bond remaining in vinyl acetate. With an increase in the content of VA, low-molecular weight substances that diffused from wood flour are better distributed in the volume and surface layers of the biocomposite (increase in the peak of benzene rings by 1590 cm^{-1}), where the oxidation reactions of phenols occur (increase in the peak of the carbonyl group of benzophenones at 1650 cm^{-1}).

To confirm the discovered patterns, the thermal stability of highly filled biocomposites was assessed by the OIT method in isothermal mode. Pure EVA begins to oxidize earlier than the biocomposite with WF (Figure 7). Moreover, the filler itself (WF) is not subject to oxidation under these experimental conditions. That is, the degradation of WF occurs at a higher temperature as a result of thermal destruction and not due to oxidation. It is noticeable that biocomposites based on EVA with a lower content of VA (15%) have a shorter period of thermal stability (about 60 min). While for EVA biocomposites with 19–28% VA, this parameter increases to 80 min. This increase in the thermal stability of biocomposites can also be associated with better diffusion and solubility of natural antioxidants in EVA with a high VA content. In addition, it is possible that the formation of hydrogen bonds preferentially occurs at a higher content of polar VA groups.

For EVA-MCC biocomposites, no increase in thermal stability was detected. Figure 8 shows the isothermal OIT curves of biocomposites based on EVA 19150 and MCC. The course of the curves based on other EVA trademarks is identical, so they are not shown in the figure. Table 5 contains the oxidation induction time data of all studied biocomposites. It can be seen that biocomposite EVA + MCC starts to oxidize at the same time as pure EVA. This confirms the assumption that there is no intermolecular interaction between MCC and EVA. Consequently, when EVA is filled with MCC, the copolymer is not stabilized, but when it is filled with WF, thermo-oxidative stabilization of EVA occurs.

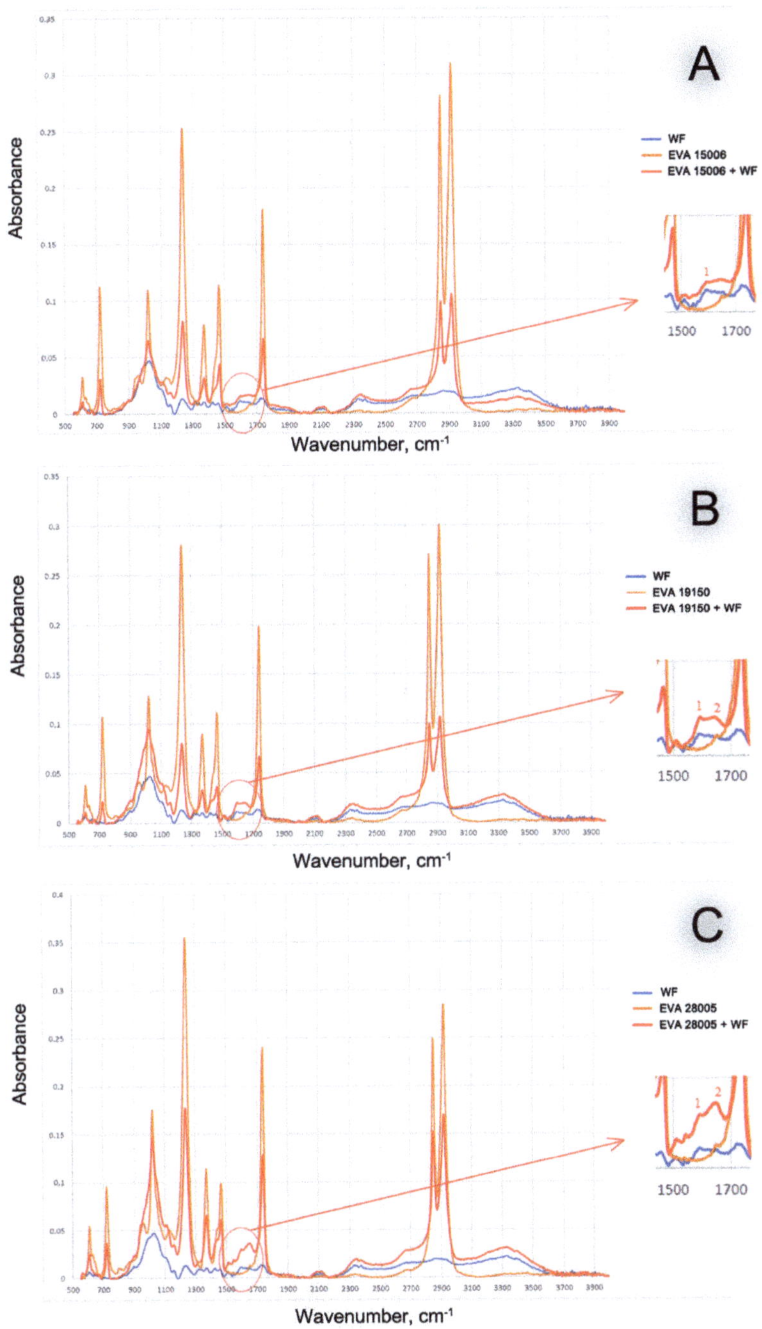

Figure 6. FTIR spectra of WF-based biocomposites. FTIR spectra of pure EVA and WF are also presented as a reference. (**A**)—EVA with 15% of VA; (**B**)—EVA with 19% of VA; (**C**)—EVA with 28% of VA.

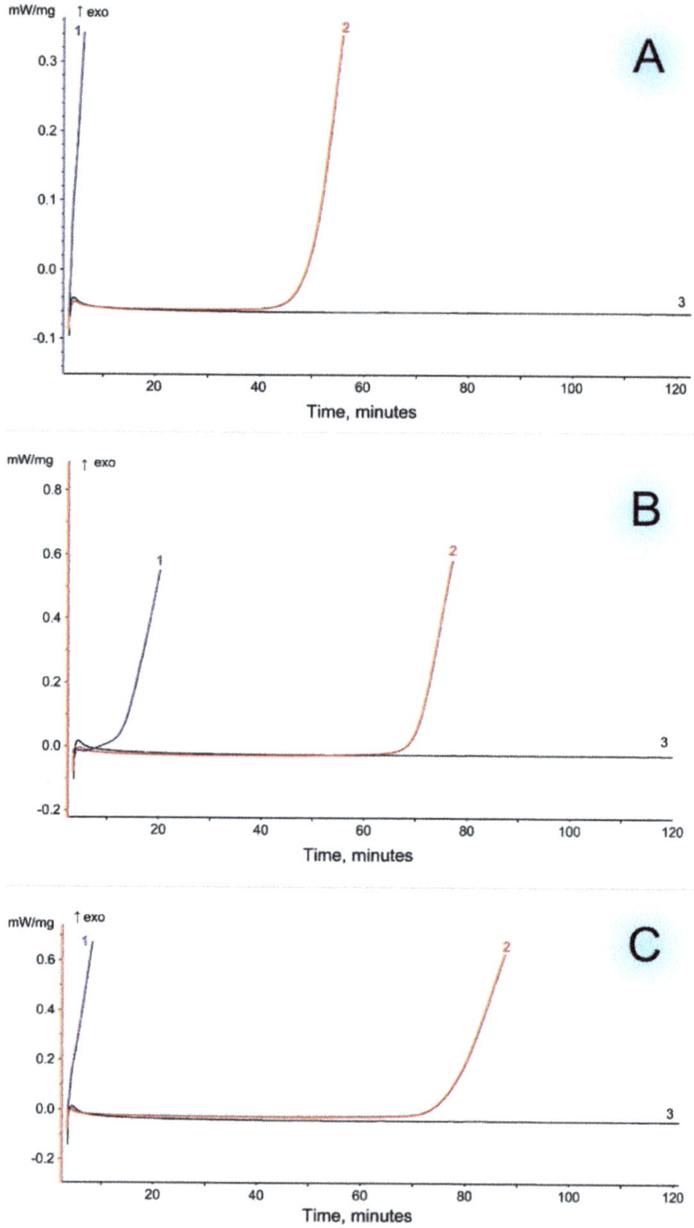

Figure 7. Oxidation induction time (isothermal OIT) of WF-based biocomposites (WF) at 200 °C. Curves: 1—EVA, 2—EVA + WF, 3—WF. (**A**)—EVA with 15% of VA; (**B**)—EVA with 19% of VA; (**C**)—EVA with 28% of VA.

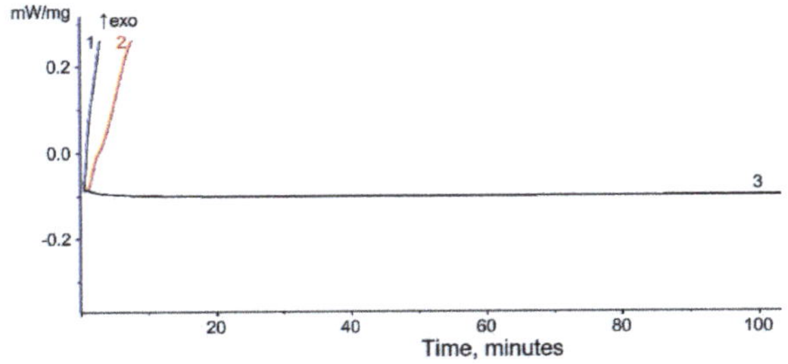

Figure 8. Oxidation induction time (isothermal OIT) of MCC-based biocomposites with EVA 28005 matrix at 200 °C. Curves: (1)—EVA, (2)—EVA + MCC, (3)—MCC.

Table 5. Oxidation induction time (isothermal OIT).

Content of VA in EVA, [wt.%]	Filler	Onset Time of Thermal-Oxidative Degradation [min]		
		EVA	Filler	EVA + Filler
15	WF	0.2	>120	46.2
19		9.8		66.8
28		0.2		73.4
15	MCC	0.2	>120	0.2
19		9.8		9.8
28		0.2		0.2

4. Conclusions

This work examined the influence of two natural fillers, WF and MCC, on the thermal stability of highly filled biocomposites based on EVA. These masterbatches were developed for the purpose of being used as an additive to the matrix of synthetic polymers. When the masterbatch additive is introduced to the main polymer, a biodegradable biocomposite is formed, which significantly reduces the decomposition period of synthetic polymers in the natural environment. The results showed that in EVA biocomposites with MCC, the filler does not have a stabilizing effect on the EVA matrix, and the thermal-oxidative degradation of EVA/MCC biocomposites occurs at the same time when the matrix (EVA) starts oxidizing. In biocomposites with WF, a significant effect of thermal-oxidative stabilization was observed, confirmed in both dynamic and isothermal modes. Moreover, it was found that this effect intensifies with an increase in the content of vinyl acetate in EVA. After addition of WF to EVA 28005, the oxidation induction time increased from 0 (for pure EVA) to 73 min (for EVA + WF biocomposite). The discovered effect of thermal stabilization can be explained by the diffusion of phenolic compounds from wood into the polymer matrix during compounding. Phenolic compounds, as natural antioxidants, can protect the polymer from oxidative degradation. Thus, biocomposites filled with WF are more heat resistant than pure matrix polymer and can withstand more cycles of recycling, while being more biodegradable and cheaper. In previous papers, it was shown that extracts from vegetable fillers might be added to the polyolefin polymer matrix and act as efficient antioxidants. In this work, it was shown that the same effect may be obtained in highly filled biocomposites without additional treatments like extraction. The discovered effect of thermal stabilization is most likely characteristic not only of wood flour, but also of other vegetable fillers that are not chemically purified, which makes such biocomposites more promising.

Supplementary Materials: The following supporting information can be downloaded at: https://www.mdpi.com/article/10.3390/polym16152103/s1, Figure S1: Thermogravimetric analysis of biocomposites based on EVA 15006; Figure S2: Thermogravimetric analysis of biocomposites based on EVA 28005.

Author Contributions: Conceptualization, P.G.S., P.V.P. and A.A.P.; methodology, P.G.S.; formal analysis, P.G.S., Z.R.A. and A.A.P.; investigation, P.G.S., P.V.P., S.V.A., A.A.M. and Z.R.A.; resources, A.A.P. and J.J.S.-A.; data curation, P.G.S., P.V.P., S.V.A., A.A.M., Z.R.A. and M.P.; writing—original draft, P.G.S. and P.V.P.; writing—review and editing, P.V.P. and M.P.; supervision, J.J.S.-A.; project administration, P.V.P. and A.A.P.; funding acquisition, J.J.S.-A. All authors have read and agreed to the published version of the manuscript.

Funding: This research received no external funding.

Institutional Review Board Statement: Not applicable.

Data Availability Statement: Data are contained within the article and supplementary materials.

Acknowledgments: Preparation of the samples and laboratory tests were carried out using the scientific and technological equipment at the Center of Shared Usage «New Materials and Technologies» at Emanuel Institute of Biochemical Physics and Joint Research Center at Plekhanov Russian University of Economics.

Conflicts of Interest: Authors Pavel Gennadievich Shelenkov and Svetlana Vladimirovna Aleshinskaya were employed by the company Metaclay Research and Development. The remaining authors declare that the research was conducted in the absence of any commercial or financial relationships that could be construed as a potential conflict of interest.

References

1. Zimmermann, M.V.G.; Turella, T.; Santana, R.M.C.; Zattera, A.J. Comparative study between poly(ethylene-co-vinyl acetate)—EVA expanded composites filled with banana fiber and wood flour. *Mater. Res.* **2014**, *17*, 1535–1544. [CrossRef]
2. Sartore, L.; D'Amore, A.; Di Landro, L. Ethylene vinyl acetate blends with cellulosic fillers and reinforcements. *Polym. Compos.* **2015**, *36*, 980–986. [CrossRef]
3. Sessini, V.; Arrieta, M.P.; Raquez, J.-M.; Dubois, P.; Kenny, J.M.; Peponi, L. Thermal and composting degradation of EVA/Thermoplastic starch blends and their nanocomposites. *Polym. Degrad. Stab.* **2019**, *159*, 184–198. [CrossRef]
4. Lightsey, G.R. Organic Fillers for Thermoplastics. In *Polymer Applications of Renewable-Resource Materials*; Springer: Boston, MA, USA, 1983; pp. 193–211. [CrossRef]
5. Buchachenko, A.L.; Emanuel, N.M. *Chemical Physics of Polymer Degradation and Stabilization*; CRC Press: Boca Raton, FL, USA, 1987.
6. Maghsoud, Z.; Rafiei, M.; Famili, M.H.N. Effect of processing method on migration of antioxidant from HDPE packaging into a fatty food simulant in terms of crystallinity. *Packag. Technol. Sci.* **2018**, *31*, 141–149. [CrossRef]
7. Kang, K.; Chang, Y.; Choi, J.C.; Park, S.; Han, J. Migration Study of Butylated Hydroxytoluene and Irganox 1010 from Polypropylene Treated with Severe Processing Conditions. *J. Food Sci.* **2018**, *83*, 1005–1010. [CrossRef] [PubMed]
8. Poletto, M. Natural oils as coupling agents in recycled polypropylene wood flour composites: Mechanical, thermal and morphological properties. *Polym. Polym. Compos.* **2020**, *28*, 443–450. [CrossRef]
9. Vorobyova, E.V.; Prykhod'ko, E.L. Stabilization of polyethylene by natural fillers and their extracts. *Chem. Plant Raw Mater.* **2019**, *2*, 213–223. [CrossRef]
10. Cerruti, P.; Malinconico, M.; Rychly, J.; Matisova-Rychla, L.; Carfagna, C. Effect of natural antioxidants on the stability of polypropylene films. *Polym. Degrad. Stab.* **2009**, *94*, 2095–2100. [CrossRef]
11. Koontz, J.L.; Marcy, J.E.; O'Keefe, S.F.; Duncan, S.E.; Long, T.E.; Moffitt, R.D. Polymer processing and characterization of LLDPE films loaded with α-tocopherol, quercetin, and their cyclodextrin inclusion complexes. *J. Appl. Polym. Sci.* **2010**, *117*, 2299–2309. [CrossRef]
12. Samper, M.D.; Fages, E.; Fenollar, O.; Boronat, T.; Balart, R. The potential of flavonoids as natural antioxidants and UV light stabilizers for polypropylene. *J. Appl. Polym. Sci.* **2013**, *129*, 1707–1716. [CrossRef]
13. Shelenkov, P.G.; Pantyukhov, P.V.; Popov, A.A. Mechanical Properties of Superconcentrates Based on Ethylene-Vinyl Acetate Copolymer and Microcrystalline Cellulose. *Mater. Sci. Forum.* **2020**, *992*, 306–310. [CrossRef]
14. Shelenkov, P.G.; Pantyukhov, P.V.; Popov, A.A. Highly filled biocomposites based on ethylene-vinyl acetate copolymer and wood flour. *IOP Conf. Ser. Mater. Sci. Eng.* **2018**, *369*, 012043. [CrossRef]
15. Shelenkov, P.G.; Pantyukhov, P.V.; Poletto, M.; Popov, A.A. Influence of Vinyl Acetate Content and Melt Flow Index of Ethylene-Vinyl Acetate Copolymer on Physico-Mechanical and Physico-Chemical Properties of Highly Filled Biocomposites. *Polymers* **2023**, *15*, 2639. [CrossRef] [PubMed]
16. Nikitin, N.I. *Chemistry of Wood and Cellulose*; Academy of Sciences of the USSR: Leningrad, Russia, 1962.
17. *ISO 11357-6:2018*; Differential Scanning Calorimetry (DSC). ISO: Geneva, Switzerland, 2018.
18. *ISO 11358-1:2022*; Plastics—Thermogravimetry (TG) of Polymers. ISO: Geneva, Switzerland, 2022.
19. Shurvell, H.F. Spectra- Structure Correlations in the Mid- and Far-Infrared. In *Handbook of Vibrational Spectroscopy*; Chalmers, J.M., Ed.; Wiley: Hoboken, NJ, USA, 2001. [CrossRef]

20. Wenwei, Z.; Xiaoguang, Z.; Li, Y.; Yuefang, Z.; Jiazhen, S. Determination of the vinyl acetate content in ethylene-vinyl acetate copolymers by thermogravimetric analysis. *Polymer* **1994**, *35*, 3348–3350. [CrossRef]
21. Pantyukhov, P.; Monakhova, T.; Popov, A.; Zykova, A. Chemical interaction of polyethylene matrix with vegetable fillers in biocomposites. *AIP Conf. Proc.* **2016**, *1736*, 020122. [CrossRef]
22. Pantyukhov, P.V.; Khvatov, A.V.; Monakhova, T.V.; Popov, A.A.; Kolesnikova, N.N. The Degradation of Materials Based on Low-Density Polyethylene and Natural Fillers. *Int. Polym. Sci. Technol.* **2013**, *40*, 55–58. [CrossRef]
23. Iyer, K.A.; Zhang, L.; Torkelson, J.M. Direct Use of Natural Antioxidant-rich Agro-wastes as Thermal Stabilizer for Polymer: Processing and Recycling, ACS Sustain. *Chem. Eng.* **2016**, *4*, 881–889. [CrossRef]

Disclaimer/Publisher's Note: The statements, opinions and data contained in all publications are solely those of the individual author(s) and contributor(s) and not of MDPI and/or the editor(s). MDPI and/or the editor(s) disclaim responsibility for any injury to people or property resulting from any ideas, methods, instructions or products referred to in the content.

polymers

MDPI

Article

Preparation and Characterization of Chloroprene Latexes Modified with Vinyl-POSS

Junhua Chen [1,2], Zhenxian Wu [1], Qingwei Wang [1], Chuanghui Yang [1], Jinlian Chen [1], He Zhang [1,2], Yinping Wu [1,2], Dong-Yu Zhu [3] and Xiangying Hao [1,2,*]

1 School of Environmental and Chemical Engineering, Zhaoqing University, Zhaoqing 526061, China; cehjchen@yeah.net (J.C.); wuzhenxian2023@163.com (Z.W.); wangqingwei0721@163.com (Q.W.)
2 Guangdong Provincial Key Laboratory of Environmental Health and Land Resource, College of Environmental and Chemical Engineering, Zhaoqing University, Zhaoqing 526061, China
3 School of Chemical Engineering and Light Industry, Guangdong University of Technology, Guangzhou 510006, China; zdy16@gdut.edu.cn
* Correspondence: xyinghao@zqu.edu.cn

Abstract: Water-based chloroprene latex is a solvent-free, environmentally friendly adhesive. Currently, its market demand is growing rapidly. However, there are problems such as a lack of heat resistance and poor mechanical properties, which limit its application. The introduction of vinyl-POSS (OVS) into the resin structure can effectively improve the thermal stability of chloroprene adhesives. In this paper, modified waterborne chloroprene latex was prepared by copolymerization of methyl methacrylate and OVS with chloroprene latex. The results showed that vinyl-POSS was successfully grafted onto the main chain of the waterborne chloroprene latex, and the modified waterborne chloroprene latex had good storage stability. With the increase in vinyl-POSS, the tensile strength of the chloroprene latex firstly increased and then decreased, the tensile property (peel strength of 20.2 kgf) was maintained well at a high temperature (100 °C), and the thermal stability of the chloroprene latex was improved. When the addition amount was 4%, the comprehensive mechanical properties were their best. This study provides a new idea for the construction of a new and efficient waterborne chloroprene latex system and provides more fields for the practical application of waterborne chloroprene latex. This newly developed vinyl-POSS modified chloroprene latex has great application potential for use in home furniture, bags, and seat cushions.

Keywords: vinyl-POSS; chloroprene latexes; tensile strength; thermal stability

check for updates

Citation: Chen, J.; Wu, Z.; Wang, Q.; Yang, C.; Chen, J.; Zhang, H.; Wu, Y.; Zhu, D.-Y.; Hao, X. Preparation and Characterization of Chloroprene Latexes Modified with Vinyl-POSS. *Polymers* **2024**, *16*, 462. https://doi.org/10.3390/polym16040462

Academic Editors: Markus Gahleitner and Changwoon Nah

Received: 23 December 2023
Revised: 31 January 2024
Accepted: 1 February 2024
Published: 7 February 2024

Copyright: © 2024 by the authors. Licensee MDPI, Basel, Switzerland. This article is an open access article distributed under the terms and conditions of the Creative Commons Attribution (CC BY) license (https://creativecommons.org/licenses/by/4.0/).

1. Introduction

Chloroprene latex, a type of water-based emulsion chloroprene rubber adhesive, is composed of polychloroprene micelles and various types of emulsifiers [1–4]. This versatile adhesive is widely utilized in bonding fabrics, leather, wood, plastics, glass, and other materials, playing a crucial role in various industries worldwide [5–8]. The environmental friendliness, safety, non-toxicity, non-flammability, and affordability of water-based chloroprene latex are notable advantages [9–13]. However, it also possesses certain drawbacks, such as slow drying, limited substrate compatibility, low initial viscosity, poor resistance to extreme temperatures, susceptibility to freezing, poor storage stability, and color change over time. Considering these limitations, the modification of waterborne chloroprene latex adhesives is often necessary to enhance their overall performance [14,15]. Various industries rely on the development of chloroprene adhesives to meet their specific bonding needs. In recent years, significant efforts have been made to improve the formulation of chloroprene latex adhesives, addressing their inherent limitations, and expanding their application range. These efforts have resulted in the development of modified chloroprene latex adhesives with improved drying times, enhanced substrate compatibility, increased initial viscosity, and better resistance to environmental factors, such as heat and cold.

Furthermore, advancements in storage stability and color retention have contributed to the broader use of chloroprene latex adhesives in diverse industrial settings. As a result, the market for modified chloroprene latex adhesives continues to grow, offering superior performance and versatility for various bonding applications.

There are several common methods for modifying chloroprene latex, including blending modification, graft modification, and copolymerization modification [16–21]. Blending modification is prone to phase separation problems, making it less suitable for certain applications. Graft modification, on the other hand, is a more useful method of polymer modification as it is easy to control various reaction parameters [22,23]. In the past, Kai Zhang et al. grafted methyl methacrylate and styrene onto polychloroprene rubber using emulsion polymerization. By incorporating the bulking agent made using this method, they achieved better mechanical properties of the contact rubber compared to simple blends [14].

Polyhedral oligosiloxanes (POSS) are small molecules that have cage-like three-dimensional structures at the nanoscale. The general formula for POSS is $(RSiO_{1.5})_n$ [24–26]. The core of POSS is made up of inert inorganic materials, such as organosilicon and oxygen ($SiO_{1.5}$). When incorporated into a polymer matrix, POSS enhances the mechanical properties and thermal stability of the polymer [27–30]. The substituents (R) attached to the silicon atoms situated at the corners of the cage can be classified as hydrogen, reactive, or inert organic groups [31,32]. These organic groups promote specific interactions and compatibility between POSS and the polymer or monomer [33,34]. By utilizing traditional chemical transformations, it is possible to replace one or more of the substituents with functional groups. These multifunctional groups, such as methacrylates, styrenes, epoxides, alcohols, and phenols, facilitate the introduction of POSS into a polymer chain or network through grafting or polymerization [35–38]. The addition of nanosized inorganic particles is an effective approach to enhance the properties of polymers while maintaining their low density and high ductility [39]. Moreover, when applied to emulsion polymerization, POSS does not suffer from the problem of alkoxy hydrolysis polycondensation that general organosilicon monomers encounter. Furthermore, POSS can form connections with multiple organic groups due to its high reactivity [40,41]. The vinyl group in OVS (organosilicon vinyl ether) allows for the conversion of OVS into valuable products through various reactions [42,43]. The objective of this study was to modify waterborne chloroprene latex using functional silicone macro-monomers with a nanoscale three-dimensional structure derived from OVS. The effects of these modifications on the stability, latex particle size, heat resistance, and T-peel strength were investigated. The modified chloroprene rubber can be widely used in home furniture, bags, seat cushions for aircraft, cars, high-speed trains, trains, ships, and other modes of transportation. Additionally, it can be employed in the bonding of leather, sponge, plastics, and wood [40,44,45].

2. Materials and Methods

2.1. Materials

Industrial-grade chloroprene latex (CRL) was purchased from Shanghai Costron Polymers Co., Ltd., Shanghai, China. The methyl methacrylate (MMA), sodium dodecyl biphenyl ether sulfonate (DSB), and vinyl-POSS (95%, OVS) used in this study were all analytical reagent grade and were purchased from Shanghai Maclin Biochemical Technology Co., Ltd., Shanghai, China. Tert-butyl hydroperoxide (TBHP) and tetraethylenepentamine (TEPA) were all analytical grade and were purchased from Shanghai Aladdin Biochemical Technology Co., Ltd., Shanghai, China. Benzene propylene emulsion was industrial reagent grade and was purchased from Shandong Haoshun Chemical Co., Ltd. (Jinan, China). Deionized water is homemade in the laboratory. All the other reagents underwent further purification before use.

2.2. Preparation of Modified Aqueous Chloroprene Latex

The grafting reaction was conducted using the seed emulsion polymerization method displayed in Scheme 1. A 250 mL three-necked round bottom flask equipped with a thermostatically heated magnetic stirrer, peristaltic pump, condenser pipe, and thermometer was used. In this flask, 25 g of CRL and DSB (1% of the total monomer weight) was added. The reaction system was then diluted with deionized water to achieve a solid content of 30% while stirring with a thermostatically heated magnetic stirrer at 50 °C. An amount of 8.25 g of mixed monomer (MMA and OVS) was added dropwise at a uniform rate over 100 min. The OVS component accounted for 0–5% of the total weight ratio of monomer and polymer. During this process, a mixture of TBHP (TBHP/TEPA = 1:1, 0.5% of the dry weight of CRL) was added, alongside the dropwise addition of aqueous TEPA, maintaining a constant temperature for 3.5–4 h. Finally, the emulsion was allowed to cool to room temperature to obtain the OVS-modified aqueous chloroprene latex.

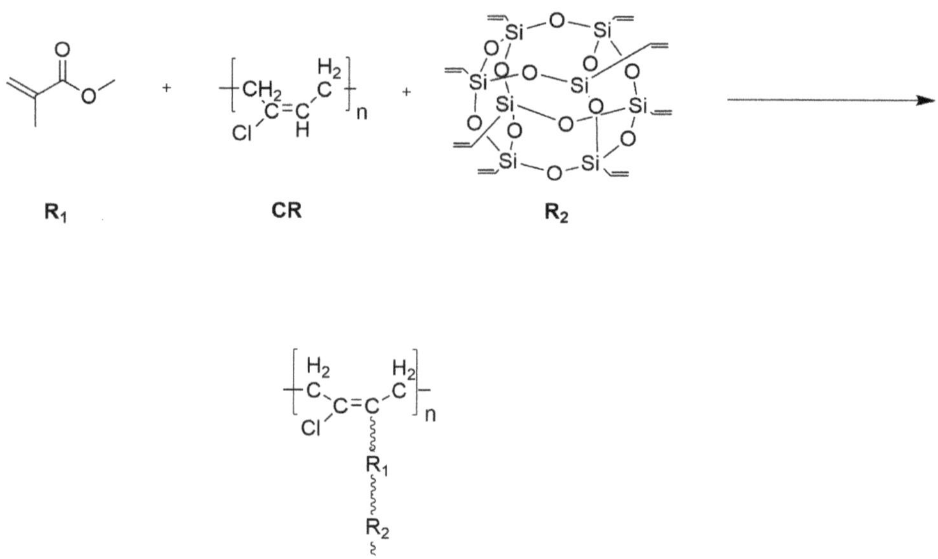

Scheme 1. OVS modified aqueous chloroprene synthetic approaches.

2.3. Characterization

Fourier transform infrared spectroscopy (FT-IR) analysis of the samples was obtained in the range of 400–4000 cm^{-1} using a Nicolet IS50-Nicolet Contin FT-IR spectrometer (Thermo Fisher Scientific Inc., Waltham, MA, USA). The viscosity of the samples was measured using a DVS+ viscometer according to the standard GB/T 2794-1995 [46]. The TG and DTG curves were obtained by heating the samples from 40~600 °C at a scanning rate of 10 °C/min under a nitrogen atmosphere using a TG209F1 thermogravimetric analyzer (NETZSCH-Gerätebau GmbH., Selb, Germany). The samples were thermally analyzed to obtain the glass transition temperature by DSC 214 with a temperature scanning range of 80–120 °C, a heating rate of 10 °C/min and a nitrogen atmosphere. The T-peeling strength of the sample was measured using a temperature-controlled tensile strength machine according to the standard of GB/T 2790-1995 [47] displayed in Scheme 2.

To test the stability of the emulsion dilution, 2 mL of emulsion was diluted in 10 mL of water. Then, after a week, the mixture was observed to see if any phenomena such as precipitation or delamination occurred. The emulsion samples were placed in 50 mL clear vials and left at room temperature (25 °C, 50% humidity) for 24 h. To test the stability of the emulsion at room temperature during the measurement period, a visual method of observation was used. In an oven set to a constant temperature of 50 °C, the emulsion samples were put. To ascertain the emulsion's stability at high temperatures, the emulsion was visually examined during the measurement period to see if flocculation or emulsion breakup occurred.

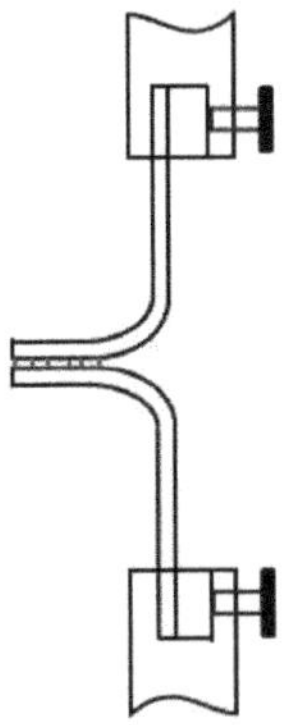

Scheme 2. Schematic diagram of T-peeling test.

The emulsion samples and phenylpropylene emulsion were mixed together before being uniformly applied to a piece of sponge. A second piece of sponge was then aligned and bonded, and after waiting for it to dry at room temperature and being placed into the oven for an hour at a high temperature, the heat resistance of the latex was assessed by testing its tensile properties both before and after exposure to high temperature. The synthesized vinyl-POSS modified neoprene emulsion samples were co-mingled with the phenylpropylene emulsion and coated uniformly on the paper skins, and the coated paper skins (15 cm × 3.5 cm) were aligned and glued together with another paper skin, rolled with a rolling machine and pressed until 48 h. In accordance with the GB/T 2909 standard [48], the T-peel strength of the sample emulsion was tested using an Al-7000S temperature-controlled tensile machine (High Speed Rail Testing Instruments Ltd., Taipei City, Taiwan, China) at a testing temperature of 40 °C and with a cross-head speed of 100 mm/min.

The mechanism of polymerization kinetics was to use a linear regression equation to find the slope *(dx/dt)* of the monomer conversion (x)–time (t) curve in the constant velocity phase, and then the rate of polymerization $R_p = \frac{dX}{dt} \times [M_0]$ was calculated according to the formula, where the formula $[M_0]$ = the amount of substance of the monomer (mol)/volume of water in the formulation (L).

3. Results and Discussion

As shown in the infrared spectra of 0% OVS-CRL and 5% OVS-CRL in Figure 1, the 5% OVS waterborne chloroprene latex, in addition to the typical characteristic peaks, has a strong and sharp ester group C=O telescoping vibration peak at 1730 cm^{-1} and a C-O telescoping vibration peak at 1246 cm^{-1}. The peaks at 1083 cm^{-1} represent the stretching vibrations of the siloxane network (Si-O-Si). This indicates that methyl methacrylate and vinyl-POSS have been successfully grafted onto the chain of waterborne chloroprene latex.

The effect of the addition of OVS on the viscosity, solid content and conversion rate of chloroprene latexes was investigated with the same amount of chloroprene, emulsifier, initiator and reducing agent added, the content of fixed MMA and other conditions being the same, and the results are shown in Table S1. As can be seen from Table S1, as the content of the monomer OVS increases, the viscosity decreases, with the first conversion increasing and then decreasing. As the conversion decreases, the latex particles decrease, the water component increases, the total surface area of the system particles decreases, making the interaction and resistance to movement between the latex particles weaken, and the Brownian motion of the latex particles becomes easier, so the viscosity decreases.

Table S2 demonstrates the stability of OVS modified chloroprene emulsions. As the content of monomer OVS increased, the samples with 0% to 3% OVS content had a good appearance without gelation, and the emulsions reacting at 4% and 5% OVS content produced a small amount of gelation. The emulsion has good dilution stability, room temperature and high temperature storage stability. Among them, Figure 2 shows the appearance state of OVS-modified waterborne chloroprene emulsion after 90 days of placement. The modified emulsion was light yellow in color with no obvious delamination and precipitation.

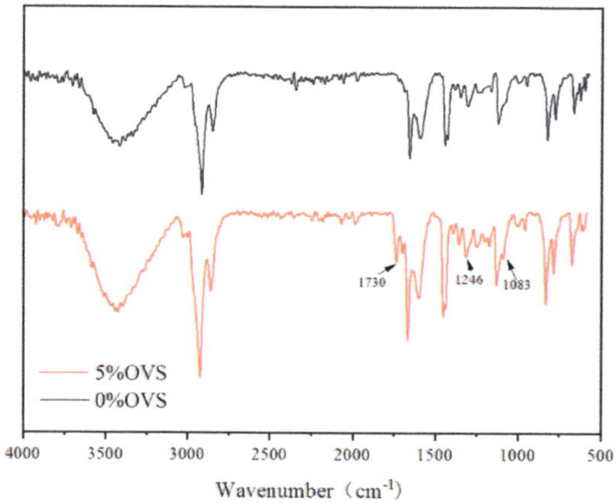

Figure 1. Infrared spectra of CRL and 5% OVS-CRL.

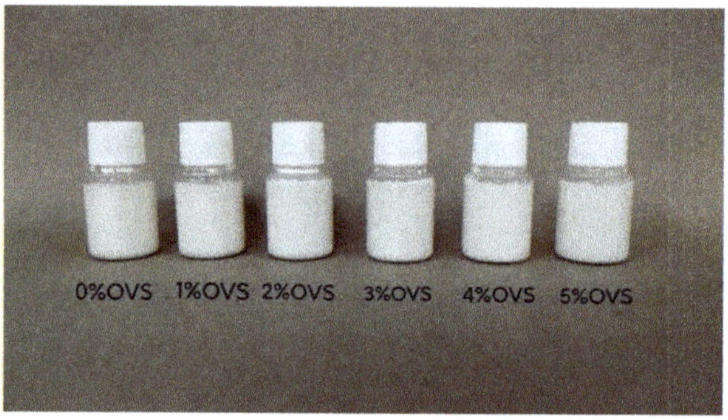

Figure 2. The appearance state of the OVS modified emulsion after 90 days of placement.

Figure 3 shows the particle size distribution of modified aqueous chloroprene latexes with 0%, 2% and 5% OVS content, respectively.

The graph reveals that the particle sizes of the modified chloroprene latexes with 0%, 2%, and 5% OVS are all single-peaked. The average particle size of the 0% OVS-modified waterborne chloroprene latex is 275.9 nm, while that of the 2% OVS-modified latex is 355.6 nm. The particle size of the 5% OVS-modified latex is the largest, at 401.5 nm. As the OVS content in the monomer increases, the average particle size gradually increases. This observation suggests that the modified chloroprene latex has been successfully grafted with MMA and OVS, leading to larger latex particles. Furthermore, there is no phase separation in the modified latex, indicating successful grafting and a more homogeneous emulsion. The increase in particle size can be attributed to the presence of OVS, which acts as a stabilizer and allows for more-controlled particle growth. The larger particle size may also be attributed to the fact that OVS contains vinyl groups that can undergo polymerization reactions, resulting in a more cross-linked and stable latex network. The grafting of MMA and OVS onto the chloroprene rubber backbone not only alters the particle size but also affects other properties of the latex, such as its mechanical properties and thermal stability.

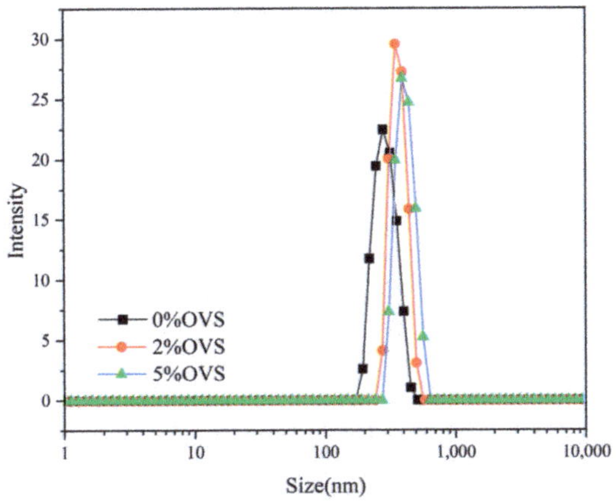

Figure 3. Particle size distribution of modified waterborne chloroprene latex.

Figure 4 presents the potential distribution of the modified waterborne chloroprene latexes containing 0%, 2%, and 5% OVS content, respectively. It can be observed that the potentials of the modified chloroprene latexes with different OVS contents are single-peaked. The potential of the 0% OVS-modified latex is −23.3 mV, while that of the 2% OVS-modified latex is −68.3 mV. The potential of the 5% OVS-modified latex is the lowest, at −71.7 mV. As the OVS content in the monomer increases, the potential of the modified waterborne chloroprene latex gradually decreases. This trend suggests that most of the CRL particles are compatible with OVS and MMA, leading to a more stable colloidal dispersion system. The potential distribution curve is a useful tool for understanding the stability and compatibility of the latex particles. A single-peaked curve indicates that the particles are relatively uniform in size and that the colloidal dispersion system is stable. The shift in potential as the OVS content increases can be attributed to changes in the surface charge and interactions between the particles. The compatibility between CRL particles and OVS/MMA is crucial for the mechanical and physical properties of the final product. A high degree of compatibility ensures that the particles remain well-dispersed and do not aggregate, leading to a more uniform final material. This uniformity in turn affects various properties such as tensile strength, tear resistance, and hardness, among others. Therefore, understanding the potential distribution and compatibility of the latex particles is essential for optimizing the final product's performance.

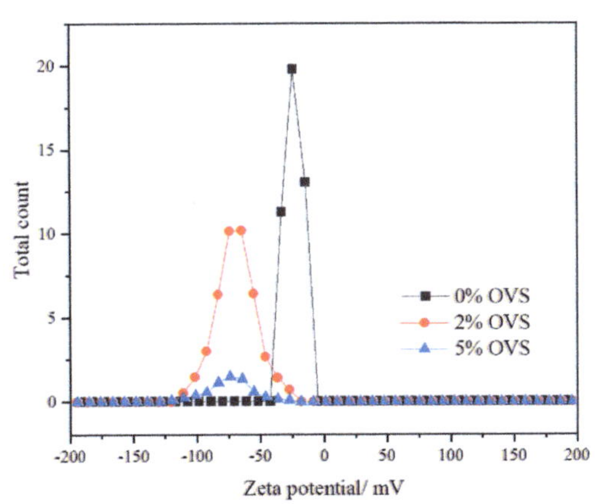

Figure 4. Potential distribution of modified waterborne chloroprene latex.

The concentration of aqueous chloroprene latex was fixed, DSB was added at 1% of the mass of the mixed monomer, TBHP-TEPA was added at 0.5% of the dry weight of CRL, the solid content of the system was controlled at 30%, the polymerization temperature was 50 °C, the concentration of MMA was fixed, and the effect of the variation of vinyl POSS concentration on the polymerization rate was investigated; the results are shown in Figure 5.

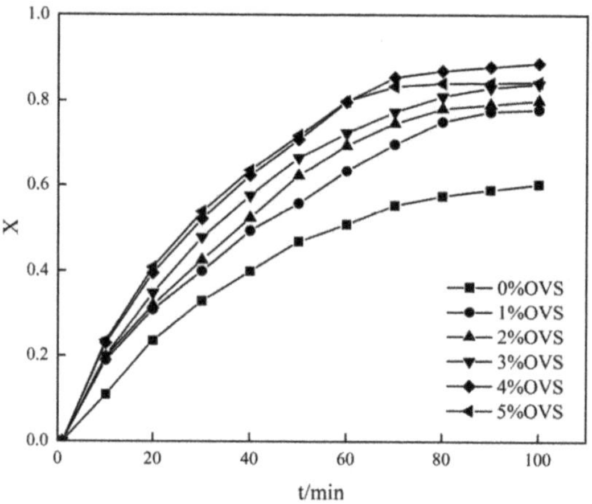

Figure 5. Effect of OVS concentration on polymerization kinetics.

As Figure 5 demonstrates, the polymerization rate of the reaction increased as the concentration of OVS increased. Initially, the polymerization rate gradually increased with the polymerization time, indicating growth in the number of polymer chains. This was followed by a period of constant rate polymerization, where the rate remained relatively constant. However, as the polymerization proceeded further, the polymerization rate gradually decreased. The increase in the polymerization rate with increasing OVS concentration can be attributed to the enhanced initiation and propagation reactions that occur due to the presence of vinyl groups on POSS. The vinyl groups act as reactive centers, facilitating the addition of monomer units to the growing polymer chains. This leads to an increase in the number of active polymer chains and ultimately a higher polymerization rate. The observed decrease in polymerization rate towards the end of the reaction can be explained by factors such as the depletion of monomer, a decrease in the availability of reactive sites on the growing polymer chains, or chain termination reactions that become more prevalent as the polymerization progresses. These factors result in a decrease in the number of active polymer chains and a corresponding reduction in the polymerization rate.

Figure 6 shows a plot of lnRp versus ln[OVS] with a linear regression with a linear slope of 1.05 and a correlation coefficient of 0.99912, which yields the equation for the relationship between the reaction rate and monomer concentration: $Rp \propto [OVS]^{1.05}$, with a correlation coefficient of 0.99912, deviating from the classical kinetic model of emulsion polymerization ($Rp \propto [OVS]^{1}$) with a monomer reaction order greater than 1. It can be correlated that at low conversions, the relationship between reaction rate and reactant concentration is $R_p = K[E]^{0.15}[I]^{0.30}[OVS]^{1.05}$ (K is a constant). The rate equation $Rp = k_p(2kd/kt1)^{0.5}[M][I_2]^{0.5}$ derived from the reaction mechanism is basically correct for the radical reaction mechanism of graft polymerization, but there are deviations in the order of reaction because the actual experimental procedure deviates from the assumptions made during the theoretical derivation.

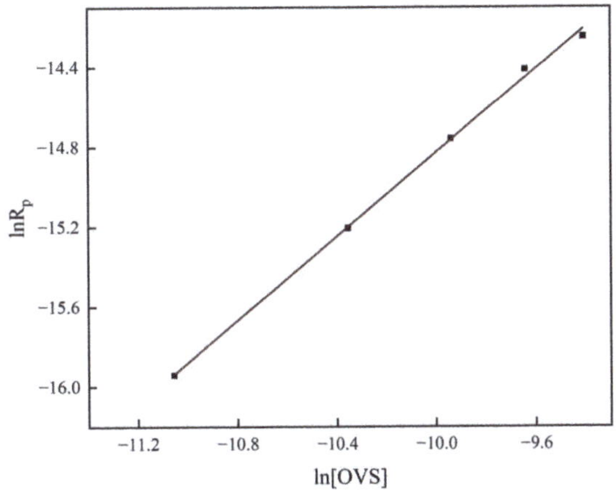

Figure 6. Relationship between lnRp and ln[OVS].

Figure 7 shows a before and after photograph of 0% and 5% OVS modified waterborne chloroprene latex tested for stretching, with no tearing of the substrate after stretching, which indicates that all have good peeling properties. When interface failure occurs between the adhesive and the substrate, the peel strength of the sample modified with 5% OVS is approximately 0.30 kgf/mm.

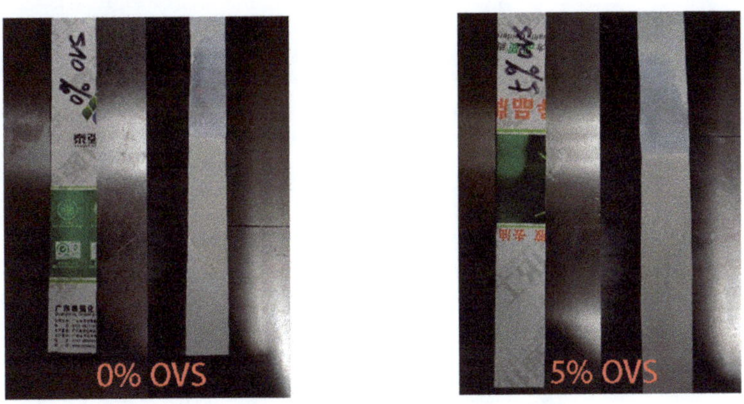

Figure 7. The surface of the adhesive sample before and after stretching.

Figure 8 illustrates that the T-peel strength of modified aqueous chloroprene latex increased and then decreased as the addition of OVS increased, which was due to the increased concentration of OVS, which increased the vinyl involved in the silicone hydrogen addition reaction in the system and increased the crosslinking density of the system, making the crosslinking point dense. The cross-linking reaction makes the molecular chain of chloroprene latex grow, and the flexible molecular chain can disperse the stress well under the action of external force; meanwhile, the cross-linking reaction enhances the force between the monomer and chloroprene latex, which improves the mechanical properties of chloroprene latex. Meanwhile, the cross-linking reaction enhances the force between the monomer and chloroprene latex, which improves the mechanical properties of chloroprene latex; moreover, due to the unique inorganic hollow cage skeleton of OVS, the chloroprene latex enhances the rigidity. When the content of OVS reaches 5%, the monomer is not uniformly dispersed in the system due to the excessive addition of OVS, resulting in agglomeration, which makes the internal bonding of the chloroprene rubber latex poorer, reduces the tensile strength, and, consequently, reduces the mechanical properties. Table S3 shows the mechanical properties of OVS-modified neoprene rubber in detail. From the table, the tensile strength and elongation at break gradually increase with the increase in OVS content. When the OVS content is 5%, the tensile strength is 1.45 MPa and the elongation at break is 523.3%. Compared with the reported mechanical properties of neoprene–montmorillonite nanocomposite, the OVS-modified neoprene rubber improved significantly [49].

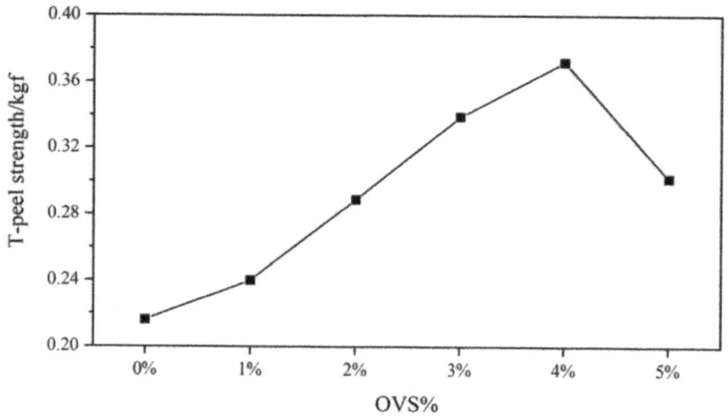

Figure 8. Effect of OVS concentration on T-peel strength.

Figure 9 reveals that the unmodified neoprene latex samples undergo rupture and debonding at the bonding site after stretching. In contrast, Figure 10 demonstrates that the OVS-modified neoprene latex sample strip exhibits breakage upon stretching. Notably, there is no debonding observed at the bonding site after stretching at room temperature. However, there is material breakdown present at other locations, indicating that the bonding performance is still satisfactory at 100 °C. These observations suggest that the addition of OVS enhances the bonding properties of the neoprene latex. The improved bonding performance can be attributed to the presence of vinyl groups on POSS, which facilitate stronger interactions between the polymer chains and the substrate. This leads to a more robust material that is better able to withstand stretching without debonding. Furthermore, the fact that the bonding performance remains good at elevated temperatures highlights the potential of OVS-modified neoprene latexes for applications where thermal stability is essential, such as in high-temperature seals or gaskets.

Figure 9. Tensile fracture cross-sectional morphology of unmodified neoprene latex specimens (fracture at the sample bond).

Figure 10. Tensile fracture of modified chloroprene latex samples at high temperatures (modified samples did not break from the bond before and after 100 °C).

After the tensile property test, the tensile fracture of sponge bonded with 4% OVS modified chloroprene latex was 16.81 kgf before high temperature treatment and 20.20 kgf after high temperature treatment. The peel strength of OVS modified chloroprene latex after high temperature treatment was larger than that before high temperature treatment, and both of them broke the material, indicating that the addition of OVS improved the heat resistance of water-based chloroprene latex.

Figure 11 displays the DSC curves of the modified waterborne chloroprene latexes containing 0%, 1%, and 5% OVS content, respectively. As observed in the figure, the glass transition temperature (T_g) of the 0% OVS-modified waterborne neoprene latex adhesive is −43.58 °C. For the 1% OVS-modified adhesive, the T_g is −42.50 °C, and for the 5% OVS-modified adhesive, the Tg is −41.55 °C. These results indicate that the addition of OVS increases the T_g of the adhesive. The T_g is influenced by various factors, including the mobility of both the main chain and side groups of the macromolecules. The observed increase in T_g suggests that the addition of OVS impairs the movement of chloroprene rubber macromolecular chain segments. This is likely due to the organosilicone resin present in OVS, which acts to hinder the flexibility of the molecular chain segments and increase their rigidity. With increasing OVS content, there is a corresponding increase in cross-linking within the molecular chain. This cross-linking leads to an increase in chain rigidity and a corresponding rise in the glass transition temperature. The more OVS that was added, the more severe the cross-linking becomes, leading to a further increase in chain rigidity and T_g.

Figures 12 and 13 show the TG curves of the modified waterborne chloroprene latexes with 0% and 5% OVS content, respectively, while the temperature of maximum mass loss rate (T_{max}) and char yield (Y_c) are summarized in Table S4.

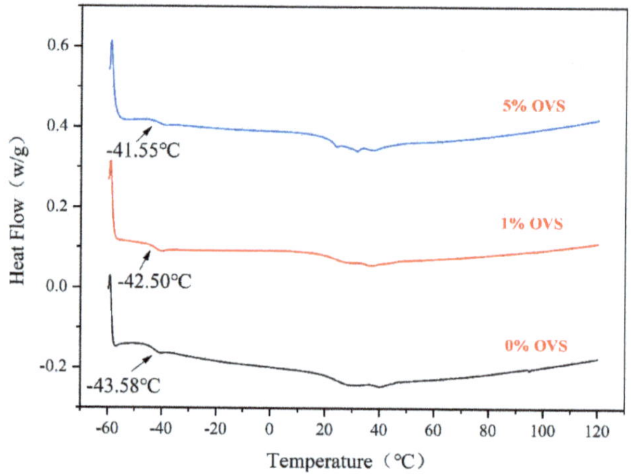

Figure 11. DSC curves of modified waterborne chloroprene latex with 0% OVS, 1% OVS and 5% OVS.

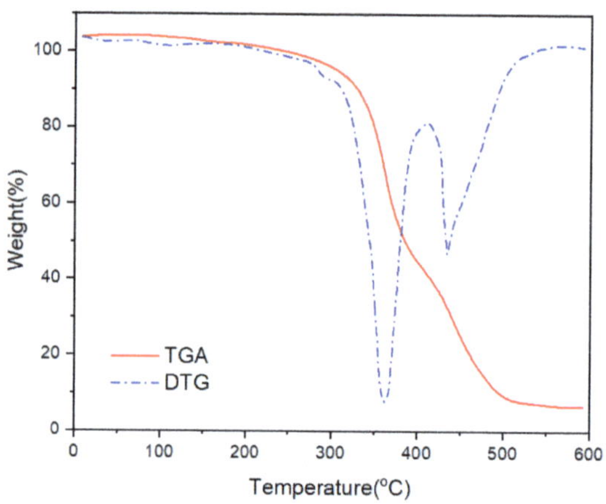

Figure 12. TGA and DTG curves of 0% OVS modified waterborne chloroprene latex.

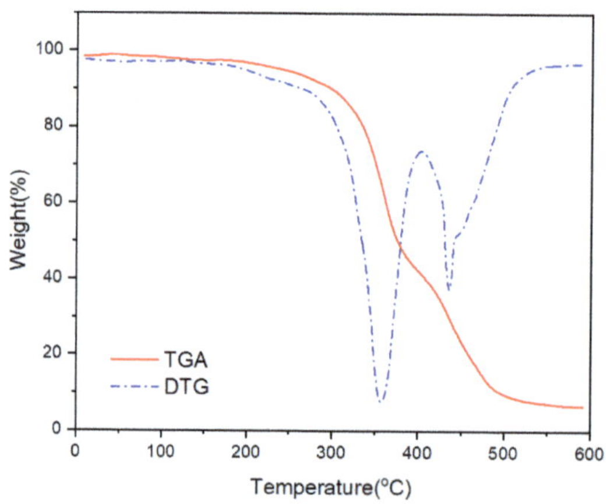

Figure 13. TGA and DTG curves of 5% OVS modified waterborne chloroprene latex.

The weight loss of all coatings was approximately 1 wt% at approximately 100 °C, probably due to the small amounts of volatile solvents and water contained in the system. The first decomposition stage occurs at 250~420 °C, mainly due to C-C and C-H degradation of the main chain. The second decomposition stage occurs in the temperature range of 350~500 °C, which may be related to the ester bonds in the structure and the C-Cl degradation structure. We found that the lowest T_{max1} was 372.6 °C and the highest T_{max2} was 450.4 °C when the OVS content was 5%, which was mainly due to the grafting of vinyl-POSS onto the chloroprene latex backbone, which increased the initial decomposition temperature of the backbone polymer but decreased the final decomposition temperature. The increase in initial decomposition temperature was attributed to the vinyl-POSS branched grafting increasing the interchain interaction force and decreasing the degree of crystallinity, thus increasing the heat resistance. The addition of vinyl-POSS improved the thermal stability of the neoprene adhesives. This is due to the formation of a three-dimensional network of the system when vinyl-POSS is added, which reduces the flexibility of the polymer chains and delays the final decomposition temperature. This is also confirmed by Table S4. The residual carbon content based on OVS-0% and OVS-5% at 600 °C is 18.58 wt% and 21.64 wt%, respectively. This is mainly due to the introduction of siloxanes with alkoxyl groups into the main chain of the neoprene polymers, resulting in a three-dimensional network crosslinked structure. Therefore, the addition of OVS improves the thermal stability of the bonding layer. Compared with the chemically modified neoprene rubber with cashew phenol grafting reported in the literature, the OVS-modified neoprene studied in this paper has a significantly higher residual carbon rate at 600 °C because of the structure of the silicone-oxygen network in the OVS, so it has a higher thermal stability and a higher T_g [50].

4. Conclusions

(1) Waterborne chloroprene latex modified with vinyl-POSS was prepared by emulsion polymerization using a redox initiator (TBHP/TEPA 0.5% of dry weight), an emulsifier (DSB 1% of total monomer) at a polymerization temperature of 50 °C, with a solid content of 30% controlled using deionized water and dropwise addition of the monomer methyl methacrylate and vinyl-POSS. The stability of the prepared aqueous chloroprene latex was good, with a small amount of gel appearing at 4% and 5% content for polymerization, and the best monomer conversion at 4% vinyl-POSS content. Infrared spectroscopic analysis demonstrated that MMA and OVS were grafted onto the chloroprene latex.

(2) The OVS grafting CRL kinetic study showed that the relationship between the monomer reaction rate and reactant concentration was $R_p = K[E]^{0.15}[I]^{0.30}[OVS]^{1.05}$ (K is a constant); the reaction mechanism deduced the rate equation $R_p = k_p(2k_d/k_{t1})^{0.5}[M][I_2]^{0.5}$ and the free radical reaction mechanism of grafting polymerization is basically correct, but there is a deviation in the order of reaction because the actual experimental procedure deviates from the assumptions made in the theoretical derivation.

(3) As the addition of vinyl-POSS increased, the tensile strength of the chloroprene latex first increased and then decreased, and the thermal stability of the chloroprene latex increased.

Supplementary Materials: The following supporting information can be downloaded at: https://www.mdpi.com/article/10.3390/polym16040462/s1, Figure S1: The TEM photograph of 5% OVS modified aqueous chloroprene latexes; Table S1: Physical properties of OVS modified waterborne neoprene emulsion; Table S2: Stability testing of OVS-modified neoprene emulsions; Table S3: The corresponding detailed data obtained from tensile measurements; Table S4: Thermal stability of OVS modified waterborne neoprene.

Author Contributions: Conceptualization, J.C. (Junhua Chen) and D.-Y.Z.; Methodology, J.C. (Junhua Chen) and Y.W.; Software, H.Z.; Validation, H.Z.; Formal analysis, Z.W. and Q.W.; Investigation, Z.W. and Q.W.; Resources, D.-Y.Z.; Data curation, C.Y. and J.C. (Jinlian Chen); Writing—original draft, J.C. (Junhua Chen); Writing—review & editing, X.H.; Visualization, J.C. (Junhua Chen), Y.W. and D.-Y.Z.; Supervision, X.H.; Project administration, X.H.; Funding acquisition, D.-Y.Z. All authors have read and agreed to the published version of the manuscript.

Funding: This study was financially supported by Youth Enhancement Project of the Natural Science Foundation of Guangdong Province (Dong-Yu Zhu, 202341515030135) and College Students' Innovative Entrepreneurial Training Plan Program (No. S202310580062).

Institutional Review Board Statement: Not applicable.

Data Availability Statement: Data are contained within the article and supplementary materials.

Conflicts of Interest: The authors declare no conflicts of interest.

References

1. Andersen, D.E.; Arnold, R.G. Aging Stability of Neoprene Latex. *Ind. Eng. Chem.* **1953**, *45*, 2727–2730. [CrossRef]
2. Sauterey, F. Production of Polychloroprene Latexes. U.S. Patent 5,322,886, 21 June 1994.
3. Maron, S.H.; Levy-Pascal, A.E. Rheology of synthetic latex: VI. The flow behavior of neoprene latex. *J. Colloid Sci.* **1955**, *10*, 494–503. [CrossRef]
4. Stewart, C.A., Jr.; Staff, U.B. Chloroprene. *Kirk-Othmer Encycl. Chem. Technol.* **2000**, 1–9. [CrossRef]
5. Yashima, H.; Mochizuki, K. Polychloroprene Latex, Process for the Production Thereof and Aqueous Adhesive Compositions. U.S. Patent 7,514,487, 7 April 2009.
6. Ozoe, S. Polychloroprene-Base Latex and Method for Producing it. U.S. Patent Application 11/946,201, 29 May 2008.
7. Guggenberger, S.K. Neoprene (polychloroprene)-based solvent and latex adhesives. In *Handbook of Adhesives*; Springer: Berlin/Heidelberg, Germany, 1990; pp. 284–306.
8. Minorikawa, N.; Takenoshita, Y.; Suzuki, E. Chloroprene Polymer Latex Composition and Use Thereof. U.S. Patent Application No. 13/512,540, 20 September 2012.
9. Ohnishi, T. Chloroprene Rubber-Based Polymer Latex Composition and Use Thereof. U.S. Patent 9,023,948, 5 May 2015.
10. Christell, L.A. High-Performance Water-Based Chloroprene Polymer Adhesive Composition. U.S. Patent No. 5,332,771, 26 July 1994.
11. Onozuka, M.; Kiyofuji, G.; Hagiwara, S. Chloroprene Rubber Latex Adhesive Composition. U.S. Patent No. 11,479,698, 25 October 2022.
12. Oba, H.; Tsuji, T.; Yamaguchi, M. Chloroprene Latex Composition, Process for Producing the Same, and Adhesive Composition Comprising the Same. U.S. Patent 6,525,132, 25 February 2003.
13. Yashima, H.; Kishi, S.; Otsuka, K.; Ishigaki, Y. Chloroprene Rubber Composition and Adhesive Composition Using Said Chloroprene Rubber Composition. U.S. Patent 9,328,228, 3 May 2016.
14. Zhang, K.; Huang, C.B.; Shen, H.F.; Chen, H.Q. Modification of Polychloroprene Rubber Latex by Grafting Polymerization and Its Application as a Waterborne Contact Adhesive. *J. Adhes.* **2012**, *88*, 119–133. [CrossRef]
15. Kobayashi, N.; Konno, M.; Nagao, D.; Mashiko, Y.; Otsu, T. Method of Producing a Chloroprene-Based Polymer, Polychloroprene Latex and Adhesive Composition. U.S. Patent No. 9,493,683, 15 November 2016.
16. Žukienė, K.; Jankauskaitė, V. The Effect of Surface Properties on the Adhesion of Modified Polychloroprene Used as Adhesive. *J. Adhes. Sci. Technol.* **2005**, *19*, 627–638. [CrossRef]
17. Zhang, K.; Huang, C.; Shen, H.; Chen, H. Grafting of methyl methacrylate and styrene onto polychloroprene latex for compatibilization of polychloroprene latex/styrene-acrylate emulsion blends. *J. Adhes.* **2015**, *91*, 419–433. [CrossRef]
18. Zhang, K.; Shen, H.; Zhang, X.; Lan, R.H.; Chen, H.Q. Preparation and properties of a waterborne contact adhesive based on polychloroprene latex and styrene-acrylate emulsion blend. *J. Adhes. Sci. Technol.* **2009**, *23*, 163–175. [CrossRef]
19. Soriano-Luna, M.; Núñez-Pérez, H.I.; Estrada, M.R. On the swelling of polychloroprene-MMT nanocomposite films. *J. Appl. Polym. Sci.* **2012**, *125*, 2986–2992. [CrossRef]
20. Rattanasom, N.; Kueseng, P.; Deeprasertkul, C. Improvement of the mechanical and thermal properties of silica-filled polychloroprene vulcanizates prepared from latex system. *J. Appl. Polym. Sci.* **2012**, *124*, 2657–2668. [CrossRef]
21. Salleh, S.; Ismail, H.; Ahmad, Z. Properties of natural rubber latex-compatibilized natural rubber/recycled chloroprene rubber blends. *J. Elastomers Plast.* **2016**, *48*, 640–655. [CrossRef]
22. Aneeq, F.; Shahid, M.; Saleem, M.; Awais, M.; Butt, M.S.; Adeel Umer, M. Enhancement in bonding strength and ageing resistance of polychloroprene solvent-base adhesives through graft copolymerization. *J. Adhes. Sci. Technol.* **2021**, *35*, 1752–1763. [CrossRef]
23. Iqbal, T.; Noon, Z.; Yasin, S.; Qaiser, A.A.; Ahmad, M.M. Temperature effects on grafting reaction to produce methyl methacrylate grafted polychloroprene adhesive. *J. Pak. Inst. Chem. Eng.* **2011**, *39*, 75–80.
24. Hussain, H.; Shah, S.M. Recent developments in nanostructured polyhedral oligomeric silsesquioxane-based materials via "controlled" radical polymerization. *Polym. Int.* **2014**, *63*, 835–847. [CrossRef]

25. Lin, X.; Deng, Y.Y.; Zhang, Q.; Han, D.; Fu, Q. Effect of POSS size on the porosity and adsorption performance of hybrid porous polymers. *Macromolecules* **2023**, *56*, 1243–1252. [CrossRef]
26. Zhang, K.; Li, B.; Zhao, Y.; Li, H.; Yuan, X. Functional POSS-containing polymers and their applications. *Prog. Chem.* **2014**, *26*, 394.
27. Chen, D.; Nie, J.; Yi, S.; Wu, W.; Zhong, Y.; Liao, J.; Huang, C. Thermal behaviour and mechanical properties of novel RTV silicone rubbers using divinyl-hexa [(trimethoxysilyl) ethyl]-POSS as cross-linker. *Polym. Degrad. Stab.* **2010**, *95*, 618–626. [CrossRef]
28. Calabrese, C.; Aprile, C.; Gruttadauria, M.; Giacalone, F. POSS nanostructures in catalysis. *Catal. Sci. Technol.* **2020**, *10*, 7415–7447. [CrossRef]
29. Imoto, H.; Sato, Y.; Naka, K. Polymers and cyclic compounds based on a side-opening type cage silsesquioxane. *J. Polym. Sci. Part A Polym. Chem.* **2019**, *57*, 2243–2250. [CrossRef]
30. Karuppasamy, K.; Prasanna, K.; Vikraman, D.; Kim, H.S.; Kathalingam, A.; Mitu, L.; Rhee, H.W. A rapid one-pot synthesis of novel high-purity methacrylic phosphonic acid (PA)-based polyhedral oligomeric silsesquioxane (POSS) frameworks via thiol-ene click reaction. *Polymers* **2017**, *9*, 192. [CrossRef] [PubMed]
31. Liu, Z.; Ma, S.; Chen, L.; Xu, J.; Ou, J.; Ye, M. Porous styryl-linked polyhedral oligomeric silsesquioxane (POSS) polymers used as a support for platinum catalysts. *Mater. Chem. Front.* **2019**, *3*, 851–859. [CrossRef]
32. Chen, D.; Yi, S.; Wu, W.; Zhong, Y.; Liao, J.; Huang, C.; Shi, W. Synthesis and characterization of novel room temperature vulcanized (RTV) silicone rubbers using Vinyl-POSS derivatives as cross linking agents. *Polymer* **2010**, *51*, 3867–3878. [CrossRef]
33. Zhao, W.; Dong, Z.; Zhao, L. Radiation synthesis of polyhedral oligomeric silsesquioxanes (POSS) gel polymers. *Radiat. Phys. Chem.* **2022**, *198*, 110251. [CrossRef]
34. Wang, Y.; Liu, F.; Xue, X. Synthesis and characterization of UV-cured epoxy acrylate/POSS nanocomposites. *Prog. Org. Coat.* **2013**, *76*, 863–869. [CrossRef]
35. Zhou, H.; Ye, Q.; Xu, J. Polyhedral oligomeric silsesquioxane-based hybrid materials and their applications. *Mater. Chem. Front.* **2017**, *1*, 212–230. [CrossRef]
36. Kausar, A. State-of-the-Art overview on polymer/POSS nanocomposite. *Polym.-Plast. Technol. Eng.* **2017**, *56*, 1401–1420. [CrossRef]
37. Pielichowski, K.; Njuguna, J.; Janowski, B.; Pielichowski, J. Polyhedral oligomeric silsesquioxanes (POSS)-containing nanohybrid polymers. *Supramol. Polym. Polym. Betains Oligomers* **2006**, 225–296. [CrossRef]
38. Ye, Q.; Zhou, H.; Xu, J. Cubic polyhedral oligomeric silsesquioxane based functional materials: Synthesis, assembly, and applications. *Chem. Asian J.* **2016**, *11*, 1322–1337. [CrossRef] [PubMed]
39. Kuo, S.W.; Chang, F.C. POSS related polymer nanocomposites. *Prog. Polym. Sci.* **2011**, *36*, 1649–1696. [CrossRef]
40. Liao, W.; Zou, L.; Zheng, S.; Zhao, L.; Huang, X.; Ye, L.; Zhong, G. Effects of colloidal silica on the properties of POSS-containing fluorinated poly (styrene–acrylate)/SiO_2 composite materials. *J. Coat. Technol. Res.* **2021**, *18*, 107–116. [CrossRef]
41. Lichtenhan, J.D.; Pielichowski, K.; Blanco, I. POSS-based polymers. *Polymers* **2019**, *11*, 1727. [CrossRef] [PubMed]
42. So, J.I.; Kim, J.B.; Jeong, H.W.; Kim, C.H.; Choi, J.; Shim, S.E.; Qian, Y. One-pot synthesis of bifunctional polyhedral oligomeric silsesquioxane: Full spectrum ratio of vinyl groups from 0 to 100%. *J. Ind. Eng. Chem.* **2022**, *113*, 502–512.
43. Zhang, L.; Abbenhuis, H.C.L.; Yang, Q.; Wang, Y.-M.; Magusin, P.C.M.M.; Mezari, B.; van Santen, R.A.; Li, C. Mesoporous organic–inorganic hybrid materials built using polyhedral oligomeric silsesquioxane blocks. *Angew. Chem.* **2007**, *119*, 5091–5094. [CrossRef]
44. Auner, N.; Weis, J. *Organosilicon Chemistry III: From Molecules to Materials*; John Wiley&Sons: Hoboken, NJ, USA, 2008.
45. Wang, X.; Chin, J.M.; He, C.; Xu, J. Highly thermally resistant polyhedral oligomeric silsesquioxanes lubricating oil prepared via a thiol-ene click reaction. *Sci. Adv. Mater.* **2014**, *6*, 1553–1561. [CrossRef]
46. *GB/T 2794-2022*; Determination of Viscosity of Adhesives, Single Cylinder Rotational Viscometer Method. Standardization Administration of China (SAC): Beijing, China, 2022.
47. *GB/T 2790-1995;* Test Method for 180° Peel Strength of Adhesives Flexure to Rigidity. Standardization Administration of China (SAC): Beijing, China, 1995.
48. *GB/T 2909-2014;* Cotton Grey Canvas for Rubber Industry. Standardization Administration of China (SAC): Beijing, China, 2015.
49. Yeh, M.H.; Hwang, W.S.; Cheng, L.R. Microstructure and mechanical proper-ties of neoprene–montmorillonite nanocomposites. *Appl. Surf. Sci.* **2007**, *253*, 4777–4781. [CrossRef]
50. Zhang, W.; Zhang, T.; Jiang, N.; Zhang, T. Chemical modification of neoprene rubber by grafting cardanol, a versatile renewable material from cashew industry. *J. Polym. Res.* **2020**, *27*, 1–7. [CrossRef]

Disclaimer/Publisher's Note: The statements, opinions and data contained in all publications are solely those of the individual author(s) and contributor(s) and not of MDPI and/or the editor(s). MDPI and/or the editor(s) disclaim responsibility for any injury to people or property resulting from any ideas, methods, instructions or products referred to in the content.

polymers

MDPI

Article

Synthesis and Characterization of Pressure-Sensitive Adhesives Based on a Naphthyl Curing Agent

Junhua Chen [1,2], Shiting Li [1], Xuan Wang [1], Lili Fang [1], Dingding Huang [1], Lin Ke [1], Jinlian Chen [1], Qingwei Wang [1], He Zhang [1,2], Yinping Wu [1,2], Dongyu Zhu [3], Chunsheng Li [1,2,*] and Xiangying Hao [1,2,*]

[1] School of Environmental and Chemical Engineering, Zhaoqing University, Zhaoqing 526061, China; cehjchen@yeah.net (J.C.); lishitingqc@163.com (S.L.); pearships@163.com (X.W.); lily_ffa@163.com (L.F.); huangdingding1116@163.com (D.H.); kelin0515@163.com (L.K.); cjlccc2003@outlook.com (J.C.); wangqingwei0721@163.com (Q.W.); zhanghe@zqu.edu.cn (H.Z.); wuyinping@zqu.edu.cn (Y.W.)

[2] Guangdong Provincial Key Laboratory of Environmental Health and Land Resource, College of Environmental and Chemical Engineering, Zhaoqing University, Zhaoqing 526061, China

[3] School of Chemical Engineering and Light Industry, Guangdong University of Technology, Guangzhou 510006, China; zdy16@gdut.edu.cn

* Correspondence: lichunsheng@zqu.edu.cn (C.L.); xyinghao@zqu.edu.cn (X.H.)

Abstract: The incorporation of a naphthyl curing agent (NCA) can enhance the thermal stability of pressure-sensitive adhesives (PSAs). In this study, a PSA matrix was synthesized using a solution polymerization process and consisted of butyl acrylate, acrylic acid, and an ethyl acrylate within an acrylic copolymer. Benzoyl peroxide was used as an initiator during the synthesis. To facilitate the UV curing of the solvent-borne PSAs, glycidyl methacrylate was added to introduce unsaturated carbon double bonds. The resulting UV-curable acrylic PSA tapes exhibited longer holding times at high temperatures (150 °C) compared to uncross-linked PSA tapes, without leaving any residues on the substrate surface. The thermal stability of the PSA was further enhanced by adding more NCA and increasing the UV dosage. This may be attributed to the formation of cross-linking networks within the polymer matrix at higher doses. The researchers successfully balanced the adhesion performance and thermal stability by modifying the amount of NCA and UV radiation, despite the peel strength declining and the holding duration shortening. This research also investigated the effects of cross-linking density on gel content, molecular weight, glass transition temperature, and other properties of the PSAs.

Keywords: pressure-sensitive adhesives; thermal analysis; glass transition temperature; UV curing

check for updates

Citation: Chen, J.; Li, S.; Wang, X.; Fang, L.; Huang, D.; Ke, L.; Chen, J.; Wang, Q.; Zhang, H.; Wu, Y.; et al. Synthesis and Characterization of Pressure-Sensitive Adhesives Based on a Naphthyl Curing Agent. *Polymers* **2023**, *15*, 4516. https://doi.org/10.3390/polym15234516

Academic Editor: Antonio Pizzi

Received: 10 October 2023
Revised: 15 November 2023
Accepted: 18 November 2023
Published: 24 November 2023

Copyright: © 2023 by the authors. Licensee MDPI, Basel, Switzerland. This article is an open access article distributed under the terms and conditions of the Creative Commons Attribution (CC BY) license (https://creativecommons.org/licenses/by/4.0/).

1. Introduction

Pressure-sensitive adhesives (PSAs) are capable of adhering to solid substrates quickly and with minimal pressure, and they can be easily removed without leaving any visible residue [1–3]. Due to their viscoelastic nature, PSAs exhibit semi-solid properties that can be evaluated through measurements of tack, peel resistance, and shear strength [4]. Acrylic PSAs are commonly employed in various industrial products, including mounting tapes, packing tapes, labels, and protective films [5]. One specific application of PSAs involves their use as a protective coating for electronic chemical materials [6–8]. The protective PSA film serves to safeguard the electronic components from external factors. The rapid advancement of integrated circuit technology necessitates the incorporation of thermal stability and enhanced reliability in the protective PSA films for microdevices [9].

The primary constituents employed in the production of PSAs through solution and emulsion polymerization methods are commonly acrylic ester monomers. Nevertheless, the inherent linear structures of the polymer constituents in synthetic PSAs impose constraints on their tensile and thermal properties [10]. In order to enhance the thermal resilience of PSAs, it becomes imperative to restrict the mobility of polymeric chains. In general, modulating the degree of cross-linking in PSAs represents a highly effective approach for minimizing molecular motion. UV curing is a widely employed cross-linking method for PSAs due to its numerous advantages, including environmentally friendly emissions, reduced power consumption, and accelerated curing speed [11–14]. The UV-curable systems used to cross-link acrylic PSAs are made up of a photoinitiator and unit with unsaturated carbon double bonds that help with subsequent cross-linking [9,15,16].

The utilization of the aromatic compound containing naphthyl groups in materials is facilitated by its thermal resistance and protective packaging capabilities [17]. This compound exhibits a higher glass transition temperature (T_g) and enhanced thermal stability. Previous studies have demonstrated that the introduction of rigid structures, such as the naphthyl group, into the molecular chains can effectively enhance thermal stability [9,18–20]. The UV-curable, acrylic pressure-sensitive adhesives (PSAs) possess a cross-linked network structure comprising acrylic polymer links and a curing-agent component [21]. The adhesive properties of the acrylic component are closely associated with its flexibility. Furthermore, it has been observed that the molecular structure of UV-curable, acrylic PSA is not affected by the curing agent [7,22]. Therefore, incorporating the naphthyl group into the polymeric chains is a viable method for obtaining a thermally stable structure. This modification results in a higher T_g, better thermal degradation and flame retardancy, and increased carbon residue compared to acrylate PSAs. Despite the benefits observed, there has been a lack of reports on the utilization of the naphthyl group for enhancing the thermal stability of PSAs.

Thermal stability is a key property of PSAs in electronic chemical materials such as packaging and protection. To improve the thermal stability of adhesion, PSAs were synthesized using an NCA in this study. The UV-curable, acrylic PSAs were then cross-linked to enhance their mechanical, adhesive, and thermal properties. The impact of the amount of NCA and UV radiation on these properties were carefully analyzed. These results demonstrate the potential of an NCA and UV irradiation in enhancing the thermal stability of PSAs, a crucial factor in electronic applications. This research is highly relevant to peers and researchers working in related fields.

2. Materials and Methods

2.1. Materials

The monomers of n-butyl acrylate (BA), acrylic acid (AA), and vinyl acetate (vAc) of chemical purity were procured from Shanghai Aladdin Reagents Co., Ltd. Benzoyl peroxide (BPO, industrial grade, Shanghai Aladdin Reagents Co., Ltd., Shanghai, China) and ethyl acetate (EAC, industrial grade, Guangzhou Chemical Reagent Factory, Guangzhou, China) were utilized as the radical initiator and solvent in radical polymerization. The acrylic copolymer was modified using glycidyl methacrylate (GMA, industrial grade, Shanghai Aladdin Reagents Co., Ltd., Shanghai, China), and the reactive vinyl group was retained. The naphthyl curing agent was synthesized using 1,5-dihydroxynaphthalene (DHNA, industrial grade), 4-dimethylaminopyridine (DMAP), and acryloyl chloride (chemical pure), which were all procured from Shanghai Aladdin Reagents Co., Ltd. 1-Hydroxycyclohexyl-phenyl ketone (CP-4, industrial grade, Tianjin Damao Chemical Reagent Co., Ltd., Tianjin, China) was employed as a photoinitiator. All the other reagents underwent further purification before use.

2.2. Preparation of Acrylic Copolymer

A polymerization process was performed to produce an acrylic copolymer containing butyl acrylate (BA, 80 g), vinyl acetate (Vac, 15 g), and acrylic acid (AA, 5 g). The monomers were combined with 2 g of benzoyl peroxide (BPO) in 185 mL of ethyl acetate. The solvent-borne acrylic copolymer was synthesized in a 500 mL flask equipped with a condenser, mechanical stirrer, and dropping funnel, with the polymerization carried out at 80 °C for a dosage time of 3 h, followed by post-reaction at 80 °C for an additional 3 h. The acrylic resin was subsequently reacted with glycidyl methacrylate (GMA) at a temperature of 50 °C for a period of 12 h, resulting in the synthesis of UV-curable acrylic copolymer.

The number-average molecular weight of the acrylic copolymer using gel permeation chromatography was determined to be 1.0 × 10^5, with a polydispersity value of 3.23.

2.3. Synthesis of the NCA

In our laboratory, the NCA was synthesized through a one-step process as depicted in Scheme 1. The procedure involved adding 20 g of DHNA, 40 mL of triethylamine, 0.5 g of 4-dimethylamineopyridine, 0.5 g of hydroquinone, and 100 mL of tetrahydrofuran into a 250 mL single-neck, round-bottomed flask which was equipped with a thermometer, stirrer, and dropping funnel. Meanwhile, acryloyl chloride (38 mL) dissolved in tetrahydrofuran (50 mL) was slowly added while stirring, maintaining the temperature at 0–5 °C. The mixture was then stirred for 4 h at room temperature and subsequently filtered. The undesired by-products were eliminated by treating the reaction mixture with a 10% dilute hydrochloric acid solution, followed by 200 mL of distilled water, and a 5% potassium carbonate solution. The crude product was filtered, washed with ethanol, and dried. Ultimately, a product with a yield of 53% was obtained after the filtration and removal of ethanol [23].

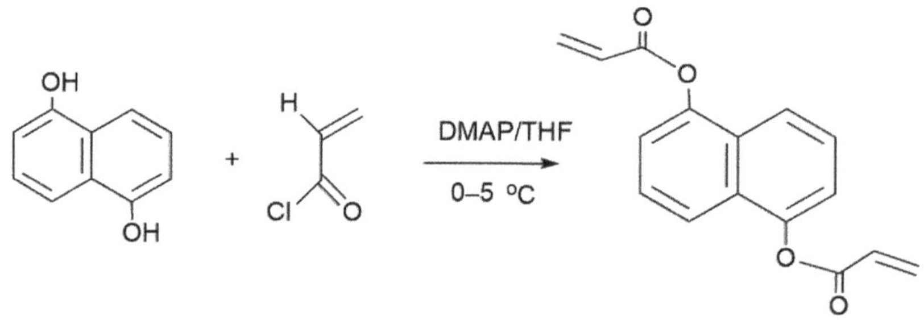

Scheme 1. Synthesis of the NCA.

2.4. Preparation of the UV-Curable, Acrylic PSA Film

The acrylic copolymer solution was supplemented with an NCA of 0–20 phr to PSA and CP-4 (5 wt% to PSA). The UV-curable, acrylic PSA was prepared by mechanically stirring the solution at room temperature. The process for creating the UV-curable, acrylic PSA resin is displayed in Schemes 2 and 3. The resulting mixture was applied onto a 90 μm thick poly(ethylene terephthalate) (PET) film using an applicator roll, ensuring a uniform coating thickness. The coated surface was then dried in an air drying oven at 120 °C for 5 min to eliminate the solvent. As a result, the acrylic PSA layer had a thickness of approximately 20 ± 3 μm using a coating thickness gauge. Finally, the acrylic PSA films were cross-linked through UV-irradiation, utilizing commercially available UV lamps.

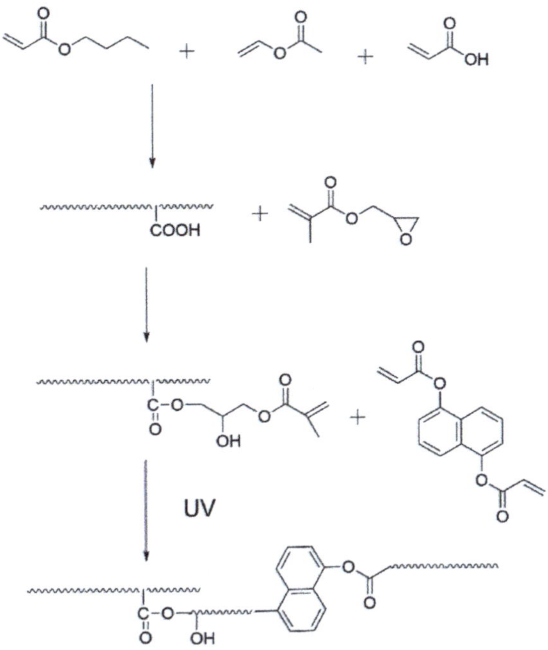

Scheme 2. Preparation of UV-curable, acrylic PSAs.

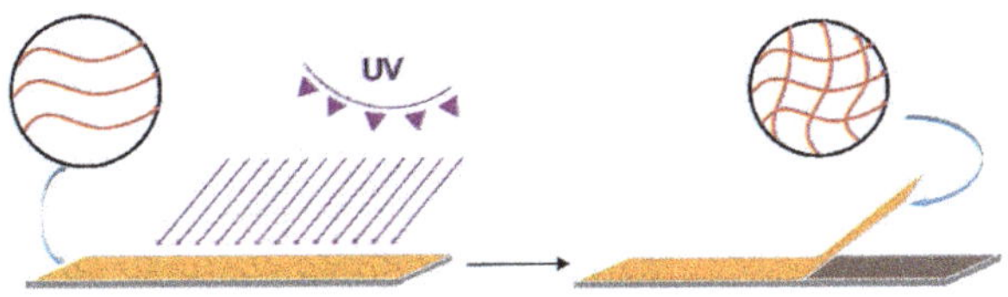

Scheme 3. Illustration of the UV-curable PSAs.

2.5. Characterization

We utilized an FT-IR analyzer (MAGNA-IR750, Nicolet, Madison, WI, USA) to acquire the Fourier-transform infrared (FT-IR) spectra of the NCA and the resulting reaction products within the frequency range of 4000–400 cm^{-1}. Meanwhile, ^{1}H-NMR of the NCA was captured on a Bruker AVANCE AV400 analyzer (Romanshorn, Switzerland) at room temperature, utilizing $CDCl_3$ as a solvent and TMS as an internal reference. After undergoing the cross-linking and UV-curing processes, the PSA polymers were synthesized. Before immersing in ethyl acetate at 60 °C for 72 h, the gel contents of the PSA samples were estimated. Subsequently, a 200-mesh-wire filtration was performed on the samples, followed by drying until a consistent weight was attained. The gel concentration of the UV-curable, acrylic PSA samples was determined utilizing the subsequent algorithm (1):

$$Gel\ content\ (\%) = (W_t/W_0) \times 100 \quad (1)$$

where W_t is the weight of the dried sample, and W_0 is the initial weight of the sample.

The molecular weights of the sticky polymer solutions were measured using gel permeation chromatography (Malvern, UK). To dissolve a small amount of dry adhesive, tetrahydrofuran (THF) was utilized, and the mixture was shaken for 24 h. The resulting solution was then purified using a 0.2 μm polytetrafluoroethylene filter. The injection volume was 0.1 mL, and the mobile-phase flow rate was 1 mL/min. Glass transition temperatures (T_g) were measured using a DSC 200 F3 instrument. Desiccated samples weighing 5 mg were packed into metal sample pans. The samples were cooled to −60 °C for 3 min before being heated to 120 °C at a rate of 10 °C/min.

Based on ASTM D3121-17 (Standard Test Method for Tack of Pressure-Sensitive Adhesives by Rolling Ball), PSA tapes with a cross-section of 100 mm × 100 mm were tested for tack using a tester (TSKL-CNS). The adhesive was applied to stainless-steel pans meticulously wiped with tissue and measured 25 mm wide. The maximum number of sizable spheres adhering to the adhesive during the descent of varied-size spheres over a sloping surface underneath it was quantified. The 180° peel strength of PSA tape was tested according to ASTM D1000-66. The adhesive tapes underwent reciprocation thrice by a traditional roller. Electronic tensile testing equipment (MIDEL KJ-1065A, Dongguan Kejian Testing Instrument Co., Ltd., Dongguan, China) operating at a velocity of 300 mm/min gauged the 180° peel strength post 30 min. The cohesive strength of PSA tapes was tested using a lasting adhesive tester (MIDELKJ-6032, Dongguan Kejian Testing Instrument Co., Ltd., Dongguan, China) according to ASTMD 3653. The stainless-steel carriers were bedded with standardized tapes measuring 25 mm × 70 mm. A customary roller facilitated reciprocation thrice for enhanced surface interaction between the PSA adhesive and the substratum. The specimens were subsequently elevated upright within the testing apparatus one hour later with a mass load of 2 kg. The pace at which the adhesive detached from the steel sheets was employed to compute the shear anchoring capacity. The temperature holding force tester (MIDEL KJ-6012, Dongguan Kejian Testing Instrument Co., Ltd., Dongguan, China) was set to 100 °C and 150 °C, which was the same as the cohesive strength of the PSA tapes, so as to evaluate the high-temperature shear resistance of the pressure-sensitive adhesive.

A Perkin-Elmer Pyris-1 TG Analyzer (Waltham, WI, USA) was leveraged to assess the thermal stability of the PSA specimens. Under nitrogen conditions, the samples underwent heating from 25 °C to 600 °C at a gradient of 10 °C/min. For another 85 min, the samples were maintained at 300 °C inside a nitrogen environment to check their isothermal durability. Cotton immersed in ethanol-derived solvent was used to clean the stainless-steel platform. Three imprints using a 2 kg roller were made on the adhesive tape samples affixed to the steel pans. After this, the adhesive tape specimens were subjected to four hours of intense heat at 150 °C in an oven. As soon as the tapes reached room temperature, the interfaces of the stainless-steel planes and the residue remnants were meticulously scrutinized. The detachment velocity from the steel sheets was calculated to determine the shear adhesiveness. The experimental procedure remained consistent with the standard shear resilience examination, the sole deviation being the increased temperature (150 °C), accomplished through the oven placement of the test specimens.

3. Results and Discussion

Figure 1 illustrates the FTIR spectra of DHNA and the NCA, where Line a represents DHNA and Line b represents the curing agent. In Line b, the significant peaks at 1250–1200 cm^{-1} and 1730 cm^{-1} corresponded to the C-O and C=O groups, respectively. Additionally, a distinctive absorption band of the C=C group was seen at 1590–1640 cm^{-1}, and the peak of the -OH group around 3000–3500 cm^{-1} disappeared in Line b, indicating the presence of C=C groups in the monomer of the curing agent. Figure 2 displays the ^{1}H-NMR spectrum of the NCA (with TMS as the standard and $CDCl_3$ as the solvent). The spectrum reveals the expected peaks, including 6.27 (-CH attached to the monomer chains), 6.59–6.71 (=CH_2 attached to the -C=C side group of the monomer), and 7.46–7.78 (-CH attached to the naphthyl group of the monomer). The signals at 2.5 and 3.3 ppm potentially indicate the presence of a hydroxyl structure with a small amount of residual DNHA. Both the FTIR and ^{1}H-NMR spectra suggest the completion of the reactions and the successful synthesis of the expected monomer.

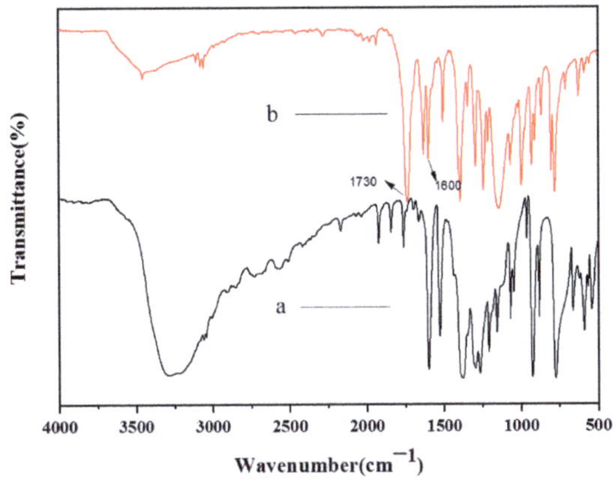

Figure 1. FTIR spectra of DHNA (**a**) and the NCA (**b**).

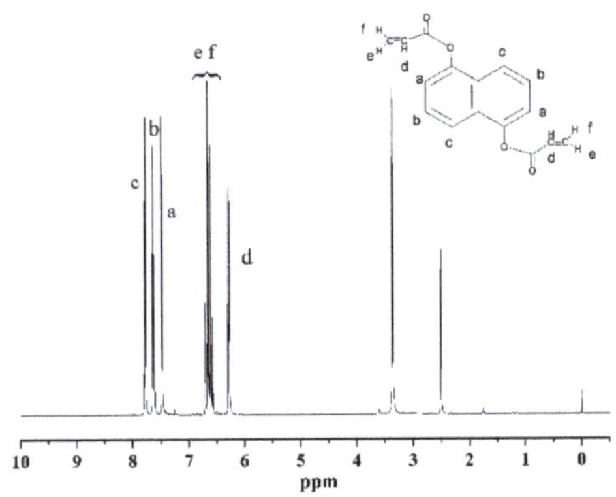

Figure 2. ^{1}H-NMR spectra of the NCA.

The determination of gel fraction is a useful method for assessing the degree of cross-linking in cross-linked or network polymers [24,25]. Figure 3 illustrates the gel content of the modified pressure-sensitive adhesives (PSAs) under UV irradiation with varying amounts of NCA and at different UV doses. The gel fraction offers insight into the amount of cross-linking in the system and better explains the impact of UV doses and NCA contents on the degree of cross-linking. While both chain scission and cross-linking can occur simultaneously following UV irradiation, we evaluate the outcome of such treatment. The gel content of PSAs increases gradually with increasing amounts of NCA, which is due to the high concentration of carbon double bonds in the composites that can form cross-links with acrylic PSAs [26]. When the NCA amount was 15 phr, the gel content reached 68% at the UV dose of 2400 mJ/cm^2. The gel content also increased with increasing UV doses, due to the enhanced potential for cross-linking between the acrylic copolymer and NCA. This cross-linking density was able to increase with the higher UV dose. Due to the tight interlocking of the unsaturated linear chain segments, the highly cross-linked network structure changed numerous PSA characteristics, including its mechanical, thermal, and adhesive capabilities [27].

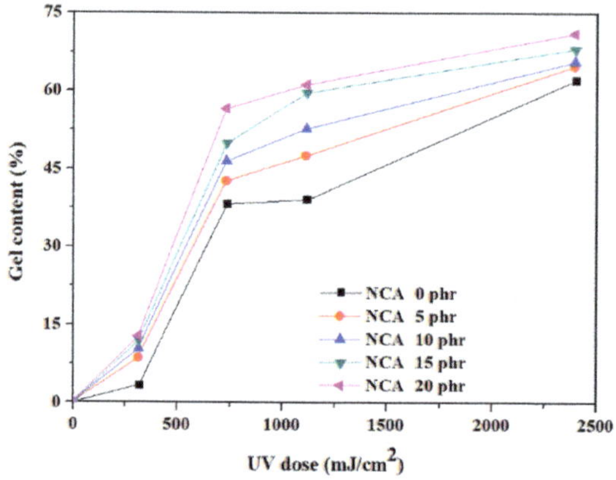

Figure 3. Variations in the gel content of PSA composites depending on NCA content and UV dose.

Figure 4 demonstrates the influence of various UV doses on the weight average molar mass (M_w) of composites with 10 phr NCA. The corresponding groups of UV dose were 0 mJ/cm^2 for UV-0, 320 mJ/cm^2 for UV-1, 737 mJ/cm^2 for UV-2, 1116 mJ/cm^2 for UV-3, and 2400 mJ/cm^2 for UV-4. The M_w was observed to increase as the UV dose increased. This phenomenon can be attributed to the initiation of radical polymerization reactions among the C=C bands in the composites, initiated by the radicals generated by the photoinitiators through UV irradiation. This ultimately resulted in a highly cross-linked structure, leading to a higher M_w. Figure 5, on the other side, shows the impact of NCA concentration on Mw when the UV dose remained constant. M_w has been identified to rise with increasing NCA content. This suggests that during polymerization, there were more cross-linking points available, allowing longer polymer chains to form. In summary, the concentration of the NCA and the UV dose played a significant role in the trend of the M_w, which exhibited a rapid increase as the concentration of the NCA and UV dose increased.

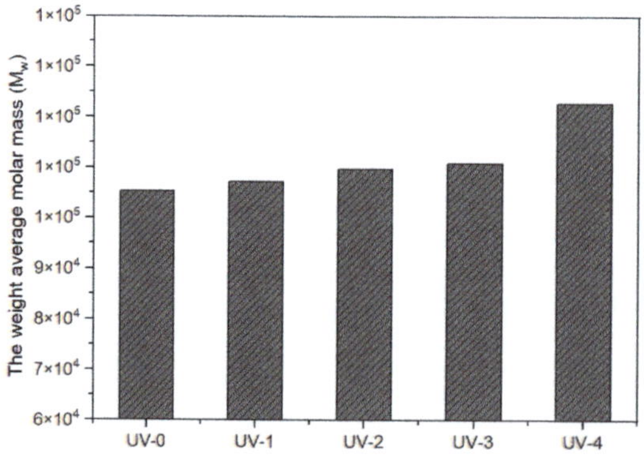

Figure 4. Effect of UV dose on M_w (the amount of addition: 10 phr).

The calculation of glass transition temperature (T_g) through the Fox equation (Equation (2)) for uncross-linked, UV-curable, acrylic pressure-sensitive adhesives (PSAs) resulted in a T_g close to −40 °C. However, the actual measured value was higher than the calculated value. One plausible explanation for this discrepancy is the formation of hydrogen bonds between the polar carboxyl groups in the polymer chains, contributing to physical cross-linking [1]. As observed in Figure 6, the T_g value increased as UV dose increased when the cross-linker content was at 10 phr. This can be attributed to the highly cross-linked networks formed in the polymer chains due to the greater extent of UV-induced cross-linking reactions, which can hinder the movement of molecular chains. Furthermore, Figure 7 shows that the T_g increased with an increase in the amount of NCA at a UV dose of 2100 mJ/cm^2.

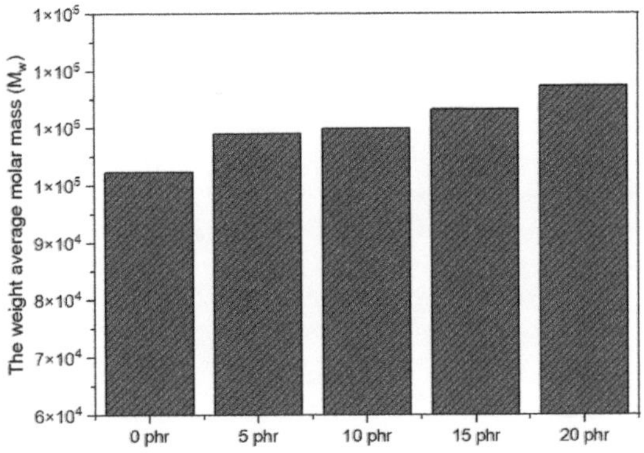

Figure 5. Effect of NCA concentration on M_w (UV dose: 737 mJ/cm^2).

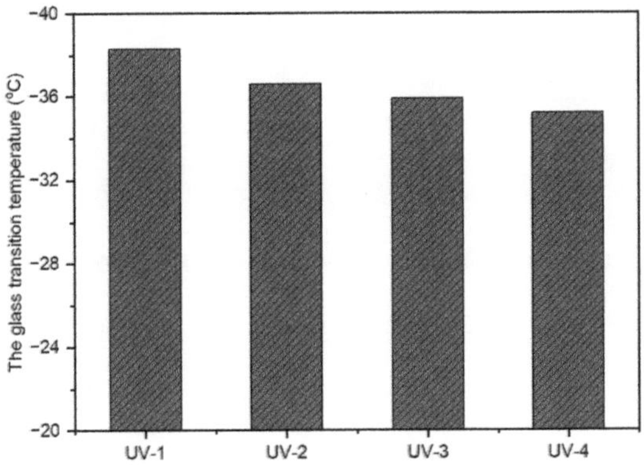

Figure 6. Effect of UV dose on T_g (the amount of addition: 10 phr).

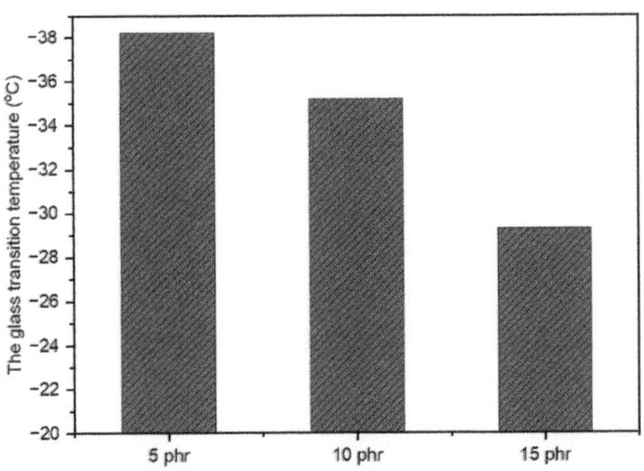

Figure 7. Effect of NCA amount on T_g (UV-4: 2100 mJ/cm^2).

The T_g of a polymer is influenced by various factors, including comonomers present in the reaction mixture. The T_g plays a critical role in determining the softness of the acrylic PSAs, which is necessary for flow and adhesion to surfaces [28]. Typically, the using comonomers like Vac and AA increase the T_g of the adhesive, enhancing its cohesive strength. The introduction of an NCA with a high T_g into the monomer mixture is expected to modify the T_g of the polymer.

$$\frac{1}{T_g} = \frac{w_1}{T_g1} + \frac{w_2}{T_g2} + \cdots \quad (2)$$

where T_g is the glass transition temperature of the copolymer, T_g1 and T_g2 are the glass transition temperatures of Component 1 and Component 2 in the copolymer, and ω_1 and ω_2 are the mass fractions of Component 1 and Component 2, respectively.

In summary, the physical and chemical properties of the polymer, including T_g, are vital considerations in the design and preparation of PSAs. The information presented in this paper sheds light on the factors that affect T_g and their implications on the adhesion properties of PSAs. Understanding these underlying mechanisms can facilitate the development of PSAs with improved performance in various industrial and biomedical applications.

In order to assess the impact of the naphthyl curing agent on the thermal stability of acrylic pressure-sensitive adhesives (PSAs), the thermal properties of UV-curable, acrylic PSAs were analyzed using TGA. The TGA thermograms of the UV-curable, acrylic PSAs with varying amounts of NCA content in a nitrogen atmosphere, after irradiation with a UV dose of 2100 mJ/cm^2, are presented in Figure 8a. The primary breakdown of the PSA samples started between 200 °C and 380 °C due to the thermal degradation of the polymer matrix [29]. The temperatures of 60% weight loss and residue content at 600 °C are presented in Table 1. The samples exhibited similar degradation trends. In comparison to the samples cross-linked without the NCA, the residual concentration and decomposition temperature of the UV-curable PSA samples significantly increased. This suggested that the enhancement of thermal stability could be attributed to the expanding cross-linking density of the films [30]. The major thermal dissociation temperature of the UV-curable, acrylic PSAs increased, primarily due to stronger covalent connections formed between the copolymer and NCA. These connections introduced the thermally stable aromatic unit to the main chains of PSAs. This incorporation of rigid structures into the acrylic PSAs was accomplished through the resonance structures of aromatic groups [31]. The introduction of aromatics was a simple and effective method for improving thermal stability. Figure 8b presents the weight loss curves of the UV-curable, acrylic PSA samples, cross-linked and kept at 300 °C for 85 min, after irradiation with a UV dose of 2100 mJ/cm^2. The isothermal stability of the samples at 300 °C was also significantly improved with increasing NCA content. This was attributed to the contribution of the thermally stable naphthyl structure introduced to the acrylic chains of the PSAs, as well as the rigid network structures formed between the acrylic copolymer and NCA.

Table 1. Thermal resistance of UV-curable PSAs with a UV dose of 2100 mJ/cm^2.

Samples	Corresponding Temperature for 60% Weight Loss (°C)	Residue Content (wt%)
0 phr	387	3.4
5 phr	399	4.3
10 phr	398	5.1
15 phr	399	6.6
20 phr	398	6.9

The adhesive attribute known as "initial tack" is measured to assess the effectiveness of PSAs, with the highest number of the largest balls that remain on the PSA surface recorded. Figure 9 illustrates the variation in tack for UV-curable, acrylic PSAs with different NCA contents and UV doses. As the NCA and UV radiation increased, the tack decreased. Given that tack is a crucial adhesion feature that allows materials to stick to substrate surfaces quickly and with minimal pressure, it is likely that it heavily relies on the mobility of molecular chains. The cross-linking parts, which decrease the mobility of molecular chains, lead to a decrease in the tack of UV-curable, acrylic PSAs. This reduction in tack can prevent the substrate from becoming wet on the surface and reduces the contact surface of the tapes with the substrate. Therefore, improving the mobility of polymer chains could enhance the efficacy of the adhesion properties of UV-curable, acrylic PSAs.

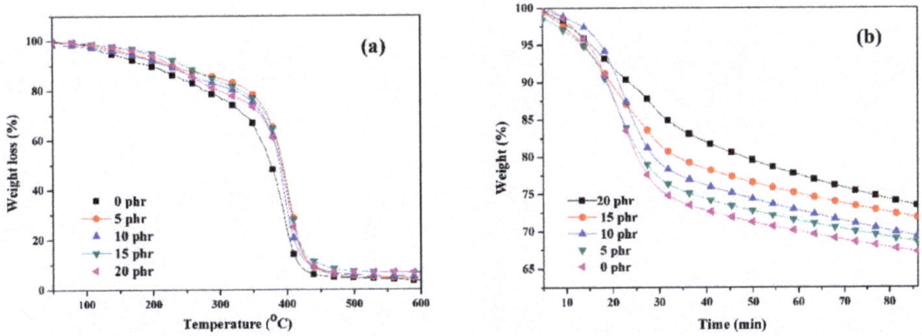

Figure 8. TGA thermograms of PSAs with various NCA content in a nitrogen atmosphere after irradiation with a UV dose of 2100 mJ/cm^2: (**a**) heating from 30 °C to 500 °C at 10 °C/min; (**b**) isothermally kept at 300 °C for 85 min.

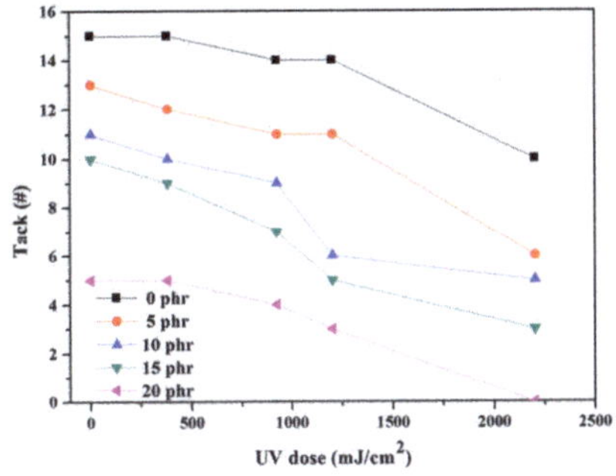

Figure 9. Variations in the tack of PSAs.

The peel strength was frequently the main consideration while analyzing the properties of PSAs. The peel strength exhibited a similar behavior to the tack, indicating that the mobility of the polymeric molecules in acrylic PSAs played a significant role in determining the peel strength. Several studies investigated the correlation between increasing PSA cross-linking levels and peel strength [32–35]. During the peel strength test, the critical features of the PSAs resulted in interfacial failure, influenced by the structure and cross-linking of the copolymer. Figure 10 shows that when the UV dose and NCA content increased, the peel strength of the PSAs considerably reduced. When the concentration of the NCA was 0 phr, the peel strength initially increased before declining. By utilizing UV light to cross-link the copolymer and the NCA, acrylic PSAs formed a complex network structure that severely limited the mobility of their polymeric chains [9]. The reduced peel strength can be primarily attributed to the restricted interfacial contact between the substrate and PSA tapes, as increased cross-linking further limits the mobility of the polymeric chains [9,18].

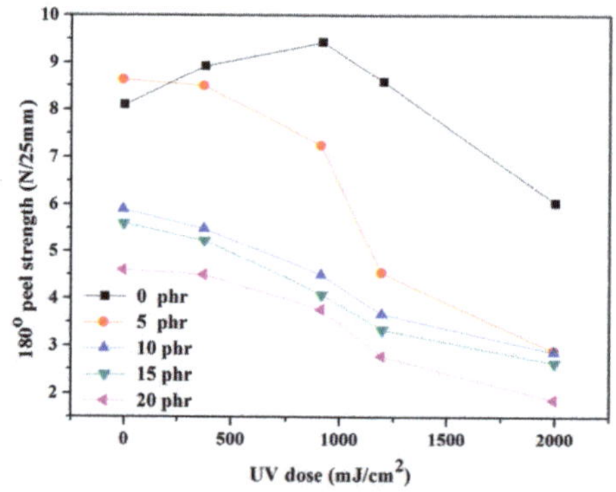

Figure 10. Variations in the peel strength of acrylic PSAs.

Figure 11 shows the shear strength of UV-curable, acrylic pressure-sensitive adhesives (PSAs) containing the NCA as a cross-linker, which was measured at room temperature. The results indicate that the shear strength of the UV-curable, acrylic PSAs was significantly affected by the amount of NCA and UV dosage used in the cross-linking process. Specifically, when the UV dosage was 380 mJ/cm^2, the holding period increased from 19 min to 1347 min with an increase in NCA content, after which the holding time rapidly declined to 449 min. Similarly, the holding time initially increased and then decreased with increasing UV dosage when the NCA content was kept constant [10]. The observed results could be attributed to the formation of polymeric networks with interconnected macromolecules and cross-linking of the linear molecular chains of the PSAs with a small quantity of NCA and UV radiation. With an increase in the cross-linker and UV dosage, the number of cross-linking sites increased, leading to the enhanced cohesiveness of the adhesives. When the NCA content reached 5 phr or the UV dose was over 380 mJ/cm^2, a highly cross-linked network structure was speculated to form. However, excessive cross-linking points could have a negative impact on the bonding, adhesive properties of PSAs.

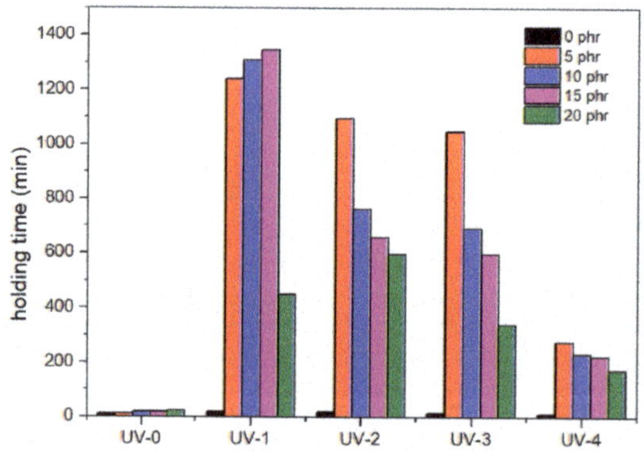

Figure 11. Variations in the holding time of acrylic PSAs (the unit of time: min; reaction temperature: 25 °C).

As demonstrated in Figure 12, the shear strength of UV-curable, acrylic PSAs increased with a higher concentration of NCA at an elevated temperature (150 °C) after UV irradiation (UV dose: 2000 mJ/cm^2). Conversely, when the NCA content exceeded 5 phr at 100 °C after irradiation, the shear strength decreased. This decline was attributed to an increased holding time caused by the creation of a cross-linking network. When the NCA was added in excess of 5 phr at 100 °C, the abundance of cross-linking points containing unsaturated bonds led to cohesion failure during high-temperature tests after a 2000 mJ/cm^2 dosage of UV exposure. Notably, when the NCA content was 20 phr, the shear strength remained constant at 150 °C for approximately 6 min, indicating a significant improvement compared to when there was no NCA present. Additionally, the average molecular weight may have an impact on cohesive strength at the cross-linking locations (M_c), which could be determined using Equation (3) as outlined in [36].

$$M_c = \frac{V_1\rho\left[\varphi^{1/3} - \varphi/2\right]}{-[\ln(1-\varphi) + \varphi + \chi\varphi^2]} \quad (3)$$

V_1: molar volume of toluene, ρ: density of the polymer, φ: volume fraction of the polymer in toluene, χ: polymer–solvent interaction parameter, M_e: the theoretical average molecular weight of the network chain.

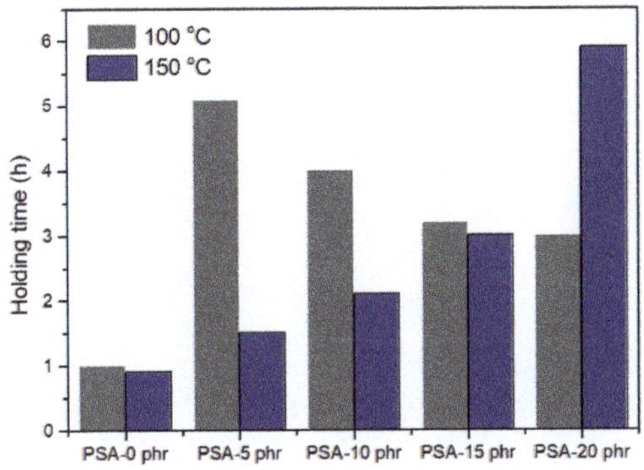

Figure 12. Effect of NCA content on the holding time at various temperatures.

A substantial cross-linking network can be created by the linear acrylic PSAs when $M_c \geq M_e$ [37]. Increasing the amount of cross-linker enhances the cohesive strength and cross-linking density of the PSAs while reducing M_c. This demonstrates an improvement in the properties and potential applications of the PSAs.

Figure 13 demonstrates the effect of NCA content on the thermoresistance properties of cross-linked PSAs after a 2000 mJ/cm^2 dosage of UV exposure. These PSAs had excellent peeling properties and left no residues on the surface of tape at room temperature. However, when the NCA content was 0 phr and the tapes were peeled at 150 °C, some evident residues were noticeable. This indicates that the degree of cross-linking was inadequate, leading to a low cohesive strength in the tapes, which resulted in ineffective debonding from the surface of steel sheets. Moreover, the adhesives were fluid, causing them to leave visible residues on the surface of the steel sheets when exposed to high temperatures. To overcome this challenge, we introduced rigid bonds to the polymeric molecules of the acrylic PSAs using UV irradiation, which subsequently underwent chemical cross-linking. The cohesive strength of the PSAs would increase during drying and UV irradiation, making them less likely to leave residues on the surface of the steel sheets. Figure 13f–j reveals that the peel properties of PSA tapes improved significantly at 150 °C. Compared to other cross-linkers, the appearance of these tapes left no residue on the surface of steel sheets, while maintaining excellent mechanical and thermal properties. Table 2 summarizes the results indicating that the thermoresistance test demonstrated that the UV-curable, acrylic PSAs had high stability at high temperatures. The incorporation of the NCA contributed to the stable configurations of the PSAs, thereby achieving superior thermoresistance performance.

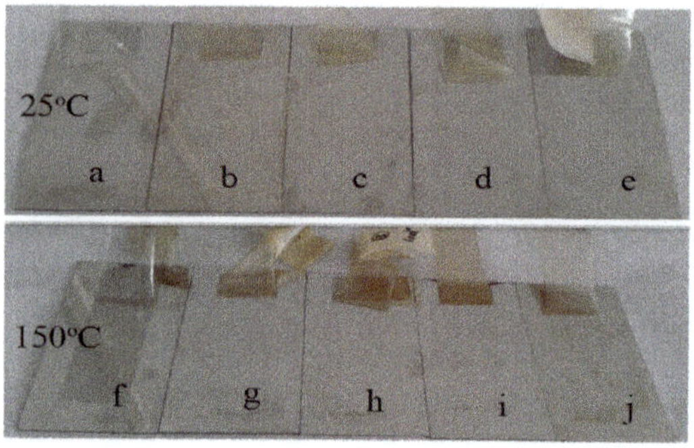

Figure 13. The thermoresistance of the tapes made of PSA cross-linked by 0 phr, 5 phr, 10 phr, 15 phr, and 20 phr of NCA were tested after UV exposure at 25 °C (**a**–**e**) and 150 °C (**f**–**j**).

Table 2. The influence of cross-linker content on the appearance of the substrate at 24 h/150 °C (additive: 20 phr).

Samples	Appearance
Polyfunctional aziridine	Large residue
2,4-tolylene diisocyanate (TDI)	Small residue
GMA	Large residue
MQ silicone resin	No residue
NCA	No residue

4. Conclusions

In this study, vinyl groups added to an acrylic copolymer by GMA modification are used to create UV-curable PSAs, as well as an NCA which introduces thermally stable aromatic structures to the main chains. The addition of the NCA and a higher UV dose was found to increase the gel fraction of the PSAs due to the formation of numerous cross-links between the components. The concentration of the NCA was identified as the main contributor to the thermal stability of PSAs, as it effectively added stiff groups and created cross-linked structures in the polymer matrix. Consequently, the T_g of PSAs was significantly improved, leading to increased thermal stability. However, the addition of the NCA also resulted in decreased tack and 180° peel strength due to limitations in polymeric chain mobility. Nonetheless, the holding duration of the PSAs after UV irradiation was significantly improved due to the cohesive strength created by the cross-linked network. Furthermore, the use of the NCA allowed for the creation of PSAs with a range of applications in electronic chemical products, such as packaging and protection. These PSAs exhibited excellent adhesive properties at elevated temperatures, thanks to the formation of cross-linked networks between the acrylic polymer matrix and the NCA. Overall, this study demonstrates the potential of using a thermally stable NCA to enhance the properties of UV-curable, acrylic PSAs, enabling their use in a diverse range of applications.

Author Contributions: Conceptualization, J.C. (Junhua Chen); methodology, J.C. (Junhua Chen) and Y.W.; software, H.Z.; validation, H.Z.; formal analysis, L.K., J.C. (Jinlian Chen) and Q.W.; investigation, S.L., X.W., L.F., D.H., L.K., J.C. (Jinlian Chen) and Q.W.; resources, D.Z.; data curation, S.L., X.W., L.F. and D.H.; writing—original draft preparation, J.C. (Junhua Chen); writing—review and editing, X.H. and C.L.; visualization, J.C. (Junhua Chen) and Y.W.; supervision, C.L.; project administration, X.H.; funding acquisition, D.Z. All authors have read and agreed to the published version of the manuscript.

Funding: This study was financially supported by the Guangdong Basic and Applied Basic Research Foundation (Chunsheng Li, 2020A1515111156), Youth Enhancement Project of the Natural Science Foundation of Guangdong Province (Dong Yu Zhu, 202341515030135).

Institutional Review Board Statement: Not applicable.

Data Availability Statement: The data presented in this study are available on request from the corresponding author.

Conflicts of Interest: The authors declare no conflict of interest.

References

1. Sun, S.; Li, M.; Liu, A. A review on mechanical properties of pressure sensitive adhesives. *Int. J. Adhes. Adhes.* **2013**, *41*, 98–106. [CrossRef]
2. Zhang, L.; Cao, Y.; Wang, L.; Shao, L.; Bai, Y. Synthesis and properties of soap-free P(2-EHA-BA) emulsion for removable pressure sensitive adhesives. *RSC Adv.* **2014**, *4*, 47708–47713. [CrossRef]
3. Sowa, D.; Czech, Z.; Byczyński, Ł. Peel adhesion of acrylic pressure-sensitive adhesives on selected substrates versus their surface energies. *Int. J. Adhes. Adhes.* **2014**, *49*, 38–43. [CrossRef]
4. Do, H.-S.; Park, Y.-J.; Kim, H.-J. Preparation and adhesion performance of UV-crosslinkable acrylic pressure sensitive adhesives. *J. Adhes. Sci. Technol.* **2006**, *20*, 1529–1545. [CrossRef]
5. Czech, Z. Solvent-based pressure-sensitive adhesives for removable products. *Int. J. Adhes. Adhes.* **2006**, *26*, 414–418. [CrossRef]
6. Chae, G.-S.; Park, H.-W.; Lee, J.-H.; Shin, S. Comparative Study on the Impact Wedge-Peel Performance of Epoxy-Based Structural Adhesives Modified with Different Toughening Agents. *Polymers* **2020**, *12*, 1549. [CrossRef] [PubMed]
7. Ebe, K.; Seno, H.; Horigome, K. UV curable pressure-sensitive adhesives for fabricating semiconductors. I. Development of easily peelable dicing tapes. *J. Appl. Polym. Sci.* **2003**, *90*, 436–441. [CrossRef]
8. Pang, B.; Ryu, C.-M.; Kim, H.-I. Improvement of thermal stability of UV curable pressure sensitive adhesive by surface modified silica nanoparticles. *Mater. Sci. Eng. B* **2013**, *178*, 1212–1218. [CrossRef]
9. Mapari, S.; Mestry, S.; Mhaske, S.T. Developments in pressure-sensitive adhesives: A review. *Polym. Bull.* **2021**, *78*, 4075–4108. [CrossRef]
10. Czech, Z. Crosslinking of pressure sensitive adhesive based on water-borne acrylate. *Polym. Int.* **2003**, *52*, 347–357. [CrossRef]
11. Czech, Z. Inhibition of radical polymerization in solvent-based systems (security of solvent-based radical polymerization of PSA-acrylic in a plant reactor). *J. Appl. Polym. Sci.* **2003**, *87*, 1354–1357. [CrossRef]
12. Yoo, Y.; Hong, G.; Hur, S.; Kim, Y.S.; Lee, S.; Kim, H.; Lee, J.H. Preparation of acrylic copolymers and crosslinking agents and properties as a film. *J. Appl. Polym. Sci.* **2009**, *112*, 1587–1594. [CrossRef]
13. Hayashi, S.; Kim, H.-J.; Kajiyama, M.; Ono, H.; Mizumachi, H.; Zufu, Z. Miscibility and pressure-sensitive adhesive performances of acrylic copolymer and hydrogenated rosin systems. *J. Appl. Polym. Sci.* **1999**, *71*, 651–663. [CrossRef]
14. Canetta, E.; Marchal, J.; Lei, C.-H.; Deplace, F.; König, A.M.; Creton, C.; Ouzineb, K.; Keddie, J.L. A Comparison of Tackified, Miniemulsion Core−Shell Acrylic Latex Films with Corresponding Particle-Blend Films: Structure−Property Relationships. *Langmuir* **2009**, *25*, 11021–11031. [CrossRef] [PubMed]
15. Mozelewska, K.; Antosik, A.K. Influence of Silicone Additives on the Properties of Pressure-Sensitive Adhesives. *Materials* **2022**, *15*, 5713. [CrossRef] [PubMed]
16. Zhang, J.; Yang, M.; Zhu, Y.; Yang, H. Synthesis and characterization of crosslinkable latex with interpenetrating network structure based on polystyrene and polyacrylate. *Polym. Int.* **2006**, *55*, 951–960. [CrossRef]
17. Wang, C.-S.; Leu, T.-S.; Hsu, K.-R. Novel bismaleimide with naphthalene side group. 1. From 1-naphthaldehyde and 2,6-dimethylaniline. *Polymer* **1998**, *39*, 2921–2927. [CrossRef]
18. Matsumoto, A.; Hasegawa, K.; Fukuda, A. Studies on modified phenolic resin. Iv: Properties of phenolic resin modified with 4-hydroxyphenyl- maleimide/n-butylacrylate copolymers. *Polym. Int.* **1993**, *30*, 65–72. [CrossRef]
19. Kaji, M.; Endo, T. Synthesis of a novel epoxy resin containing naphthalene moiety and properties of its cured polymer with phenol novolac. *J. Polym. Sci. Part A Polym. Chem.* **1999**, *37*, 3063–3069. [CrossRef]
20. Chen, J.H.; Zhao, Q.; Wang, Y.; Luo, F.; Shen, L.; Wu, K.; Liang, L.Y.; Lu, M. Improvement in mechanical properties and thermal stability of solvent-based pressure-sensitive adhesives based on triazine heterocyclic monomer. *J. Appl. Polym. Sci.* **2016**, *133*, 43281. [CrossRef]
21. Pang, B.; Ryu, C.-M.; Kim, H.-I. Effect of NCA having thermally stable structure on properties of UV-cured pressure sensitive adhesive. *J. Ind. Eng. Chem.* **2014**, *20*, 3195–3200. [CrossRef]
22. Bartkowiak, M.; Czech, Z.; Mozelewska, K.; Nowak, M. Influence of thermal reactive crosslinking agents on the tack, peel adhesion, and shear strength of acrylic pressure-sensitive adhesives. *Polym. Test.* **2020**, *90*, 106603. [CrossRef]
23. Sathyan, A.; Kurtz, I.; Rathore, P.; Emrick, T.; Schiffman, J.D. Using Catechol and Zwitterion-Functionalized Copolymers to Prevent Dental Bacterial Adhesion. *ACS Appl. Bio. Mater.* **2023**, *6*, 2905–2915. [CrossRef] [PubMed]
24. Joo, H.-S.; Park, Y.-J.; Do, H.-S.; Kim, H.-J.; Song, S.-Y.; Choi, K.-Y. The curing performance of UV-curable semi-interpenetrating polymer network structured acrylic pressure-sensitive adhesives. *J. Adhes. Sci. Technol.* **2007**, *21*, 575–588. [CrossRef]
25. Lim, D.-H.; Do, H.-S.; Kim, H.-J.; Bang, J.-S.; Yoon, G.-H. Preparation of SIS/SBS-based UV-cross-linkable pressure-sensitive adhesives using the thiolene reaction. *J. Adhes. Sci. Technol.* **2007**, *21*, 589–603. [CrossRef]
26. Pang, B.; Ryu, C.-M.; Jin, X.; Kim, H.-I. Preparation and properties of UV curable acrylic PSA by vinyl bonded graphene oxide. *Appl. Surf. Sci.* **2013**, *285*, 727–731. [CrossRef]

27. Moussa, K.; Decker, C. Semi-interpenetrating polymer networks synthesis by photocrosslinking of acrylic monomers in a polymer matrix. *J. Polym. Sci. Part A Polym. Chem.* **1993**, *31*, 2633–2642. [CrossRef]
28. Kajtna, J.; Krajnc, M. Solventless UV crosslinkable acrylic pressure sensitive adhesives. *Int. J. Adhes. Adhes.* **2011**, *31*, 822–831. [CrossRef]
29. Noel, M.; Santhanam, R. Electrochemistry of graphite intercalation compounds. *J. Power Sources* **1998**, *72*, 53–65. [CrossRef]
30. Asif, A.; Hu, L.; Shi, W. Synthesis, rheological, and thermal properties of waterborne hyperbranched polyurethane acrylate dispersions for UV curable coatings. *Colloid Polym. Sci.* **2009**, *287*, 1041–1049. [CrossRef]
31. Wang, C.-S.; Lee, M.-C. Synthesis and modification of a naphthalene-containing trifunctional epoxy resin for electronic applications. *J. Appl. Polym. Sci.* **1998**, *70*, 1907–1921. [CrossRef]
32. Ryu, C.-M.; Pang, B.-L.; Han, J.-H.; Kim, H.-I. Effect of Photo-crosslinking on Clean Debonding of Acrylic Pressure Sensitive Adhesives from Silicon Wafer. *J. Photopolym. Sci. Technol.* **2012**, *25*, 705–712. [CrossRef]
33. Asahara, J.; Hori, N.; Takemura, A.; Ono, H. Crosslinked acrylic pressure-sensitive adhesives. I. Effect of the crosslinking reaction on the peel strength. *J. Appl. Polym. Sci.* **2003**, *87*, 1493–1499. [CrossRef]
34. Czech, Z. New copolymerizable photoinitiators for radiation curing of acrylic PSA. *Int. J. Adhes. Adhes.* **2007**, *27*, 195–199. [CrossRef]
35. Czech, Z.; Urbala, M. UV-crosslinked acrylic pressure-sensitive adhesive systems containing unsaturated ethers. *Polimery* **2007**, *52*, 438–442. [CrossRef]
36. Seok, W.C.; Leem, J.T.; Kang, J.H.; Kim, Y.J.; Lee, S.; Song, H.J. Change of Characterization and Film Morphology Based on Acrylic Pressure Sensitive Adhesives by Hydrophilic Derivative Ratio. *Polymers* **2020**, *12*, 1504. [CrossRef]
37. Tobing, S.D.; Klein, A. Molecular parameters and their relation to the adhesive performance of acrylic pressure-sensitive adhesives. *J. Appl. Polym. Sci.* **2001**, *79*, 2230–2244. [CrossRef]

Disclaimer/Publisher's Note: The statements, opinions and data contained in all publications are solely those of the individual author(s) and contributor(s) and not of MDPI and/or the editor(s). MDPI and/or the editor(s) disclaim responsibility for any injury to people or property resulting from any ideas, methods, instructions or products referred to in the content.

polymers

MDPI

Article

A Study of the Phosphorylcholine Polymer Coating of a Polymethylpentene Hollow Fiber Membrane

Feihua Ye [1,2,3], Zhisheng Chen [1], Chunsheng Li [1,2], Junhua Chen [1,2,*] and Guobin Yi [3,*]

1 School of Environmental and Chemical Engineering, Zhaoqing University, Zhaoqing 526061, China; yefeihua@zqu.edu.cn (F.Y.); 15089670136@163.com (Z.C.); lichunsheng@zqu.edu.cn (C.L.)
2 Guangdong Provincial Key Laboratory of Environmental Health and Land Resource, Zhaoqing University, Zhaoqing 526061, China
3 School of Chemical Engineering and Light Industry, Guangdong University of Technology, Guangzhou 510006, China
* Correspondence: cehjchen@yeah.net (J.C.); yigb@gdut.edu.cn (G.Y.)

Abstract: A phosphorylcholine polymer (poly(MPC–co–BMA–co–TSMA), PMBT) was prepared by free radical polymerization and coated on the surface of the polymethylpentene hollow fiber membrane (PMP–HFM). ATR–FTIR and SEM analyses showed that the PMBT polymer containing phosphorylcholine groups was uniformly coated on the surface of the PMP–HFM. Thermogravimetric analysis showed that the PMBT had the best stability when the molar percentage of MPC monomer in the polymer was 35%. The swelling test and static contact angle test indicated that the coating had excellent hydrophilic properties. The fluorescence test results showed that the coating could resist dissolution with 90% (*v*/*v*%) ethanol solution and 1% (*w*/*v*%) SDS solution. The PMBT coating was shown to be able to decrease platelet adherence to the surface of the hollow fiber membrane, and lower the risk of blood clotting; it had good blood compatibility in tests of whole blood contact and platelet adhesion. These results show that the PMBT polymer may be coated on the surface of the PMP–HFM, and is helpful for improving the blood compatibility of membrane oxygenation.

Keywords: phosphorylcholine; polymer; thermal decomposition; hollow fiber membrane; surface modification

check for updates

Citation: Ye, F.; Chen, Z.; Li, C.; Chen, J.; Yi, G. A Study of the Phosphorylcholine Polymer Coating of a Polymethylpentene Hollow Fiber Membrane. *Polymers* **2023**, *15*, 2881. https://doi.org/10.3390/polym15132881

Academic Editors: Francesco Galiano and Mohamed Khayet

Received: 4 May 2023
Revised: 20 June 2023
Accepted: 26 June 2023
Published: 29 June 2023

Copyright: © 2023 by the authors. Licensee MDPI, Basel, Switzerland. This article is an open access article distributed under the terms and conditions of the Creative Commons Attribution (CC BY) license (https://creativecommons.org/licenses/by/4.0/).

1. Introduction

During COVID-19, membrane oxygenators achieved remarkable success in the life support of patients with pulmonary failure [1]. The core component of a membrane oxygenator is the hollow fiber membrane (HFM) material, the performance of which directly determines the performance of the oxygenator. Commonly used hollow fiber membrane materials are polypropylene (PP–HFM), cellulose acetate (CA–HFM), polyvinylidene fluoride (PVDF–HFM), and polymethylpentene (PMP–HFM) [2,3]. PP–HFM is the most widely used membrane material in clinical practice; however, directly prepared PP–HFM suffers problems such as a short life span and a tendency to fracture. CA–HFM has the characteristics of low cost and biodegradability, but it is susceptible to microbial corrosion and chemical attacks during utilization. PVDF–HFM has excellent oxidation resistance and mechanical strength, but its air permeability is average. PMP–HFM is a new type of membrane material that has been developed recently and has received a lot of attention because of its strong thermal stability, high mechanical strength, good air permeability, and chemical resistance [4]. However, no matter which hollow fiber membrane material is used, the surface is prone to protein adsorption, platelet adhesion, and thrombus deposition after prolonged contact with blood, leading to clinical adverse effects such as blood damage and inflammation, which often do not meet the requirements of prolonged medical treatment [5,6]. Therefore, the modification of membrane material surfaces via physical or chemical methods, in order to prepare hollow fiber membrane composites with excellent hemocompatibility, is currently a hot research topic.

Numerous efforts have been made to improve the blood compatibility on the surface of hollow fiber membrane materials, among which the construction of blood–compatible bionic coatings on the surface of fiber membranes is one of the most attractive solutions [7]. Using polymers containing phosphorylcholine (PC) groups to construct coatings on the surface of fibrous membranes can help to form amphiphilic ionic surfaces that mimic the cell membrane structure and can significantly improve the blood compatibility on the surface of fibrous membrane materials; 2-formyloxyethyl phosphorylcholine (MPC) is considered to be one of the most suitable monomers for the synthesis of phosphorylcholine polymers [8–10]. Phosphorylcholine polymer coatings have promising applications in extracorporeal circulation devices such as membrane oxygenators, stents, and blood purification devices [11–13].

Ishihara et al. [14–16] reported that the CA–HFM was coated with a copolymer (poly (MPC–co–BMA), PMB) to improve its performance with respect to solute and water permeability. The application of modified fiber membranes in blood purification devices and liver–assisted bioreactors has also been studied. Myers et al. [17] modified the surface of PP–HFM with poly(MPC–co–dodecyl methacrylate) and formed a polymer coating that could significantly reduce the blood circulation process of platelet activation. Teotia et al. [18] studied the in situ application of a phosphorylcholine polymer coating on PS–HFM to improve the blood compatibility of the membrane surface. Nishigochi et al. [19] coated the PMB polymer on PVDF–HFM to form a PMB coating, and found that the modified hollow fiber membrane could resist the adsorption of bovine serum protein. Wang et al. [20] used a dip–coating method to coat PMBT polymer on PP–HFM, and showed that the coating improved the hemocompatibility of the fiber membrane without hindering the gas exchange performance of the hollow fiber membrane. The authors of the paper coated poly(MPC–co–LMA–co–TSMA), PMLT) on PMP–HFM and found that the PMLT coating also showed a positive effect, improving the hemocompatibility of the fibrous membrane surface [21]. In addition, on the basis of earlier research, we found some new interesting phenomena. When different hydrophobic monomers, such as lauryl methacrylate (LMA) or *N*–butyl methacrylate (BMA), are used in the preparation of phosphorylcholine polymers, the differences in the hydrophobicity, chain length, and rigidity of the different monomers may lead to different bonding and contact angles between the polymer and the hollow fiber membrane surface, which in turn can affect the other related properties of the polymer coating. For this reason, we plan to further investigate the effect of phosphorylcholine polymers synthesized with different hydrophobic monomers on the modification of PMP–HFM.

Herein, we synthesized a phosphorylcholine polymer (poly(MPC–co–BMA–co–TSMA), PMBT) via a free radical polymer reaction, and coated the polymer on the surface of PMP–HFM to successfully prepare hollow fiber membrane composites with a PMBT coating on the surface. The morphology, thermal stability, hydrophilic properties, and chemical stability of the composite surface coating were tested via ATR–FTIR, SEM, TGA, contact angle, and fluorescence microscopy, respectively. Further, the blood compatibility of the coatings was analyzed using a platelet adhesion test and whole blood contact test. The results of the study can be used to prepare PMP–HFM bionic polymer coatings to improve the hemocompatibility of membrane oxygenators.

2. Experimental

2.1. Materials

2-Methacryloxyethyl phosphorylcholine (MPC, 97%) was purchased from Sigma–Aldrich, Sigma–Aldrich Trading Co., Ltd., Shanghai, China. *N*–butyl methacrylate (BMA, 99%), 3–(Trimethoxysilyl)propyl methacrylate (TSMA, 97%) and rhodamine 6G were supplied by Shanghai Aladdin Biochemical Technology Co., Ltd., Shanghai, China. The 2, 2′–Azoisobutyronitrile (AIBN, 99%) was recrystallized from methanol before being used. Platelet–rich plasma (PRP) was prepared using whole pig blood provided by Shanghai

Yudo Biotechnology Co., Ltd., Shanghai, China. A polymethylpentene hollow fiber membrane (PMP–HFM, inner diameter 200 μm) was purchased from 3M Deutschland Gmbh, Seefeld, Germany. The sodium dodecyl sulfate (SDS), triethylamine (TEA, 99%), and other reagents were of analytical grade.

2.2. Synthesis of PMBT Polymer

The PMBT polymer was created via free radical polymerization, and the process of its making is shown in Figure 1 and Table 1. The necessary quantities of MPC, BMA, TSMA, and AIBN were weighed and then mixed together in ethanol, before being put into a funnel under constant pressure. In order to remove oxygen, nitrogen was bubbled into the reaction flask containing the solvent ethanol for 30 min. The flask was then heated to 83 °C. Drop by drop, the mixed reaction solution was added. The nitrogen gas flow was stopped once the dropwise addition was halted. After 24 h, the reaction was allowed to proceed before the heat source was cut off. The warm product solution was poured into a large amount of petroleum ether, and the polymer was precipitated while removing unreacted monomers to obtain a white flocculent precipitate. The precipitate was dissolved in absolute ethanol and precipitated again with petroleum ether. To produce a pure PMBT polymer, this procedure was performed three times. Finally, after being vacuum–dried at room temperature for 24 h, the polymer was then dissolved in a methanol solution (9:1 methanol to water volume ratio) to create a PMBT polymer solution with a specific concentration.

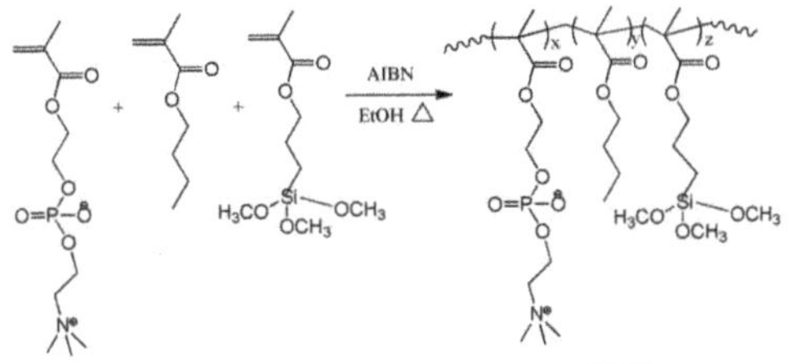

Figure 1. Synthesis route of the PMBT polymer.

Table 1. Monomer feed ratio of PMBT polymer.

Sample	MPC (mol%)	BMA (mol%)	TSMA (mol%)
PMBT25	25	65	10
PMBT35	35	55	10
PMBT45	45	45	10

2.3. Preparation of the PMBT Coating

Coating [18–22], blending [23,24], and self–assembly [25,26] can be used to modify the surface of hollow fiber membranes with phosphorylcholine polymers. Among them, the coating method is widely used because of its convenience and simplicity. In this study, a coating method was used to prepare PMBT polymer coatings on the PMP–HFM of membrane oxygenators. The specific operation is as follows. The PMP–HFM was cut to a size of 2 cm × 2 cm, washed with ethanol and deionized water, respectively, and then dried under a vacuum at room temperature for 24 h. The clean PMP–HFM was immersed into the PMBT solution at room temperature for a certain time. Then, the coated PMBT film was crosslinked in an environment of 10% (*v*/*v*%) triethylamine aqueous solution for at least three days after the PMP–HFM was removed from the solution. Finally, the cross–linked phospholipid polymer surface was then completely washed with deionized water before being vacuum–dried at 30 °C.

2.4. Polymer Coating Analysis

The bare and PMBT35–coated PMP–HFM were characterized using ATR–FTIR via an IS 50R Fourier transform infrared (FTIR) spectrometer (Thermo Fisher Scientific, Waltham, MA, USA). The surface coating morphology of PMP–HFM was observed using an SU 8220 field emission scanning electron microscope (Hitachi, Tokyo, Japan).

2.5. Thermal Stability Test

The PMBT polymer thermal weight loss was tested using a TGA analyzer (Netzsch, Serb, Germany). The sample weight was about 6 mg, the heating rate was 20 °C/min, and the nitrogen protection was carried out. The test range was 25~800 °C.

2.6. Swelling Degree Test

The dried PMBT polymer was weighed and placed in distilled water for soaking, and the water on the surface was quickly blotted out gently with filter paper after being removed at certain time intervals, and the mass was weighed and recorded. In this way, the water content (*WC*) of the polymer was measured with the changing pattern of dissolution time. Each sample was tested three times, and the *WC* was calculated as follows.

$$WC = \frac{m_t - m_0}{m_t} \times 100\% \quad (1)$$

In the equation, t is the dissolution time of PMBT polymer in water, m_t is the mass of PMBT polymer at the moment of dissolution t, and m_0 is the mass of dry PMBT polymer before dissolution.

2.7. Static Contact Angle Measurements

The water contact angle of the bare and PMBT35–coated PMP–HFM surfaces was measured using an OCA 100 optical contact angle-measuring instrument (DataPhysics, Filderstadt, Germany). To verify reproducibility, a minimum of ten samples were examined in each instance. In addition, the PMBT–coated PMP–HFM was immersed in 90% (*v/v*%) ethanol solution for 12 h or 1% (*w/v*%) SDS solution for 30 min, removed, and vacuum dried at 30 °C for 24 h. The static contact angle of the PMP–HFM surface was tested before and after treatment. The hydrophilic properties and stability of the PMBT coating were analyzed by comparing the change in the contact angle after ethanol soaking.

2.8. Fluorescence Properties of the Polymer Coating

The bare and PMBT35–coated PMP–HFM were submerged in a 200 ppm rhodamine 6 G solution in water for 30 s before undergoing two cycles of washing in clear water. Before examining with a 1X73 inverted fluorescence microscope (Olympus, Tokyo, Japan), the PMP–HFM was allowed to dry for 4 h in the open. To enable precise positioning, the PMP–HFM was viewed under normal lighting. After that, the microscope was set to fluorescence mode. The exposure time was 10 ms. The fluorescence excitation wavelength was 530~550 nm.

To further investigate the stability of the PMBT polymer coatings in ethanol and SDS solutions, the coated PMP–HFM was immersed in 90% (*v/v*%) ethanol solution for 12 h or 1% (*w/v*%) SDS solution for 30 min. The PMP–HFM was removed and stained with rhodamine 6 G solution, and the fluorescence performance of the coating was observed after treatment with a fluorescence microscope.

2.9. Platelet Adhesion

Pigs' citrated whole blood was centrifuged at 1000 RPM for 10 min to create platelet-rich plasma (PRP). The bare and PMBT35 coated PMP–HFM were immersed in 0.01 mol/L PBS solution at pH 7.4 for 2 h until swelling equilibrium was reached. To get rid of the weak adhering platelets, the PMP–HFM was rinsed five times with PBS after being submerged in PRP for 1 h at 37 °C. To immobilize the blood components on the PMP–HFM, it was then submerged in 2.5 wt% glutaraldehyde and maintained at 4 °C for 12 h. The PMP–HFM was then freeze-dried at −50 °C after being repeatedly washed with deionized water. Using SEM, the samples' surface was examined.

2.10. Whole Blood Contact

PBS was used to equilibrate the surfaces of the bare and PMBT35–coated PMP–HFM for two hours at room temperature. The samples were then submerged in the same porcine whole blood that had been citrated, and to balance off the anticoagulant effects of the citrate, a small amount of $CaCl_2$ solution (0.2 mol/L) was added to the solution. The PMP–HFM was extracted, washed with deionized water, and fixed with 2.5 wt% glutaraldehyde at 4 °C for 12 h, after shaking for 2 h at a frequency of 70 cycles/min at 37 °C. SEM was used to examine the samples' surface.

3. Results and Discussion

3.1. PMBT Coating Analysis

It can be seen from Figure 2 that compared with the bare PMP–HFM, the PMBT35–coated PMP–HFM exhibits C=O absorption peaks at 1727 cm^{-1}, and O–P=O absorption peaks at 1240 cm^{-1} and 1080 cm^{-1}. As a result of infrared analysis, the surface of PMP–HFM has phosphorylcholine groups after being treated with PMBT solution.

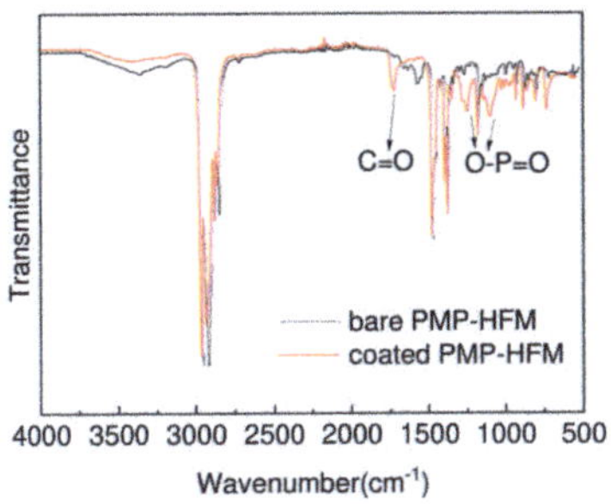

Figure 2. ATR-FTIR spectrum of bare and PMBT35-coated PMP-HFM.

Scanning electron microscope was used to observe the surface morphology changes of the PMBT35–coated PMP–HFM. The surface of bare PMP–HFM (Figure 3a) is significantly different from coated PMP–HFM (Figure 3b); the coated PMP–HFM surface is attached with a uniform PMBT polymer coating. When the polymer PMBT solution comes into contact with the PMP–HFM surface, the hydrophobic unit BMA in the polymer interacts with the fibrous membrane, which also has a hydrophobic surface, allowing the polymer to quickly adhere to the surface of the fibrous membrane via physical action. In addition, the cross–linkable unit TSMA in the polymer can form a network structure that further firmly encapsulates the PMBT polymer on the surface of the hollow fiber membrane, thus forming a stable PMBT coating.

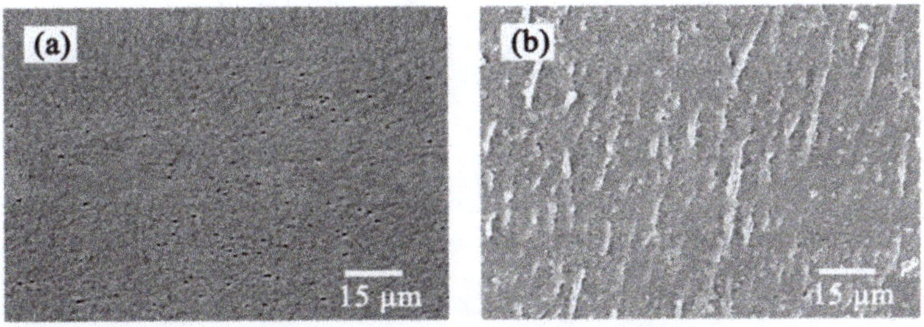

Figure 3. SEM images of bare (**a**) and PMBT35-coated (**b**) PMP-HFM.

3.2. Thermal Stability Analysis of PMBT Polymer

Thermogravimetric analysis was used to test the thermal decomposition stability of PMBT polymer. $T_{(5\%)}$ is the temperature corresponding to the polymer mass loss of 5%, representing the initial decomposition temperature of the polymer, which can be used to judge the thermal stability of the polymer with different MPC contents. Combined with Figure 4 and Table 2, it can be seen that with the increase in the amount of MPC, the thermal decomposition temperature of the PMBT polymer increases first and then decreases, among which the initial decomposition temperature of the PMBT35 polymer is the highest (252 °C) and has the best stability.

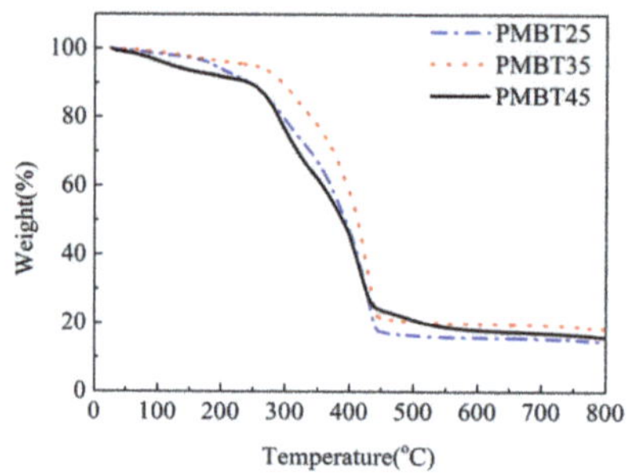

Figure 4. TGA curves of PMBT polymers.

Table 2. TGA data of different MPC contents of PMBT polymers.

Samples	$T_{(5\%)}$/°C	$T_{(10\%)}$/°C	$T_{(50\%)}$/°C
PMBT25	191	245	393
PMBT35	252	295	412
PMBT45	123	244	389

The thermal decomposition of the polymer is mainly related to chemical bond energy and molecular structure. The MPC unit of PMBT polymer contains a bond energy of C–N (305 kJ/mol), which is lower than that of C–C (332 kJ/mol) and Si–O (460 kJ/mol), and the fracture decomposition occurs first in the process of thermal decomposition, resulting in mass loss of the polymer. In addition, the −OH produced by the hydrolysis of the TSMA unit in the polymer will form an intramolecular hydrogen bond with the O atom in the MPC unit, thus enhancing the thermal stability of the polymer. Therefore, when the amount of MPC is initially increased, the intramolecular hydrogen bond in the polymer increases the thermal decomposition temperature. When the amount of MPC reached a certain level, a large number of C–N bonds began to break in the process of polymer thermal decomposition, the mass loss of the polymer accelerated, and the thermal decomposition temperature decreased.

3.3. Swelling Degree Analysis of PMBT Polymer

The study of the swelling behavior of PMBT polymer in water and the presence of water in the polymer is important for determining the hydrophilic properties and hemocompatibility of the polymer. As seen in Figure 5, the PMBT polymer rapidly absorbs water and reaches saturation within a few minutes. Subsequently, the water content within the PMBT polymer decreases slightly, and this particular phenomenon of water content rising and then falling during the swelling process is called over–swelling. The possible reason for this phenomenon is that the phosphorylcholine group (PC) contained in the PMBT polymer has a strong hydrophilic ability, which makes the polymer absorb water too quickly at the beginning of the swelling, meaning the molecular chain segments do not have time to spring back and a hysteresis effect occurs. In addition, the figure shows that the swelling of the PMBT polymer increases as the MPC concentration rises. This is due to the fact that each PC group can bind 12 water molecules, and the higher the MPC content, the more water molecules are bound to the PMBT polymer, which leads to a greater swelling of the polymer.

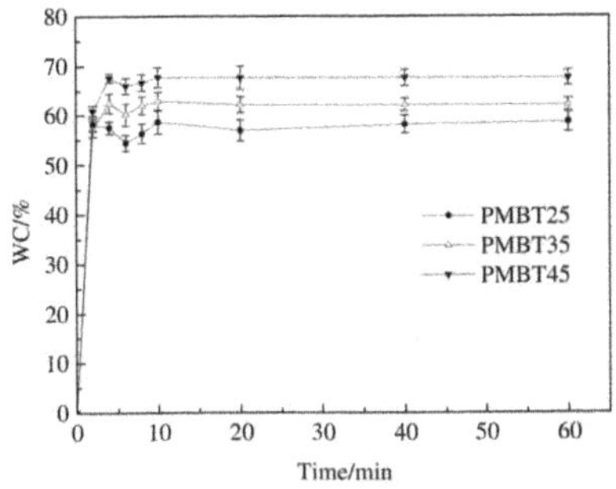

Figure 5. The relationship between *WC* and soaking time of PMBT polymers.

3.4. Static Contact Angle Analysis

Contact angle is an effective method used to observe the structure of the polymer surface and its changes, and a convenient means to measure the wetting degree of the polymer surface liquid. As seen from Table 3, the surface of bare PMP–HFM is hydrophobic. After coating with the PMBT polymer solution, the surface of PMP–HFM became hydrophilic. The hydrophilic surface was beneficial for improving the compatibility of the hollow fiber membrane with blood.

Table 3. Static contact angles of the bare and PMBT35-coated PMP-HFM.

Samples	Bare PMP-HFM	PMBT35 Coated PMP-HFM	Treated with Ethanol Solution	Treated with SDS Solution
θ (deg)	106 ± 0.6 *	79 ± 0.8 *	77 ± 1.3 *	76 ± 1.1 *

* $p < 0.001$ vs. bare.

Furthermore, the coated PMP–HFM was immersed in 90% (*v/v*%) ethanol solution for 12 h or 1% (*w/v*%) SDS solution for 30 min, and the static contact angle on the surface of the PMP–HFM was tested. It was found that the contact angle was almost unchanged. This result shows that the PMBT coating can stably adhere to the surface of PMP–HFM. The stability of the PMBT coating was related to the introduction of the TSMA monomer during polymer synthesis. When the PMBT–coated PMP–HFM was immersed in the aqueous solution of triethylamine (TEA), the coating component, TSMA, first hydrolyzed to form a hydroxyl group attached to silicon. This hydroxyl group was very active and dehydrated with other hydroxyl groups to form ether bonds and cross–linking. The network structure allowed the PMBT polymer to tightly coat the surface of the PMP–HFM to form a stable polymer coating.

3.5. Analysis of the Fluorescence Properties of the Coatings

Rhodamine 6 G could interact with the PC group in the phosphorylcholine polymer and emit fluorescence under the action of specific excitation light [27]. Observing the fluorescence characteristics of the coating could be one of the methods used to evaluate the continuity and stability of the surface coating of the hollow fiber membranes [28]. Figure 6 shows a microscope image of bare (Figure 6a) and coated (Figure 6b) PMP–HFM under visible light. Because the PMP–HFM material was white and opaque, it appeared as a black tube under an inverted fluorescence microscope. After switching to fluorescence mode, the bare PMP–HFM (Figure 6c) had no fluorescence emission, showing a dark image, while the coated PMP–HFM (Figure 6d) emitted bright orange–yellow fluorescence, and the hollow fiber membrane could be clearly seen. Through continuous observation of the moving stage, it was found that the entire hollow fiber membrane surface emits fluorescent light, indicating that the PMP–HFM was uniformly covered by the PMBT polymer coating.

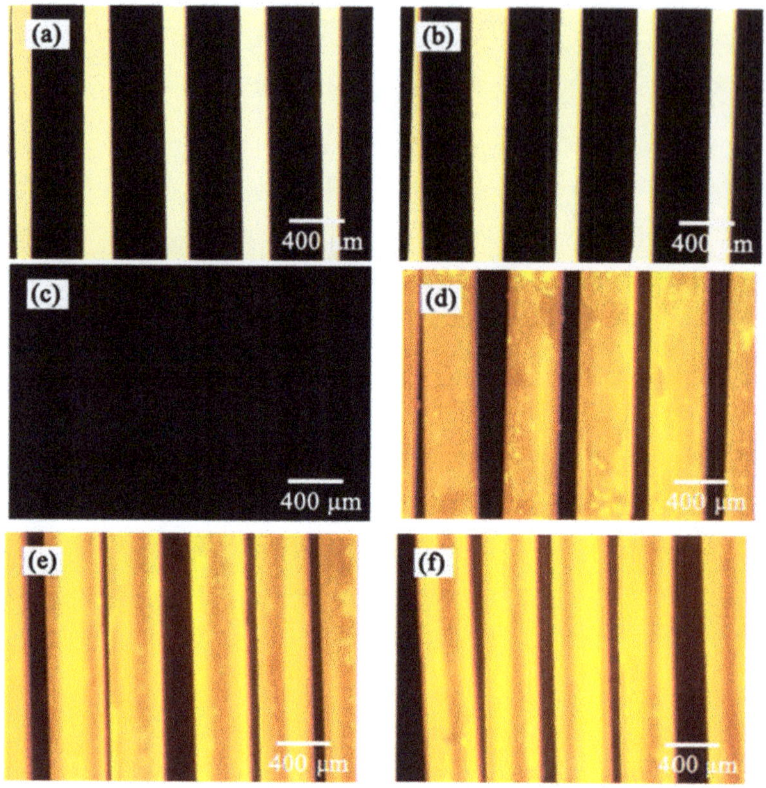

Figure 6. Microscope spectrum of bare and PMBT35-coated PMP-HFM. Underlit microscope images of bare (**a**) and coated (**b**) PMP-HFM; (**c**,**d**) are the same images viewed in fluorescence mode; (**e**,**f**) are fluorescence images of coated PMP-HFM after treatment with 90% (*v*/*v*%) ethanol solution or 1% (*w*/*v*%) SDS solution.

To further investigate the stability of the coating, the coated PMP–HFM was either immersed in 90% (*v*/*v*%) ethanol solution (Figure 6e) and left for 12 h, or was immersed in 1% (*w*/*v*%) SDS solution (Figure 6f) with 30 min of ultrasonic treatment. After drying, the sample underwent a dyeing treatment, and then the fluorescent effect of the coating on the surface of the hollow fiber membrane was observed. The results show that the fluorescence characteristics of the PMBT polymer coating on the surface of the PMP–HFM after treatment are still obvious. This result shows that the PMBT coating is stable in its adherence to the PMP–HFM, and can withstand ethanol or SDS solution dissolution.

3.6. Biocompatibility Evaluation

The adhesion and activation of platelets on the surface of medical materials is an important factor leading to coagulation. Therefore, studying the adhesion behavior of platelets on the surface of materials is an important method to evaluate the anticoagulant properties and blood compatibility of materials. It can be seen from Figure 7 that after the platelet adhesion experiment, there were many platelets adhered to the surface of the bare PMP–HFM (Figure 7a), and some of the platelets adhered thereon were aggregated and deformed. In contrast, the surface platelet adhesion of the PMP–HFM (Figure 7b) with the PMBT coating was significantly reduced, indicating that the PMBT coating can improve the blood compatibility of the PMP–HFM surface. The PMBT coating has a structure that mimics cell membranes. On the one hand, when in contact with blood, the phospholipids in blood will preferentially adsorb on the coating surface over proteins and platelets, forming a bionic surface and reducing protein adsorption; on the other hand, the polymer surface maintains a high free water content, and the contact between proteins and the polymer surface is reversible, which does not cause protein conformation changes and avoids protein adsorption. The biocompatibility of the PMBT35–coated PMP–HFM is shown schematically in Figure 8.

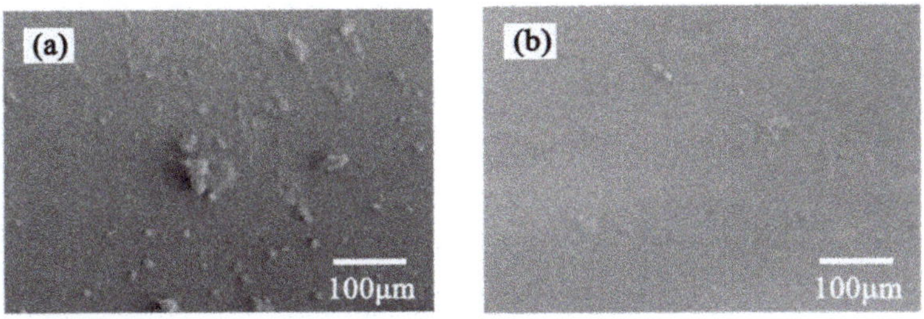

Figure 7. SEM images of platelet adhesion on the bare (**a**) and PMBT35-coated (**b**) PMP-HFM.

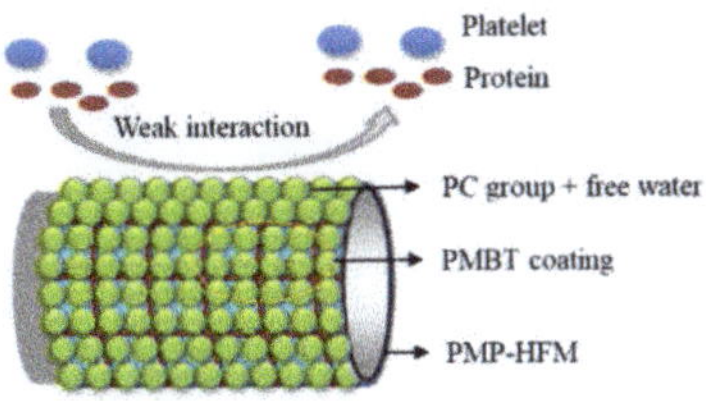

Figure 8. Schematic diagram of platelet and protein interaction on the coated PMP-HFM surfaces.

The whole blood contact test results showed that after the hollow fiber membrane was immersed in the blood for two hours, the surface of the bare PMP–HFM (Figure 9a) was adhered to by more thrombi, while the surface of the coated PMP–HFM (Figure 9b) was very clean, with no visible thrombus adhesion. The PMBT coating has an excellent antithrombotic effect; as a result of the amphiphilic phospholipid polymer coating's ability to mimic the outer membrane of cell, it exhibits significant resistance to platelet attachment and protein absorption. Therefore, the blood compatibility of the material surface can be significantly improved by coating the surface of the PMP–HFM with PMBT coating.

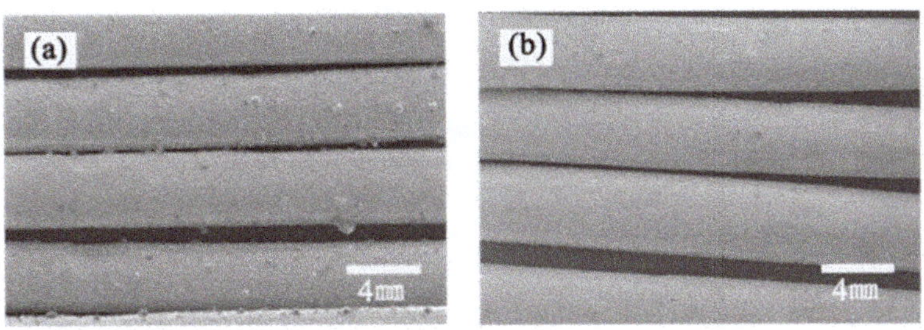

Figure 9. SEM images of whole blood contact on the bare (**a**) and PMBT35-coated (**b**) PMP-HFM.

4. Conclusions

We have successfully synthesized a phosphorylcholine polymer PMBT and applied it to the surface of PMP–HFM, uniformly, via a dip–coating method, to form a stable biocompatible coating. ATR–FTIR and SEM results confirmed that a PMBT coating was formed on the surface of PMP–HFM, and that the coating was uniformly encapsulated on the surface of the fiber membrane. The thermogravimetric test results showed that with the increase in the amount of MPC, the thermal decomposition temperature of the PMBT polymer increases first and then decreases, among which the initial decomposition temperature of PMBT35 polymer is the highest. The swelling test results indicate that the coating had excellent hydrophilic properties. Static contact angle and fluorescence microscopy results showed that the network–like structure formed in the polymer due to the introduction of the cross–linkable monomer TSMA enables the PMBT coating to remain stable in extreme environments, exhibiting good chemical stability. The results of platelet adhesion and whole blood contact experiments showed that the phosphorylcholine polymer PMBT could significantly improve the blood compatibility of PMP–HFM surface, thus prolonging the contact time between the hollow fiber membrane and blood, and providing a new solution idea for the clinical application of PMP–HFM. Encouragingly, the polymer coating is very simple to prepare, and the coating is uniform and stable; this is promising for application in the mass production of membrane oxygenators, which will create more blood–compatible membrane oxygenators with a longer service life.

Author Contributions: Methodology and data curation, F.Y. and C.L.; software and validation, C.L. and Z.C.; formal analysis, J.C.; investigation, J.C.; resources, G.Y.; data curation and writing—original draft preparation, F.Y.; writing—review and editing, G.Y.; visualization, J.C. All authors have read and agreed to the published version of the manuscript.

Funding: This research was funded by the University Teacher Characteristic Innovation Research Project, (No. 2021JNHB15), Guangdong Province Enterprise Science and Technology Special Representative (No. GDKTP2021023000), Zhaoqing University Science Fund (No. QN202229 and QN202006), Guangdong University Innovation Project (No. 2021KQNCX106), the Fund of Guangdong Provincial Key Laboratory of Environmental Health and Land Resource (No. 2020121201014), and Guangdong Basic and Applied Basic Research Foundation (No. 2020A1515111156).

Institutional Review Board Statement: Not applicable.

Data Availability Statement: The data presented in this study are available on request from the corresponding author.

Acknowledgments: We appreciate Dong Ren and Jing Zhang's generous assistance with sample preparation. We appreciate the expert lab help provided by Zilun Tang and Jianyu Wu.

Conflicts of Interest: The authors declare no conflict of interest.

References

1. Fiore, A.; Piscitelli, M.; Adodo, D.K.; Thomas, C.; Dessap, A.M.; Bagate, F.; Folliguet, T. Successful Use of Extracorporeal Membrane Oxygenation Postpartum as Rescue Therapy in a Woman with COVID-19. *J. Cardiothorac. Vasc. Anesth.* **2021**, *35*, 2140–2143. [CrossRef] [PubMed]
2. Teber, O.O.; Altinay, A.D.; Mehrabani, S.A.N.; Tasdemir, R.S.; Zeytuncu, B.; Genceli, E.A.; Dulekgurgen, E.; Pekkan, K.; Koyuncu, İ. Polymeric hollow fiber membrane oxygenators as artificial lungs: A review. *Biochem. Eng. J.* **2022**, *180*, 108340. [CrossRef]
3. Ogawa, T.; Uemura, T.; Matsuda, W.; Sato, M.; Ishizuka, K.; Fukaya, T.; Kinoshita, N.; Nakamoto, T.; Ohmagari, N.; Katano, H.; et al. SARS-CoV-2 Leakage from the Gas Outlet Port During Extracorporeal Membrane Oxygenation for COVID-19. *Asaio J.* **2021**, *67*, 511–516. [CrossRef] [PubMed]
4. Li, Y.K.; Hei, F.L. Improvement and new development of hollow fiber membrane materials for membrane artificiall lung. *Chin. J. Tissue Eng. Res.* **2022**, *26*, 2608–2612.
5. Zhi, Z.; Wang, W.; Xin, H.; Fan, W.; Lei, L.; Zhi, Z.; Wang, W.; Xin, H.; Fan, W.; Lei, L. Surface modification of polysulfone hollow fiber membrane for extracorporeal membrane oxygenator using low-temperature plasma treatment. *Plasma Process. Polym.* **2017**, *15*, 1700122.

6. Xin, H.; Wang, W.; Zhi, Z.; Fan, W.; Mao, C.; Shi, J.; Lei, L. Surface monofunctionalized polymethyl pentene hollow fiber membranes by plasma treatment and hemocompatibility modification for membrane oxygenators. *Appl. Surf. Sci.* **2016**, *362*, 355–363.
7. Garimella, N.; Wu, Z.; Griffith, B. Biocompatible Coating Composition. Patent US20150352265, 10 December 2015.
8. Ishihara, K. *Phospholipid Polymers*; John Wiley & Sons, Inc.: Hoboken, NJ, USA, 2012.
9. Iwasaki, Y.; Ishihara, K. Cell membrane-inspired phospholipid polymers for developing medical devices with excellent biointerfaces. *Sci. Technol. Adv. Mater.* **2012**, *13*, 064101. [CrossRef]
10. Santin, M.; Phillips, G. History of Biomimetic, Bioactive and Bioresponsive Biomaterials. In *Biomimetic, Bioresponsive, and Bioactive Materials*; Wiley: Hoboken, NJ, USA, 2012; pp. 1–34.
11. Goda, T.; Ishihara, K.; Miyahara, Y. Review: Critical Update on 2-Methacryloyloxyethyl Phosphorylcholine (MPC) Polymer Science. *J. Appl. Polym. Sci.* **2015**, *132*, 41766. [CrossRef]
12. Wei, Y.; Jing-xun, Z.; Ying, J.; Jian, J. REDV/Rapamycin-loaded polymer combinations as a coordinated strategy to enhance endothelial cells selectivity for a stent system. *Colloids Surf. B.* **2015**, *136*, 1166–1173. [CrossRef]
13. Xie, L.; Gong, L.; Zhang, J.; Han, L.; Xiang, L.; Chen, J.; Liu, J.; Yan, B.; Zeng, H. Wet Adhesion Strategy Via Synergistic Cation–π and Hydrogen Bonding Interactions of Antifouling Zwitterions and Mussel-Inspired Binding Moiety. *J. Mater. Chem. A* **2019**, *7*, 21944–21952. [CrossRef]
14. Sang, H.Y.; Watanabe, J.; Takai, M.; Iwasaki, Y.; Ishihara, K. High functional hollow fiber membrane modified with phospholipid polymers for a liver assist bioreactor. *Biomaterials* **2006**, *27*, 1955–1962.
15. Sang, H.Y.; Watanabe, J.; Takai, M.; Iwasaki, Y.; Ishihara, K. Design of functional hollow fiber membranes modified with phospholipid polymers for application in total hemopurification system. *Biomaterials* **2005**, *26*, 5032–5041.
16. Sang-Ho, Y.; Arazawa, D.T.; Yang, Z.; Venkat, S.; Malkin, A.D.; Kimmel, J.D.; Gamble, L.J.; Kazuhiko, I.; Federspiel, W.J.; Wagner, W.R. Hollow fiber membrane modification with functional zwitterionic macromolecules for improved thromboresistance in artificial lungs. *Langmuir* **2015**, *31*, 2463–2471.
17. Myers, G.; Johnstone, D.; Swyer, W.; McTeer, S.; Maxwell, S.; Squires, C.; Ditmore, S.; Power, C.; Mitchell, L.; Ditmore, J.; et al. Evaluation of Mimesys® phosphorylcholine (PC)–coated oxygenators during cardiopulmonary bypass in adults. *J. Extra Corpor. Technol.* **2003**, *35*, 6–12. [PubMed]
18. Teotia, R.S.; Dahe, G.J.; Bellare, J. In-Situ Coating of 2-Methacryloyloxyethyl Phosphorylcholine Polymer on Polysulfone Hollow Fiber Membranes for Hemodialysis. *J. Comput. Theor. Nanosci.* **2014**, *20*, A105. [CrossRef]
19. Nishigochi, S.; Ishigami, T.; Maruyama, T.; Hao, Y.; Ohmukai, Y.; Iwasaki, Y.; Matsuyama, H. Improvement of Antifouling Properties of Polyvinylidene Fluoride Hollow Fiber Membranes by Simple Dip Coating of Phosphorylcholine Copolymer via Hydrophobic Interactions. *Ind. Eng. Chem. Res.* **2014**, *53*, 2491–2497. [CrossRef]
20. Wang, Y.B.; Gong, M.; Yang, S.; Nakashima, K.; Gong, Y.K. Hemocompatibility and film stability improvement of crosslinkable MPC copolymer coated polypropylene hollow fiber membrane. *J. Membr. Sci.* **2014**, *452*, 29–36. [CrossRef]
21. Ye, F.H.; Yi, G.B. Cross-linkable phosphonylcholine polymer modified polymethylpentene hollow fiber membrane. *Acta Mater. Compos. Sinica.* **2021**, *38*, 479–486.
22. Wang, Y.B.; Shi, K.H.; Jiang, H.L.; Gong, Y.K. Significantly reduced adsorption and activation of blood components in a membrane oxygenator system coated with crosslinkable zwitterionic copolymer. *Acta Biomater.* **2016**, *40*, 153–161. [CrossRef]
23. Hasegawa, T.; Iwasaki, Y.; Ishihara, K. Preparation of blood-compatible hollow fibers from a polymer alloy composed of polysulfone and 2-methacryloyloxyethyl phosphorylcholine polymer. *J. Biomed. Mater. Res.* **2002**, *63*, 333–341. [CrossRef]
24. Asanuma, Y.; Inoue, Y.; Yusa, S.; Ishihara, K. Hybridization of poly(2-methacryloyloxyethyl phosphorylcholine-block-2-ethylhexyl methacrylate) with segmented polyurethane for reducing thrombogenicity. *Colloids Surf. B.* **2013**, *108*, 239–245. [CrossRef] [PubMed]
25. Sugihara, S.; Blanazs, A.; Armes, S.P.; Ryan, A.J.; Lewis, A.L. Aqueous Dispersion Polymerization: A New Paradigm for in Situ Block Copolymer Self-Assembly in Concentrated Solution. *J. Am. Chem. Soc.* **2011**, *133*, 15707–15713. [CrossRef]
26. Goda, T.; Kjall, P.; Ishihara, K.; Richter-Dahlfors, A.; Miyahara, Y. Biomimetic Interfaces Reveal Activation Dynamics of C-Reactive Protein in Local Microenvironments. *Adv. Healthc. Mater.* **2014**, *3*, 1733–1738. [CrossRef] [PubMed]
27. Lewis, A.L.; Tolhurst, L.A.; Stratford, P.W. Analysis of a phosphorylcholine-based polymer coating on a coronary stent pre- and post-implantation. *Biomaterials* **2002**, *23*, 1697–1706. [CrossRef] [PubMed]
28. Wang, J.H.; Bartlett, J.D.; Dunn, A.C.; Small, S.; Willis, S.L.; Driver, M.J.; Lewis, A.L. The use of rhodamine 6G and fluorescence microscopy in the evaluation of phospholipid-based polymeric biomaterials. *J. Microsc.* **2005**, *217*, 216–224. [CrossRef] [PubMed]

Disclaimer/Publisher's Note: The statements, opinions and data contained in all publications are solely those of the individual author(s) and contributor(s) and not of MDPI and/or the editor(s). MDPI and/or the editor(s) disclaim responsibility for any injury to people or property resulting from any ideas, methods, instructions or products referred to in the content.